Springer Undergraduate Mathematics Series

D1198484

More information about this series at http://www.springer.com/series/3423

Gregory T. Lee

Abstract Algebra

An Introductory Course

 Springer

Gregory T. Lee
Department of Mathematical Sciences
Lakehead University
Thunder Bay, ON
Canada

ISSN 1615-2085   ISSN 2197-4144   (electronic)
Springer Undergraduate Mathematics Series
ISBN 978-3-319-77648-4   ISBN 978-3-319-77649-1   (eBook)
https://doi.org/10.1007/978-3-319-77649-1

Library of Congress Control Number: 2018935845

Mathematics Subject Classification (2010): 20-01, 16-01, 12-01

Printed on acid-free paper

This Springer imprint is published by the registered company Springer International Publishing AG
part of Springer Nature
The registered company address is: Gewerbestrasse 11, 6330 Cham, Switzerland

In memory of my father

Preface

This book is intended for students encountering the beautiful subject of abstract algebra for the first time. My goal here is to provide a text that is suitable for you, whether you plan to take only a single course in abstract algebra, or to carry on to more advanced courses at the senior undergraduate and graduate levels. Naturally, I wish to encourage you to study the subject further and to ensure that you are prepared if you do so.

At many universities, including my own, abstract algebra is the first serious proof-based course taken by mathematics majors. While it is quite possible to get through, let us say, a course in calculus simply by memorizing a list of rules and applying them correctly, without really understanding why anything works, such an approach would be disastrous here. To be sure, you must carefully learn the definitions and the statements of theorems, but that is nowhere near sufficient. In order to master the material, you need to understand the proofs and then be able to prove things yourself. This book contains hundreds of problems, and I cannot stress strongly enough the need to solve as many of them as you can. Do not be discouraged if you cannot get all of them! Some are very difficult. But try to figure out as many as you can. You will only learn by getting your hands dirty.

As different universities have different sequences of courses, I am not assuming any prerequisites beyond the high school level. Most of the material in Part I would be covered in a typical course on discrete mathematics. Even if you have had such a course, I urge you to read through it. In particular, you absolutely must understand equivalence relations and equivalence classes thoroughly. (In my experience, many students have trouble with these concepts.) From time to time, throughout Parts II and III, some examples involving matrices or complex numbers appear. These can be bypassed if you have not studied linear algebra or complex numbers, but in any case, the material you need to know is not difficult and is discussed in the appendices. In Part IV, it is necessary to know some linear algebra, but all of the theorems used are proved in the text.

The fundamental results about groups are covered in Chaps. 3 and 4, those about rings are in Chaps. 8 and 9, and the introductory theorems concerning fields and polynomials are found in Chap. 11. I think that these chapters are essential in any course. Beyond that, there is a fair amount of flexibility in the choice of topics.

I confess my first encounter with abstract algebra was a joyous experience. I found (and still find!) the subject fascinating, and I will consider the time I put into this book well spent if you emerge with an appreciation for the field.

I would like to thank Lynn Brandon and Anne-Kathrin Birchley-Brun at Springer for their help in making this book a reality. Also, thanks to the reviewers for their many useful suggestions. I thank my wife and family for their ongoing support. Finally, thanks to my teacher, Prof. Sudarshan Sehgal, both for his advice concerning this book and for all of his help over the years.

Thunder Bay, ON, Canada

Gregory T. Lee

Contents

Part I
Preliminaries

Chapter 1
Relations and Functions

We begin by introducing some basic notation and terminology. Then we discuss relations and, in particular, equivalence relations, which we shall see several times throughout the book. In the final section, we talk about various sorts of functions.

1.1 Sets and Set Operations

A **set** is a collection of objects. We will see many sorts of sets throughout this course. Perhaps the most common will be sets of numbers. For instance, we have the set of **natural numbers**,

$$\mathbb{N} = \{1, 2, 3, \ldots\},$$

the set of **integers**,

$$\mathbb{Z} = \{\ldots, -2, -1, 0, 1, 2, \ldots\}$$

and the set of **rational numbers**

$$\mathbb{Q} = \left\{ \frac{a}{b} : a, b \in \mathbb{Z}, b \neq 0 \right\}.$$

We also write $\mathbb{R}$ for the set of **real numbers** and $\mathbb{C}$ for the set of **complex numbers**.

But sets do not necessarily consist of numbers. Indeed, we can consider the set of all letters of the alphabet, the set of all polynomials with even integers as coefficients or the set of all lines in the plane with positive slope.

The objects in a set are called its **elements**. We write $a \in S$ if a is an element of a set S. Thus, $-3 \in \mathbb{Z}$ but $-3 \notin \mathbb{N}$. The set with no elements is called the **empty set**, and denoted $\varnothing$. Any other set is said to be **nonempty**.

If S and T are sets, then we say that S is a **subset** of T, and write $S \subseteq T$, if every element of S is also an element of T. Of course, $S \subseteq S$. We say that S is a **proper** subset of T, and write $S \subsetneq T$, if $S \subseteq T$ but $S \neq T$. Thus, it is certainly true that $\mathbb{N} \subseteq \mathbb{Z}$, but we can be more precise and write $\mathbb{N} \subsetneq \mathbb{Z}$.

For any two sets S and T, their **intersection**, $S \cap T$, is the set of all elements that lie in S and T simultaneously.

Example 1.1. Let $S = \{1, 2, 3, 4, 5\}$ and $T = \{2, 4, 6, 8, 10\}$. Then $S \cap T = \{2, 4\}$.

We can extend this notion to the intersection of an arbitrary collection of sets. If I is a nonempty set and, for each $i \in I$, we have a set T_i, then we write $\bigcap_{i \in I} T_i$ for the set of elements that lie in all of the T_i simultaneously.

Example 1.2. For each $q \in \mathbb{Q}$, let $T_q = \{r \in \mathbb{R} : r < 2^q\}$. Then $\bigcap_{q \in \mathbb{Q}} T_q = \{r \in \mathbb{R} : r \leq 0\}$.

Also, for any sets S and T, their **union**, $S \cup T$, is the set of all elements that lie in S or T (or both).

Example 1.3. Using the same S and T as in Example 1.1, we have

$$S \cup T = \{1, 2, 3, 4, 5, 6, 8, 10\}.$$

Furthermore, if I is a nonempty set and we have a set T_i for each $i \in I$, then we write $\bigcup_{i \in I} T_i$ for the union of all of the T_i; that is, the set of all elements that lie in at least one of the T_i.

Example 1.4. If we use the same sets T_q as in Example 1.2, we have $\bigcup_{q \in \mathbb{Q}} T_q = \mathbb{R}$.

In addition, for any two sets S and T, the **set difference** (or **relative complement**) is the set $S \backslash T = \{a \in S : a \notin T\}$.

Example 1.5. Once again using S and T as in Example 1.1, we have $S \backslash T = \{1, 3, 5\}$.

We will need one more definition. The following construction is named after René Descartes.

Definition 1.1. Let S and T be any sets. Then the **Cartesian product** $S \times T$ is the set of all ordered pairs (s, t), with $s \in S$ and $t \in T$.

Example 1.6. Let $S = \{1, 2, 3\}$ and $T = \{2, 3\}$. Then

$$S \times T = \{(1, 2), (1, 3), (2, 2), (2, 3), (3, 2), (3, 3)\}.$$

There is also a Cartesian product of finitely many sets. For any sets $T_1, T_2, \ldots, T_n$, we let $T_1 \times T_2 \times \cdots \times T_n$ be the set of all ordered n-tuples $(t_1, t_2, \ldots, t_n)$, with $t_i \in T_i$ for all i.

Example 1.7. Let $T_1 = \{1, 2\}$, $T_2 = \{a, b\}$ and $T_3 = \{2, 3\}$. Then $T_1 \times T_2 \times T_3$ is the set

$$\{(1, a, 2), (1, a, 3), (1, b, 2), (1, b, 3), (2, a, 2), (2, a, 3), (2, b, 2), (2, b, 3)\}.$$

Exercises

1.1. Let $S = \{1, 2, 3\}$ and $T = \{3, 4\}$. Find $S \cap T$, $S \cup T$, $S \setminus T$, $T \setminus S$ and $S \times T$.

1.2. Let $R = \{a, b, c\}$, $S = \{a, c, d\}$ and $T = \{c, e, f\}$. Find $R \cap S$, $R \cap (S \setminus T)$, $S \cup T$, $S \cap (R \cup T)$ and $R \times S$.

1.3. Let R, S and T be sets with $R \subseteq S$. Show that $R \cup T \subseteq S \cup T$.

1.4. Let $S = \{1, 2, \ldots, n\}$, for some positive integer n. Show that S has 2^n subsets.

1.5. Let R, S and T be any sets. Show that $R \cup (S \cap T) = (R \cup S) \cap (R \cup T)$.

1.6. For each positive integer n, let $T_n = \{\frac{a}{n} : a \in \mathbb{Z}\}$.

1. What is $\bigcup_{n=1}^{\infty} T_n$?
2. What is $\bigcap_{n=1}^{\infty} T_n$?

1.2 Relations

We are going to use relations (in particular, the equivalence relations and functions that we will see in the next two sections) quite a few times in this course.

Definition 1.2. Let S and T be sets. Then a **relation** from S to T is a subset ρ of $S \times T$. If $s \in S$ and $t \in T$, then we write $s\rho t$ if $(s, t) \in \rho$; otherwise, we write $s \not\rho t$. In particular, a **relation** on S is a relation from S to S.

Example 1.8. Let $S = \{1, 2, 3\}$ and $T = \{1, 2, 3, 4\}$. Define a relation ρ from S to T via $s\rho t$ if and only if $st^2 \leq 4$. Then $\rho = \{(1, 1), (1, 2), (2, 1), (3, 1)\}$. In particular, $3\rho 1$ but $1 \not\rho 3$.

We will focus on relations on a set. Let us discuss a few properties enjoyed by some relations.

Definition 1.3. Let ρ be a relation on S. We say that ρ is **reflexive** if $a\rho a$ for all $a \in S$.

Example 1.9. On $\mathbb{Z}$, the relation $\leq$ is reflexive, but $<$ is not. Indeed, $a \leq a$ for all integers a, but 1 is not less than 1.

Definition 1.4. A relation ρ on a set S is **symmetric** if $a\rho b$ implies $b\rho a$.

Example 1.10. On $\mathbb{Z}$, neither $\leq$ nor $<$ is symmetric, as $1 < 2$ but 2 is not less than 1 (and similarly for $\leq$). Define ρ via $a\rho b$ if and only if $|a - b| \leq 10$. Then ρ is symmetric. Indeed, if $a\rho b$, then $|a - b| \leq 10$, and so $|b - a| = |a - b| \leq 10$; thus, $b\rho a$.

Definition 1.5. Let ρ be a relation on a set S. We say that ρ is **transitive** if, whenever $a\rho b$ and $b\rho c$, we also have $a\rho c$.

Example 1.11. On $\mathbb{Z}$, the relations $\leq$ and $<$ are both transitive. (If $a \leq b$ and $b \leq c$, then $a \leq c$.) However, the relation ρ from Example 1.10 is not, since $1\rho 8$ and $8\rho 13$, but $1 \not\rho 13$.

These three properties lead us directly to the next section.

Exercises

1.7. Let $S = \{1, 2, 3\}$ and $T = \{3, 4, 5, 6, 7, 8\}$. Define a relation ρ from S to T via $a\rho b$ if and only if $|a^2 - b| \leq 1$. Find all pairs $(a, b) \in S \times T$ such that $a\rho b$.

1.8. Define a relation ρ on $\mathbb{Z}$ via $a\rho b$ if and only if ab is even. Is ρ reflexive? Symmetric? Transitive?

1.9. Define a relation ρ on $\mathbb{R}$ via $a\rho b$ if and only if $a - b \in \mathbb{Q}$. Is ρ reflexive? Symmetric? Transitive?

1.10. Define a relation ρ on $\mathbb{R}$ via $a\rho b$ if and only if $a - b \in \mathbb{N}$. Is ρ reflexive? Symmetric? Transitive?

1.11. 1. How many relations are there on $\{1, 2, 3\}$?
2. How many of these relations are symmetric?

1.12. For each of the eight subsets of {reflexive, symmetric, transitive}, find a relation on $\{1, 2, 3\}$ that has the properties in that subset, but not the properties that are not in the subset.

1.3 Equivalence Relations

Definition 1.6. An **equivalence relation** on a set S is a relation that is reflexive, symmetric and transitive.

We will use the symbol $\sim$ to denote an equivalence relation.

Example 1.12. On $\mathbb{Z}$, let us say that $a \sim b$ if and only if $a + b$ is even. We claim that $\sim$ is an equivalence relation. If $a \in \mathbb{Z}$, then $a + a$ is certainly even, so $a \sim a$, and $\sim$ is reflexive. If $a \sim b$, then $a + b$ is even. But this also means that $b + a$ is even, and hence $b \sim a$. Thus, $\sim$ is symmetric. Finally, suppose that $a \sim b$ and $b \sim c$. Then $a + b$ and $b + c$ are both even. This means that their sum, $a + 2b + c$, is even. As $2b$ is even, we see that $a + c$ is even, and hence $a \sim c$. That is, $\sim$ is transitive.

Example 1.13. On the set $S = \{a \in \mathbb{Z} : 1 \le a \le 20\}$, let $a \sim b$ if and only if $a = 2^m b$ for some $m \in \mathbb{Z}$. Let us verify that this is an equivalence relation. Reflexivity: Note that $a = 2^0 a$, and hence $a \sim a$. Symmetry: If $a \sim b$, say $a = 2^m b$, then $b = 2^{-m} a$, and hence $b \sim a$. Transitivity: If $a \sim b$ and $b \sim c$, say $a = 2^m b$ and $b = 2^n c$, then $a = 2^{m+n} c$, and therefore $a \sim c$.

Example 1.14. On $\mathbb{R}$, let us say that $a \sim b$ if and only if $a - b \in \mathbb{Z}$. Let us check that it is an equivalence relation. Reflexivity: If $a \in \mathbb{R}$, then $a - a = 0 \in \mathbb{Z}$, and hence $a \sim a$. Symmetry: Let $a \sim b$. Then $a - b \in \mathbb{Z}$, and hence $b - a = -(a - b) \in \mathbb{Z}$. Thus, $b \sim a$. Transitivity: Suppose that $a \sim b$ and $b \sim c$. Then $a - b, b - c \in \mathbb{Z}$, and hence $a - c = (a - b) + (b - c) \in \mathbb{Z}$. That is, $a \sim c$.

Let us try something slightly more complicated.

Example 1.15. Let $S = \mathbb{Z} \times (\mathbb{Z} \setminus \{0\})$. Define $\sim$ on S via $(a, b) \sim (c, d)$ if and only if $ad = bc$. We must verify that $\sim$ is an equivalence relation. Reflexivity: As $ab = ba$, we have $(a, b) \sim (a, b)$ for all integers a and nonzero integers b. Symmetry: Suppose that $(a, b) \sim (c, d)$. Then $ad = bc$, and this also tells us that $(c, d) \sim (a, b)$. Transitivity: Let $(a, b) \sim (c, d)$ and $(c, d) \sim (e, f)$. Then $ad = bc$ and $cf = de$. Thus, $adf = bcf = bde$. Since we are assuming that $d \neq 0$, this means that $af = be$. Therefore, $(a, b) \sim (e, f)$.

Equivalence relations are very special.

Definition 1.7. Let $\sim$ be an equivalence relation on a set S. If $a \in S$, then the **equivalence class** of a, denoted $[a]$, is the set $\{b \in S : a \sim b\}$.

Why are equivalence classes so interesting? We need another definition.

Definition 1.8. Let S be a set, and let T be a set of nonempty subsets of S. We say that T is a **partition** of S if every $a \in S$ lies in exactly one set in T.

Example 1.16. Let $S = \{1, 2, 3, 4, 5, 6, 7\}$ and $T = \{\{1, 3, 4, 6\}, \{2, 7\}, \{5\}\}$. Then T is a partition of S.

What is the connection between these concepts?

Theorem 1.1. *Let S be a set, and $\sim$ an equivalence relation on S. Then the equivalence classes with respect to $\sim$ form a partition of S. In particular, if $a \in S$, then $a \in [a]$ and, furthermore, $a \in [b]$ if and only if $[a] = [b]$.*

Proof. As $\sim$ is reflexive, $a \sim a$, and hence $a \in [a]$ for every $a \in S$. In particular, the equivalence classes are not empty, and every element of S is in at least one of them. Suppose that $d \in [a] \cap [c]$. We must show that $[a] = [c]$. If $e \in [a]$, then $a \sim e$. Also, $d \in [a]$ means that $a \sim d$, and hence $d \sim a$ by symmetry. Also, $c \sim d$. By transitivity, $c \sim a$, and then $c \sim e$. Thus, $e \in [c]$, and therefore $[a] \subseteq [c]$. By the same argument, $[c] \subseteq [a]$, and hence $[a] = [c]$. Thus, the equivalence classes do indeed form a partition. To prove the final statement of the theorem, note that if $a \in [a] \cap [b]$, then $[a] = [b]$ and, conversely, if $[a] = [b]$, then $a \in [a] = [b]$. $\square$

So, the equivalence classes break the set down into subsets having no elements in common. It is important to note that, unless there is only one element in an equivalence class, the representative chosen for that class is not unique. That is, if $b \in [a]$, then we could just as easily write $[b]$ instead of $[a]$. They are the same class. This complicates matters a bit when we define operations on equivalence classes, as we will find ourselves doing throughout the course. We must make sure that our operations are well-defined; that is, that they do not depend upon the particular representative of the class that we use.

Let us discuss the equivalence classes determined by the relations in our earlier examples. The plan is always the same. We know that each element of the set is in exactly one class. Thus, we will keep looking for elements of the set that are not in any classes we have constructed, and obtain new classes in this way.

Example 1.17. In Example 1.12, let us start with 0. We know that $a \sim 0$ if and only if a is even. Thus,

$$[0] = \{\ldots, -6, -4, -2, 0, 2, 4, 6, \ldots\}.$$

(Note that we would have obtained the same class had we started, for instance, with 14. Since $14 \in [0]$, we have $[0] = [14]$.) We have not yet found 1, so we note that $a \sim 1$ if and only if $a + 1$ is even; that is, if and only if a is odd. Therefore,

$$[1] = \{\ldots, -5, -3, -1, 1, 3, 5, \ldots\}.$$

(Again, we could just as easily have used $[-3]$.) We have now found all elements of $\mathbb{Z}$. Thus, there are only two equivalence classes, $[0]$ and $[1]$.

Example 1.18. In Example 1.13, we may as well start with 1. We have

$$[1] = \{1, 2, 4, 8, 16\}.$$

As we have not yet found 3,

$$[3] = \{3, 6, 12\}.$$

We still do not have 5, and thus we take

$$[5] = \{5, 10, 20\}.$$

Similarly, we obtain

$$[7] = \{7, 14\}, \ [9] = \{9, 18\}, \ [11] = \{11\},$$
$$[13] = \{13\}, [15] = \{15\}, [17] = \{17\}, \text{ and } [19] = \{19\}.$$

Once again, we could have used $[8]$ in place of $[1]$, for instance.

The other two examples are a bit trickier, since there are infinitely many equivalence classes. But we can attempt to describe them.

Example 1.19. In Example 1.14, we see that $b \in [a]$ if and only if the difference between a and b is an integer. Thus, for instance,

$$[23.86] = \{\dots, -2.14, -1.14, -0.14, 0.86, 1.86, 2.86, \dots\}.$$

Listing the classes is an impossible task. How, then, to describe them? We note that for any real number a, there is certainly an integer k such that $0 \le a - k < 1$. Now, $a \sim (a - k)$, and hence every element of $\mathbb{R}$ is in a class $[b]$, for some $0 \le b < 1$. Furthermore, if $0 \le b, c < 1$, then $0 \le |b - c| < 1$ and therefore $b - c$ can only be an integer if $b = c$. That is, if $0 \le b, c < 1$ and $b \ne c$, then $[b] \ne [c]$. Thus, the equivalence classes are precisely

$$\{[b] : b \in \mathbb{R}, \ 0 \le b < 1\}.$$

Example 1.20. What about Example 1.15? We note that $(c, d) \in [(a, b)]$ if and only if $ad = bc$. Another way to say this is that $\frac{a}{b} = \frac{c}{d}$. Thus, $[(a, b)]$ consists of all ordered pairs (c, d), with $c, d \in \mathbb{Z}$ and $d \ne 0$, such that $\frac{a}{b} = \frac{c}{d}$. This is, in fact, exactly how the rational numbers are constructed! We need to ensure that $\frac{2}{3}$ and $\frac{4}{6}$ are treated as the same fraction, and these equivalence classes make that happen. We obtain one equivalence class for each fraction $\frac{a}{b}$. For instance,

$$[(2, 3)] = \{\dots, (-6, -9), (-4, -6), (-2, -3), (2, 3), (4, 6), (6, 9), \dots\}.$$

Exercises

1.13. Define a relation $\sim$ on $\mathbb{N}$ via $a \sim b$ if and only if $a - b = 3k$, for some $k \in \mathbb{Z}$. Is $\sim$ an equivalence relation? If so, what are the equivalence classes?

1.14. Define a relation $\sim$ on $\{1, 2, 3, 4, 5, 6, 7\}$ via $a \sim b$ if and only if a and b are both even or both odd. Is $\sim$ an equivalence relation? If so, what are the equivalence classes?

1.15. Define a relation $\sim$ on $\mathbb{Z}$ via $a \sim b$ if and only if $|a| = |b|$. Is $\sim$ an equivalence relation? If so, what are the equivalence classes?

1.16. Define a relation $\sim$ on $\mathbb{Z}$ via $a \sim b$ if and only if $ab > 0$. Is $\sim$ an equivalence relation? If so, what are the equivalence classes?

1.17. Let S be the set of all subsets of $\mathbb{Z}$. Define a relation $\sim$ on S via $T \sim U$ if and only if $T \subseteq U$. Is $\sim$ an equivalence relation? If so, what are the equivalence classes?

1.18. Let S be the set of all subsets of $\mathbb{Z}$. Define a relation $\sim$ on S via $T \sim U$ if and only if $T \setminus U$ and $U \setminus T$ are both finite. Show that $\sim$ is an equivalence relation and describe $[\{1, 2, 3\}]$ and $[\{\dots, -4, -2, 0, 2, 4, \dots\}]$.

1.19. On the plane $\mathbb{R}^2$, define a relation $\sim$ via $(a, b) \sim (c, d)$ if and only if $3a - b = 3c - d$. Show that $\sim$ is an equivalence relation, and describe $[(4, 2)]$.

1.20. Let S be a nonempty set. Show that for any partition of S, there is an equivalence relation on S having the sets in the partition as its equivalence classes.

1.21. Find an equivalence relation on $\mathbb{N}$ having exactly two equivalence classes, one of which contains exactly three elements.

1.22. Suppose there is a relation ρ on a set S, such that ρ is both reflexive and transitive. Define $\sim$ on S via $a \sim b$ if and only if $a\rho b$ and $b\rho a$. Show that $\sim$ is an equivalence relation.

1.4 Functions

Let us give two equivalent definitions of a function. Formally, if S and T are sets, then a function from S to T is a relation ρ from S to T such that, for each $s \in S$, there is exactly one $t \in T$ such that $s\rho t$. In practice, nobody really thinks of functions in this way. The working definition follows.

Definition 1.9. Let S and T be any sets. Then a **function** $\alpha : S \to T$ is a rule assigning, to each $s \in S$, an element $\alpha(s)$ of T.

Readers who have studied calculus will no doubt be familiar with functions from $\mathbb{R}$ to $\mathbb{R}$.

Example 1.21. We can define a function $\alpha : \mathbb{R} \to \mathbb{R}$ via $\alpha(a) = 5a^3 - 4a^2 + 7a + 3$ for all $a \in \mathbb{R}$.

But we do not need to go from $\mathbb{R}$ to $\mathbb{R}$.

Example 1.22. We can define a function $\alpha : \mathbb{Z} \to \mathbb{Q}$ via $\alpha(a) = (-2)^a$ for all $a \in \mathbb{Z}$.

In fact, the sets involved do not have to be sets of numbers.

Example 1.23. Let S be the set of all English words and T the set of letters in the alphabet. We can define $\alpha : S \to T$ by letting $\alpha(w)$ be the first letter of the word w, for every $w \in S$.

A few properties enjoyed by certain functions are important.

Definition 1.10. A function $\alpha : S \to T$ is **one-to-one** (or **injective**) if $\alpha(s_1) = \alpha(s_2)$ implies $s_1 = s_2$, for all $s_1, s_2 \in S$.

Putting this another way, a one-to-one function sends different elements to different places.

Example 1.24. Define functions α and β from $\mathbb{R}$ to $\mathbb{R}$ via $\alpha(a) = a^2$ and $\beta(a) = a^3$, for all $a \in \mathbb{R}$. Then α is not one-to-one, since $\alpha(1) = \alpha(-1)$, but β is one-to-one, since if $a^3 = b^3$, then taking the cube root of both sides, we have $a = b$, for any $a, b \in \mathbb{R}$.

Definition 1.11. A function $\alpha : S \to T$ is **onto** (or **surjective**) if, for every $t \in T$, there exists at least one $s \in S$ such that $\alpha(s) = t$.

Example 1.25. Define α and β as in Example 1.24. Then α is not onto, since there is no $a \in \mathbb{R}$ such that $\alpha(a) = -1$. However, if $b \in \mathbb{R}$, then $\beta(\sqrt[3]{b}) = b$; thus, β is onto.

We should not get the idea that one-to-one and onto always occur together.

Example 1.26. Define $\alpha : \mathbb{R} \to \mathbb{R}$ via $\alpha(a) = 2^a$. Then α is one-to-one, for if $2^a = 2^b$, then taking the base 2 logarithm of both sides, we see that $a = b$. On the other hand, there is no $a \in \mathbb{R}$ such that $2^a = -1$, so α is not onto.

However, it is nice when we can combine the two properties.

Definition 1.12. A function $\alpha : S \to T$ is **bijective** if it is one-to-one and onto.

An equivalent way of expressing this property is that for each $t \in T$, there is exactly one $s \in S$ such that $\alpha(s) = t$. There must be such an s, since α is onto, but if $\alpha(s_1) = \alpha(s_2) = t$, for some $s_1, s_2 \in S$, then since α is one-to-one, $s_1 = s_2$. For this reason, a bijective function is also known as a **one-to-one correspondence**.

Example 1.27. Combining Examples 1.24 and 1.25, we see that $\alpha : \mathbb{R} \to \mathbb{R}$ given by $\alpha(a) = a^3$ is bijective.

Let us discuss how to combine functions.

Definition 1.13. Let R, S and T be sets, and let $\alpha : R \to S$ and $\beta : S \to T$ be functions. Then the **composition**, $\beta \circ \alpha$, or simply $\beta\alpha$, is the function from R to T given by $(\beta\alpha)(r) = \beta(\alpha(r))$ for all $r \in R$.

Note that when we write $\beta\alpha$, we are applying α first, then β. The order is important! Indeed, depending upon the sets involved, it is possible that applying β first, then α, would not make sense. But even if it did make sense, the result would not necessarily be the same.

Example 1.28. Define functions α and β from $\mathbb{R}$ to $\mathbb{R}$ via $\alpha(a) = a^3 + 1$ and $\beta(a) = a^2$, for all $a \in \mathbb{R}$. Then $(\beta\alpha)(a) = \beta(a^3 + 1) = a^6 + 2a^3 + 1$, whereas $(\alpha\beta)(a) = \alpha(a^2) = a^6 + 1$, for all $a \in \mathbb{R}$. That is, $\beta\alpha$ and $\alpha\beta$ are different functions.

We can list a few important properties of the composition of functions.

Theorem 1.2. *Let $\alpha : R \to S$, $\beta : S \to T$ and $\gamma : T \to U$ be functions. Then*

1. *$(\gamma\beta)\alpha = \gamma(\beta\alpha)$;*
2. *if α and β are one-to-one, then so is $\beta\alpha$;*

3. *if α and β are onto, then so is βα; and*
4. *if α and β are bijective, then so is βα.*

Proof. (1) Take any $r \in R$. Then $((\gamma\beta)\alpha)(r) = (\gamma\beta)(\alpha(r)) = \gamma(\beta(\alpha(r)))$. Similarly, $(\gamma(\beta\alpha))(r) = \gamma((\beta\alpha)(r)) = \gamma(\beta(\alpha(r)))$.

(2) Suppose that $(\beta\alpha)(r_1) = (\beta\alpha)(r_2)$ for some $r_1, r_2 \in R$. Then $\beta(\alpha(r_1)) = \beta(\alpha(r_2))$. Since β is one-to-one, $\alpha(r_1) = \alpha(r_2)$. Since α is one-to-one, $r_1 = r_2$.

(3) Take any $t \in T$. Since β is onto, there exists an $s \in S$ such that $\beta(s) = t$. But α is also onto, so there exists an $r \in R$ such that $\alpha(r) = s$. Thus, $(\beta\alpha)(r) = \beta(\alpha(r)) = \beta(s) = t$.

(4) Combine (2) and (3). □

The following additional property of bijective functions can be useful.

Theorem 1.3. *Let $\alpha : S \to T$ be a bijective function. Then there exists a bijective function $\beta : T \to S$ such that $(\beta\alpha)(s) = s$ for all $s \in S$ and $(\alpha\beta)(t) = t$ for all $t \in T$.*

Proof. Since α is bijective, for any $t \in T$, there is a unique $s \in S$ such that $\alpha(s) = t$. Define $\beta : T \to S$ via $\beta(t) = s$. By definition, we have $(\beta\alpha)(s) = \beta(\alpha(s)) = s$, for all $s \in S$. Also, if $t \in T$, then choosing s such that $\alpha(s) = t$, we have $\beta(t) = s$, and therefore $(\alpha\beta)(t) = \alpha(\beta(t)) = \alpha(s) = t$, as required. It remains to show that β is bijective. But if $\beta(t_1) = \beta(t_2)$, then

$$t_1 = (\alpha\beta)(t_1) = \alpha(\beta(t_1)) = \alpha(\beta(t_2)) = (\alpha\beta)(t_2) = t_2,$$

so β is one-to-one. Furthermore, if $s \in S$, then $\beta(\alpha(s)) = (\beta\alpha)(s) = s$, and hence β is onto. □

Example 1.29. Let $\alpha : \mathbb{R} \to \mathbb{R}$ be given by $\alpha(a) = a^3$ for all a. Example 1.27 showed us that α is bijective. It is easily checked that if we let $\beta : \mathbb{R} \to \mathbb{R}$ be given by $\beta(a) = \sqrt[3]{a}$ for all a, then $(\alpha\beta)(a) = (\beta\alpha)(a) = a$ for all a.

We close the chapter by defining two special types of functions.

Definition 1.14. A **permutation** of a set S is a bijective function from S to S.

Example 1.30. By Example 1.27, the function $\alpha : \mathbb{R} \to \mathbb{R}$ given by $\alpha(a) = a^3$ is a permutation of $\mathbb{R}$.

Example 1.31. Let $S = \{1, 2, 3, 4\}$. Define $\alpha : S \to S$ via $\alpha(1) = 3$, $\alpha(2) = 2$, $\alpha(3) = 4$ and $\alpha(4) = 1$. Then α is a permutation of S.

As this last example illustrates, a permutation is simply a rearrangement of the elements of S.

Definition 1.15. Let S be a set. Then a **binary operation** on S is a function from $S \times S$ to S.

Example 1.32. We can define a binary operation $*$ on $\mathbb{R}$ via $a * b = 2a^2b - 3b^4 + 5$, for all $a, b \in \mathbb{R}$. (Putting this in terms of functions, we could write $\alpha((a, b)) = 2a^2b - 3b^4 + 5$ for all $a, b \in \mathbb{R}$.)

Note that in order to obtain a binary operation, we must stay within our original set. For instance, we would not get a binary operation on $\mathbb{N}$ if we tried to let $a * b = \frac{a}{b}$, for the simple reason that $1 * 2 = \frac{1}{2} \notin \mathbb{N}$.

Exercises

1.23. Define $\alpha : \{1, 2, 3, 4\} \to \{1, 2, 3, 4, 5, 6, 7\}$ via $\alpha(a) = 2a - 1$. Is this function one-to-one? Is it onto?

1.24. Define $\alpha : \mathbb{R} \to \mathbb{R}$ via $\alpha(a) = \sqrt[3]{a + 1} - 2$. Is this function one-to-one? Is it onto?

1.25. Let S be the set of real numbers and T the set of positive real numbers. Define $\alpha : S \to T$ via $\alpha(a) = 2^{3a-5}$. Show that α is a bijection and find $\beta : T \to S$ such that $(\beta\alpha)(a) = a$ for all $a \in S$.

1.26. Define $\alpha : \mathbb{R} \to \mathbb{R}$ via

$$\alpha(a) = \begin{cases} 4a - 3, & a \leq 1 \\ a^2, & a > 1. \end{cases}$$

Show that α is bijective and find $\beta : \mathbb{R} \to \mathbb{R}$ such that $(\beta\alpha)(a) = a$ for all $a \in \mathbb{R}$.

1.27. Which of the following are binary operations on $\mathbb{N}$?

1. $a * b = ab$
2. $a * b = a - b$
3. $a * b = 3$ for all a and b

1.28. Let S be a finite set, and suppose that $\alpha : S \to S$ is a one-to-one function. Show that α is a permutation of S. Construct an explicit counterexample to show that this need not be true if S is infinite.

1.29. Let $\alpha : R \to S$ and $\beta : S \to T$ be functions, and suppose that $\beta\alpha$ is onto. Must α be onto? Must β?

1.30. Let $\alpha : R \to S$ and $\beta : S \to T$ be functions, and suppose that $\beta\alpha$ is one-to-one. Must α be one-to-one? Must β?

1.31. Let S be a set with m elements and T a set with n elements, for some positive integers m and n.

1. How many functions are there from S to T?
2. How many of these functions are one-to-one?

1.32. Let S and T be sets and $\alpha : S \to T$ a function. Show that there exist a set R and functions $\beta : S \to R$ and $\gamma : R \to T$ such that β is onto, γ is one-to-one and $\alpha = \gamma\beta$.

Chapter 2
The Integers and Modular Arithmetic

In this chapter, we begin with a discussion of mathematical induction. Next, we examine a number of properties of the integers, with an emphasis on divisibility and prime factorization. We conclude by introducing modular arithmetic.

2.1 Induction and Well Ordering

We begin with an important property of the set of natural numbers.

Property 2.1 (Well Ordering Axiom). If S is a nonempty set of positive integers, then S has a smallest element.

This seems so obvious, but it is actually a rather special property of $\mathbb{N}$. Indeed, $\mathbb{Z}$ has no smallest element; neither, for that matter, does the set of positive real numbers.

There is an equivalent form of the Well Ordering Axiom that is especially useful. To state it, we need a definition. A **proposition** is a statement that is either true or false. For instance, "Ottawa is the capital of Canada" is a true proposition, and "There are only finitely many even integers" is a false one. We avoid statements having no truth value, such as "This statement is false" as well as statements that are a matter of opinion, such as "*Xena: Warrior Princess* was a great television program"[1]. What we would like to do is define a sequence of propositions, $P(1)$, $P(2)$, $P(3)$ and so on, and prove that all of them are true at once. This is where induction comes in.

Theorem 2.1 (Principle of Mathematical Induction). *Suppose that, for each positive integer n, we have a proposition $P(n)$. Further suppose that*

---

[1]Of course, any reasonable person would agree with this statement, but in principle, it is a matter of opinion.

© Springer International Publishing AG, part of Springer Nature 2018
G. T. Lee, *Abstract Algebra*, Springer Undergraduate Mathematics Series,
https://doi.org/10.1007/978-3-319-77649-1_2

1. *P(1) is true; and*
2. *for each n ∈ ℕ, if P(n) is true, then so is P(n + 1).*

Then P(n) is true for every positive integer n.

Proof. Suppose the theorem is false, and let S be the set of all positive integers n such that $P(n)$ is false. Then S is a nonempty subset of $\mathbb{N}$. By the Well Ordering Axiom, S has a smallest element k. Now, we are assuming that $P(1)$ is true, so $k > 1$. Then $k - 1 \notin S$, and hence $P(k - 1)$ is true. By our assumption, $P(k)$ is true as well, giving us a contradiction and completing the proof. ☐

Induction is a powerful tool! We can prove infinitely many propositions in just two steps. Here is a simple example.

Example 2.1. We claim that for every positive integer n, we have

$$1^2 + 2^2 + \cdots + n^2 = \frac{n(n + 1)(2n + 1)}{6}.$$

We proceed by induction. For each $n \in \mathbb{N}$, the proposition $P(n)$ is the statement

$$1^2 + 2^2 + \cdots + n^2 = \frac{n(n + 1)(2n + 1)}{6}.$$

First, we must prove $P(1)$. But it states that

$$1^2 = \frac{1(1 + 1)(2 \cdot 1 + 1)}{6},$$

which is obvious. Now, we assume $P(n)$ and prove $P(n + 1)$. But

$$1^2 + 2^2 + \cdots + n^2 + (n + 1)^2 = \frac{n(n + 1)(2n + 1)}{6} + (n + 1)^2,$$

by our inductive hypothesis, $P(n)$. Simplifying, we have

$$
\begin{aligned}
1^2 + 2^2 + \cdots + (n + 1)^2 &= \frac{(n + 1)(n(2n + 1) + 6(n + 1))}{6} \\
&= \frac{(n + 1)(2n^2 + 7n + 6)}{6} \\
&= \frac{(n + 1)(n + 2)(2n + 3)}{6} \\
&= \frac{(n + 1)((n + 1) + 1)(2(n + 1) + 1)}{6}.
\end{aligned}
$$

But this is precisely $P(n + 1)$. Thus, the proof is complete.

There is another result that we can prove by induction, and which we will need later. A bit of notation is required. For any positive integer n, we define $n!$ (read "n factorial") via $n! = n(n-1)(n-2)\cdots(2)(1)$. Also, $0! = 1$. If n and k are integers, with $n \geq k \geq 0$, then we define $\binom{n}{k}$ (read "n choose k") via $\binom{n}{k} = \frac{n!}{(n-k)!k!}$.

Example 2.2. We have $5! = 5 \cdot 4 \cdot 3 \cdot 2 \cdot 1 = 120$ and $\binom{6}{2} = \frac{6!}{4!2!} = \frac{720}{24 \cdot 2} = 15$.

Theorem 2.2 (Binomial Theorem). *Let a and b be real numbers and n a positive integer. Then*

$$(a+b)^n = a^n + \binom{n}{1}a^{n-1}b + \binom{n}{2}a^{n-2}b^2 + \cdots + \binom{n}{n-1}ab^{n-1} + b^n.$$

Proof. Let us proceed by induction on n. When $n = 1$, both sides of the equation are $a + b$, so there is nothing to do. Assume the result for n, and prove it for $n + 1$. But

$$(a+b)^{n+1} = (a+b)^n(a+b)$$
$$= \left(a^n + \binom{n}{1}a^{n-1}b + \cdots + \binom{n}{n-1}ab^{n-1} + b^n\right)(a+b),$$

by our inductive hypothesis.

When we expand this product, we obtain a sum of terms consisting of a coefficient multiplied by $a^{n+1-k}b^k$, where $0 \leq k \leq n+1$. The coefficients of a^{n+1} and b^{n+1} are clearly 1, whereas if $0 < k < n+1$, then the coefficient of $a^{n+1-k}b^k$ is $\binom{n}{k} + \binom{n}{k-1}$, since these terms arise from $(\binom{n}{k}a^{n-k}b^k)a$ and $(\binom{n}{k-1}a^{n-(k-1)}b^{k-1})b$. However,

$$\binom{n}{k} + \binom{n}{k-1} = \frac{n!}{(n-k)!k!} + \frac{n!}{(n-k+1)!(k-1)!}$$
$$= \frac{n!(n-k+1) + n!k}{(n-k+1)!k!}$$
$$= \binom{n+1}{k}.$$

That is,

$$(a+b)^{n+1} = a^{n+1} + \binom{n+1}{1}a^n b + \cdots + \binom{n+1}{n}ab^n + b^{n+1},$$

and the proof is complete. $\square$

Sometimes, a slightly different form of induction is required.

Theorem 2.3 (Strong Induction). *Suppose that, for each positive integer n, we have a proposition $P(n)$. Further suppose that*

1. *P(1) is true; and*
2. *for each integer $n > 1$, if $P(k)$ is true for every $k < n$, then $P(n)$ is true.*

Then $P(n)$ is true for every positive integer n.

Proof. Suppose that the theorem is false, and let S be the set of positive integers n such that $P(n)$ is false. Then S is a nonempty subset of $\mathbb{N}$. By the Well Ordering Axiom, it has a smallest element j. As $P(1)$ is true, $j > 1$. But then by the minimality of j, we see that $P(k)$ is true whenever $k < j$. Thus, $P(j)$ is true, giving us a contradiction.   □

As before, we must prove the first proposition. But after that, instead of just assuming that the previous case is true, we assume that all prior cases are true. This can give us more to work with.

Example 2.3. Define a sequence via $a_1 = 1$, $a_2 = 3$, $a_3 = 7$ and, for each $n \geq 4$, $a_n = a_{n-1} + a_{n-2} + a_{n-3}$. We claim that $a_n < 2^n$ for all $n \in \mathbb{N}$. We need strong induction here, because when we consider a_n, we require information not just about a_{n-1}, but about the terms before it as well. When $n = 1$, there is nothing to do. Assume that $n > 1$ and that the claim is true for smaller values of n. If $n = 2$ or 3, again, the result is obvious, so assume that $n \geq 4$. Then $a_n = a_{n-1} + a_{n-2} + a_{n-3} < 2^{n-1} + 2^{n-2} + 2^{n-3}$, by our inductive hypothesis. However, $2^{n-1} + 2^{n-2} + 2^{n-3} = 7 \cdot 2^{n-3} < 2^n$. We are done.

Exercises

2.1. Show that for every positive integer n,

$$1 + 2 + \cdots + n = \frac{n(n+1)}{2}.$$

2.2. Show that for every positive integer n,

$$1 \cdot 2 + 2 \cdot 3 + \cdots + n(n+1) = \frac{n(n+1)(n+2)}{3}.$$

2.3. Show that for every positive integer n, the following two identities hold.

1.

$$\binom{n}{0} + \binom{n}{1} + \binom{n}{2} + \cdots + \binom{n}{n} = 2^n$$

2.

$$\binom{n}{0} - \binom{n}{1} + \binom{n}{2} - \binom{n}{3} + \cdots + (-1)^n \binom{n}{n} = 0$$

2.4. In the plane $\mathbb{R}^2$, let us draw n lines, no two of which are parallel and no three of which meet at a point. Into how many regions do they divide the plane?

2.5. Show that for all integers $n \geq 2$, we have

1. $(1+a)^n > 1 + na$, for all positive real numbers a; and
2. $\sqrt[n]{n} < 2 - \frac{1}{n}$.

2.6. Show that $\binom{2n}{n}$ is less than 4^{n-1} for all positive integers $n \geq 5$.

2.7. We define the Fibonacci sequence via $f_1 = f_2 = 1$, and if $n > 2$, then $f_n = f_{n-1} + f_{n-2}$. Show that, for every positive integer n, $f_n \leq (7/4)^{n-1}$.

2.8. With f_n as in the preceding exercise, show that for every positive integer n,

$$f_n = \frac{\left(\frac{1+\sqrt{5}}{2}\right)^n - \left(\frac{1-\sqrt{5}}{2}\right)^n}{\sqrt{5}}.$$

2.9. A bar of chocolate is a rectangular array consisting of r rows and c columns of unit square chocolate pieces, with thin lines separating the rows and columns. A single action consists of taking one bar, and breaking it along a line separating two rows or two columns, producing two smaller bars. Show that it will take precisely $rc - 1$ such actions to turn the bar into rc square pieces. (This can be done using strong induction, or with no induction at all.)

2.10. Show that for every positive integer n, there exist a positive integer k, and integers $a_i \in \{0, 1\}$, such that $n = a_0 + 2a_1 + 2^2 a_2 + 2^3 a_3 + \cdots + 2^k a_k$.

2.2 Divisibility

The following theorem simply formalizes the usual division process in the integers.

Theorem 2.4 (Division Algorithm). *Let $a, b \in \mathbb{Z}$ with $b > 0$. Then there exist unique integers q and r such that $a = bq + r$, with $0 \leq r < b$.*

Proof. We will prove the existence of q and r first, and then worry about their uniqueness. Let $S = \{a - bt : t \in \mathbb{Z}, a - bt \geq 0\}$. If $0 \in S$, then $a - bq = 0$ for some $q \in \mathbb{Z}$, and hence $a = bq + 0$, as desired. Therefore, we may assume that $S \subseteq \mathbb{N}$. We claim that S is nonempty. Let $t = -|a|$. Then $a - bt = a + |a|b$. If $a \geq 0$, then $a + |a|b \geq 0$, since $b > 0$. If $a < 0$, then $a + |a|b = a(1 - b)$. But $a < 0$ and since $b \geq 1$, we have $1 - b \leq 0$. Thus, $a(1 - b) \geq 0$. Either way, the claim is proved.

In view of the Well Ordering Axiom, S has a least element, say $r = a - bq$. By definition, $r \geq 0$. If $r \geq b$, then $0 \leq r - b < r$, but also $r - b = a - bq - b = a - b(q + 1)$, and therefore $a - b(q + 1)$ is a smaller element of S than r, contradicting the choice of r. Thus, $a = bq + r$, with $0 \leq r < b$.

As to uniqueness, suppose that $a = bq_1 + r_1 = bq_2 + r_2$, with $q_i, r_i \in \mathbb{Z}$ and $0 \leq r_i < b$. Then $b(q_1 - q_2) = r_2 - r_1$. In particular, $b|q_1 - q_2| = |r_2 - r_1|$. If $q_1 \neq q_2$, then $b|q_1 - q_2| \geq b$. But $0 \leq r_1, r_2 < b$, so $|r_2 - r_1| < b$, which is impossible. Therefore, $q_1 = q_2$. But then $r_1 = r_2$ as well. $\square$

We call q and r in the preceding theorem the **quotient** and **remainder** respectively.

Example 2.4. Using $b = 5$, we have $68 = 5(13) + 3$ and $-21 = 5(-5) + 4$.

The case in which the remainder is 0 is of particular interest.

Definition 2.1. Let a and b be integers. We say that a **divides** b (or b is a **multiple** of a) if there exists an integer c such that $b = ac$. In this case, we write $a|b$.

Example 2.5. As $84 = 6(14)$ and $84 = -3(-28)$, we write $6|84$ and $-3|84$. On the other hand, $10 \nmid 84$.

Here are a few basic properties of divisibility, the proofs of which are left as Exercise 2.14.

Lemma 2.1. *Let $a, b, c \in \mathbb{Z}$. Then*

1. *if $a|b$ and $b|c$, then $a|c$;*
2. *if $a|b$ and $b \neq 0$, then $a \leq |b|$; and*
3. *if $a|b$ and $a|c$, then $a|(bu + cv)$ for any $u, v \in \mathbb{Z}$.*

Definition 2.2. Let a and b be integers, not both 0. Then the **greatest common divisor** (or **gcd**) of a and b, written (a, b), is the largest positive integer g such that $g|a$ and $g|b$.

Example 2.6. We have $(60, 170) = 10$ and $(42, -55) = 1$.

Note that the gcd must always exist. As 1 divides everything, a and b must have a common divisor. Also, by Lemma 2.1, if $a \neq 0$, then $(a, b) \leq |a|$. Thus, only the numbers from 1 to $|a|$ need to be considered. We specifically exclude the case $a = b = 0$, since everything divides 0.

Let us mention a couple of easy facts about gcds.

Lemma 2.2. *Take any integers a and b with $a \neq 0$. Then*

1. *$(a, b) = (-a, b)$; and*
2. *$(a, 0) = |a|$.*

Proof. (1) Any divisor of a also divides $-a$, and vice versa.

(2) Clearly $|a|$ divides both a and 0, and Lemma 2.1 shows that no larger integer can do so. $\square$

One particular case is important.

Definition 2.3. Let $a, b \in \mathbb{Z}$, not both 0. Then we say that a and b are **relatively prime** if $(a, b) = 1$.

Example 2.7. By Example 2.6, 60 and 170 are not relatively prime, but 42 and -55 are.

Why is the gcd so significant? The following theorem gives us an idea.

Theorem 2.5. *Let a and b be integers, not both 0. Then there exist $u, v \in \mathbb{Z}$ such that $(a, b) = au + bv$. Furthermore, (a, b) is the smallest positive integer that can be written in this way.*

Proof. Let $S = \{ax + by : x, y \in \mathbb{Z}, ax + by > 0\}$. Clearly $S \subseteq \mathbb{N}$. Without loss of generality, we may assume that $a \neq 0$. Then $a^2 + b(0) = a^2 \in S$, and hence S is not empty. By the Well Ordering Axiom, S has a least element, say $g = au + bv$. We claim that $g = (a, b)$. This will complete the proof.

Suppose that $c|a$ and $c|b$. By Lemma 2.1, $c|g$ and, hence, $c \leq g$. It remains only to show that g divides both a and b. Using the division algorithm, write $a = gq + r$, where q and r are integers and $0 \leq r < g$. Then

$$r = a - gq = a - (au + bv)q = a(1 - uq) + b(-vq).$$

Thus, if $r > 0$, then $r \in S$. But $r < g$, contradicting the minimality of g. Therefore, $r = 0$ and $g|a$. By the same argument, $g|b$. □

The following is an immediate consequence.

Corollary 2.1. *Let $a, b \in \mathbb{Z}$, not both 0. Then a and b are relatively prime if and only if there exist integers u and v such that $au + bv = 1$.*

We can now prove a couple of useful results for relatively prime numbers.

Corollary 2.2. *Let $a, b, c \in \mathbb{Z}$ with a and b not both 0. If $(a, b) = 1$ and $a|bc$, then $a|c$.*

Proof. By the preceding corollary, we may write $au + bv = 1$, for some u, $v \in \mathbb{Z}$. Then $acu + bcv = c$. But $a|a$ and $a|bc$ hence, by Lemma 2.1, $a|(acu + bcv) = c$. □

Corollary 2.3. *Let $a, b \in \mathbb{Z}$, not both 0. If a and b are relatively prime, and for some integer n, we have $a|n$ and $b|n$, then $ab|n$.*

Proof. See Exercise 2.18. □

Be careful not to apply the last two corollaries if a and b are not relatively prime! For instance, $6|4 \cdot 3$, but $6 \nmid 4$ and $6 \nmid 3$. Also, $4|12$ and $6|12$ but $24 \nmid 12$.

What we have not yet discussed is how to find (a, b) and the numbers u and v from Theorem 2.5. We could certainly list the common divisors of a and b and see which one is largest, but if the numbers are large, this would be rather time-consuming. It would also give us no insight into finding u and v. Happily, there is a better way. The following technique is attributed to the ancient Greek mathematician Euclid.

Theorem 2.6 (Euclidean Algorithm). *Let a and b be integers, with b positive. If* $b|a$, *then* $(a, b) = b$. *Otherwise, apply the division algorithm repeatedly. Let*

$$a = bq_1 + r_1$$
$$b = r_1 q_2 + r_2$$
$$r_1 = r_2 q_3 + r_3$$
$$\vdots$$
$$r_{k-2} = r_{k-1} q_k + r_k$$
$$r_{k-1} = r_k q_{k+1} + 0,$$

where $q_i, r_i \in \mathbb{Z}$ *for all i and* $0 < r_k < r_{k-1} < \cdots < r_1 < b$. *Then* $(a, b) = r_k$.

Proof. If $b|a$, then b is a common divisor of a and b. In view of Lemma 2.1, it is the largest possible common divisor. Assume that $b \nmid a$. Note that we will only apply the division algorithm finitely many times, as each $r_{i+1} < r_i$, and all are positive. Suppose that $c|a$ and $c|b$. By Lemma 2.1, $c|(a - bq_1) = r_1$. Thus, every common divisor of a and b is also a common divisor of b and r_1. But if $d|b$ and $d|r_1$, then $d|(bq_1 + r_1) = a$. That is, the common divisors of a and b are precisely the same as those of b and r_1. In particular, $(a, b) = (b, r_1)$. But by exactly the same argument,

$$(a, b) = (b, r_1) = (r_1, r_2) = (r_2, r_3) = \cdots = (r_k, 0) = r_k,$$

by Lemma 2.2. $\qquad\square$

We do require b to be positive in the Euclidean algorithm, but we can use the fact that $(a, b) = (-a, b)$ if neither a nor b is positive.

In fact, the Euclidean algorithm is doubly useful, because if we start with the penultimate equation and work our way backwards, we can find integers u and v such that $(a, b) = au + bv$. Indeed, we have

$$(a, b) = r_k = r_{k-2}(1) + r_{k-1}(-q_k),$$

and so (a, b) is a multiple of r_{k-2} plus a multiple of r_{k-1}. But then

$$r_{k-1} = r_{k-3} - r_{k-2} q_{k-1}$$

and substitution yields

$$(a, b) = r_{k-2}(1) + (r_{k-3} - r_{k-2} q_{k-1})(-q_k) = r_{k-2}(1 + q_{k-1} q_k) + r_{k-3}(-q_k).$$

That is, (a, b) is a multiple of r_{k-3} plus a multiple of r_{k-2}. Eventually, we will write it in the desired form.

Example 2.8. Let $a = 45$ and $b = 33$. Applying the Euclidean algorithm, we have

$$45 = 33(1) + 12$$
$$33 = 12(2) + 9$$
$$12 = 9(1) + 3$$
$$9 = 3(3) + 0.$$

Thus, $(a, b) = 3$. Let us find u and v such that $au + bv = 3$. We have

$$3 = 12(1) + 9(-1)$$
$$= 12(1) + (33(1) + 12(-2))(-1)$$
$$= 12(3) + 33(-1)$$
$$= (45(1) + 33(-1))(3) + 33(-1)$$
$$= 45(3) + 33(-4).$$

That is, $(a, b) = 3a - 4b$.

Exercises

2.11. In each case, use the Euclidean algorithm to find (a, b).

1. $a = 57, b = 20$
2. $a = 117, b = 51$

2.12. For each of the two parts of the preceding problem, find integers u and v such that $(a, b) = au + bv$.

2.13. Let a and b be integers such that $a|b$ and $b|a$. Show that $a \in \{b, -b\}$.

2.14. Prove Lemma 2.1.

2.15. Show that if a, b and c are positive integers, with $(a, b) = 1$, and $c|a$, then $(c, b) = 1$.

2.16. Show that $n^5 - n$ is divisible by 5 for every positive integer n.

2.17. Let a and n be positive integers. Show that there exists an integer u such that $n|(au - 1)$ if and only if a and n are relatively prime.

2.18. Let $a, b \in \mathbb{Z}$, not both 0. If a and b are relatively prime, and for some integer n, we have $a|n$ and $b|n$, show that $ab|n$.

2.19. Take f_n as in Exercise 2.7. Show that $3|f_n$ if and only if $4|n$.

2.20. Take f_n as in Exercise 2.7. Show that $4|f_n$ if and only if $6|n$.

2.3   Prime Factorization

Prime numbers will have a special importance throughout the course.

Definition 2.4. A natural number $p > 1$ is said to be **prime** if its only positive divisors are 1 and p. Otherwise, it is **composite**.

Note that 1 is neither prime nor composite.

Example 2.9. The first few primes are $2, 3, 5, 7, 11, 13, 17, \ldots$.

An equivalent way of defining a prime number is given in the following result due to Euclid.

Theorem 2.7 (Euclid's Lemma). *Let $p > 1$ be a positive integer. Then the following are equivalent:*

1. *p is prime; and*
2. *if a and b are integers such that $p|ab$, then $p|a$ or $p|b$.*

Proof. Suppose that p is prime and $p|ab$. Now, $(p, a)|p$, so $(p, a) = 1$ or p. If $(p, a) = p$, then since $(p, a)|a$, we have $p|a$. Otherwise, by Corollary 2.1, there exist integers u and v such that $pu + av = 1$. But then $pbu + abv = b$. Now, $p|p$ and $p|ab$, so by Lemma 2.1, $p|b$.

On the other hand, if p is composite, then let $p = cd$, where $1 < c, d < p$. In this case, $p|cd$, but by Lemma 2.1, $p \nmid c$ and $p \nmid d$.   □

Corollary 2.4. *Let p be a prime number and $a_1, \ldots, a_n \in \mathbb{Z}$. If $p|a_1 a_2 \cdots a_n$, then $p|a_i$, for some i.*

Proof. Exercise 2.24.   □

In fact, every positive integer larger than 1 can be written as a product of primes, called its **prime factorization**.

Theorem 2.8 (Fundamental Theorem of Arithmetic). *If $a \in \mathbb{N}$ and $a > 1$, then there exist primes $p_1, \ldots, p_n$ (not necessarily distinct) such that $a = p_1 p_2 \cdots p_n$. Furthermore, this product is unique up to order. That is, if $a = q_1 q_2 \cdots q_m$, for some primes q_i, then $m = n$ and, after rearranging the primes, $p_i = q_i$ for all i.*

Proof. Let us prove the existence of the prime factorization and then handle the uniqueness. We will prove the result by strong induction on a. We are excluding the case $a = 1$, so start with $a = 2$. There is nothing to do here, since 2 is prime. Thus, let $a > 2$ and assume that the theorem is true for smaller numbers. If a is prime, there is nothing to do. Otherwise, we can write $a = bc$, with $1 < b, c < a$. But then by our inductive hypothesis, b and c are both products of primes, and hence a is a product of primes.

Now let us prove the uniqueness. Suppose that

$$a = p_1 \cdots p_n = q_1 \cdots q_m,$$

for some primes p_i and q_i. Without loss of generality, say $n \leq m$. Now, $p_1 | a$. Thus, by Corollary 2.4, $p_1 | q_i$, for some i. Rearranging the primes as needed, we may assume that $p_1 | q_1$. But q_1 is prime, so $p_1 = 1$ or q_1. As 1 is not prime, $p_1 = q_1$. Cancelling p_1 and q_1 from the two sides of our equation, we have

$$p_2 \cdots p_n = q_2 \cdots q_m.$$

Now do the same for p_2 and repeat. We find that, after rearranging, $p_i = q_i$, $1 \leq i \leq n$. If $m = n$, we are done. Otherwise, we are left with $1 = q_{n+1} \cdots q_m$. But then $q_m | 1$, which is impossible, as $q_m > 1$. ☐

Example 2.10. We can write $1400 = 2 \cdot 2 \cdot 2 \cdot 5 \cdot 5 \cdot 7$, and there is no other way to write 1400 as a product of primes, except by rearranging (for instance, $2 \cdot 5 \cdot 7 \cdot 2 \cdot 2 \cdot 5$).

Note that this gives us one good reason not to consider 1 as a prime: we would have to abandon uniqueness, as we could multiply by 1 as many times as we wanted.

We can use the existence of prime factors to prove a handy fact.

Corollary 2.5. *Let a, b and n be integers, with $n \neq 0$. If $(a, n) = (b, n) = 1$, then $(ab, n) = 1$.*

Proof. If $(ab, n) > 1$, then by Theorem 2.8, there exists a prime p dividing (ab, n). Since $p | ab$, Theorem 2.7 tells us that $p | a$ or $p | b$. But $p | n$ as well; thus, $(a, n) \geq p$ or $(b, n) \geq p$. Either way, we have a contradiction. ☐

Exercises

2.21. Factor each of the following numbers into a product of primes: 3528, 30030 and 220000.

2.22. Show that for every prime $p > 3$, there exists a positive integer k such that $p = 6k + 1$ or $p = 6k - 1$.

2.23. Let p be a prime and n an integer. Show that either $p | n$ or $(p, n) = 1$.

2.24. Use induction to prove Corollary 2.4.

2.25. Let $p_1, \ldots, p_k$ be any primes. Show that for each i, $p_i \nmid (p_1 p_2 \cdots p_k + 1)$.

2.26. Use the preceding exercise to show that there are infinitely many primes.

2.27. Let p be a prime and $a, n \in \mathbb{N}$. Suppose that $p | a^n$. Show that $p^n | a^n$.

2.28. Let $p_1, \ldots, p_k$ be distinct primes, and let m_i, n_i be nonnegative integers. Find the gcd of $p_1^{m_1} \cdots p_k^{m_k}$ and $p_1^{n_1} \cdots p_k^{n_k}$.

2.4　Properties of the Integers

This section may seem a tad underwhelming. Indeed, there are no proofs at all and we will not really learn any new facts about the integers. The whole point is to establish some terminology that we will see many times in different settings. While our discussion will take place in $\mathbb{Z}$, it is worth noting that we could just as easily use $\mathbb{Q}$, $\mathbb{R}$ or $\mathbb{C}$.

First, we observe that $\mathbb{Z}$ is **closed** under addition and multiplication. That is,

$$a + b, ab \in \mathbb{Z}$$

for all $a, b \in \mathbb{Z}$.

Next, addition and multiplication on $\mathbb{Z}$ are both **associative**. This means that

$$(a + b) + c = a + (b + c) \text{ and } (ab)c = a(bc)$$

for all $a, b, c \in \mathbb{Z}$. In particular, we can write $a + b + c$ and abc without fear of ambiguity.

Furthermore, addition and multiplication are both **commutative** on $\mathbb{Z}$. In other words,

$$a + b = b + a \text{ and } ab = ba$$

for all $a, b \in \mathbb{Z}$.

We also have the **distributive law**. Specifically,

$$a(b + c) = ab + ac$$

for all $a, b, c \in \mathbb{Z}$.

The numbers 0 and 1 are rather special. We call 0 the **additive identity** for $\mathbb{Z}$ and 1 the **multiplicative identity**. This is because

$$a + 0 = a \text{ and } a \cdot 1 = a$$

for all $a \in \mathbb{Z}$.

Finally, if $a \in \mathbb{Z}$, then $-a$ is its **additive inverse**. This means that

$$a + (-a) = 0.$$

It is important to note that we do not have **multiplicative inverses** for all integers; that is, if $a \in \mathbb{Z}$, it does not follow that there exists a $b \in \mathbb{Z}$ such that $ab = 1$. In fact, this only happens if a is 1 or -1. (The sets $\mathbb{Q}$, $\mathbb{R}$ and $\mathbb{C}$ are a bit different on this last point. Every element other than 0 has a multiplicative inverse in these sets. For instance, in $\mathbb{Q}$, the multiplicative inverse of $\frac{2}{9}$ is $\frac{9}{2}$.)

Exercises

2.29. For each of the following binary operations on $\mathbb{Z}$, decide if it is commutative; that is, do we have $a * b = b * a$ for all $a, b \in \mathbb{Z}$?

1. $a * b = ab + 1$
2. $a * b = a + b + ab$
3. $a * b = a$

2.30. For each part of the preceding exercise, are the operations associative? That is, do we have $(a * b) * c = a * (b * c)$ for all $a, b, c \in \mathbb{Z}$?

2.31. For parts (1) and (2) from Exercise 2.29, decide if $*$ has an identity; that is, does there exist an $e \in \mathbb{Z}$ such that $a * e = e * a = a$ for all $a \in \mathbb{Z}$?

2.32. Define a binary operation $*$ on $\mathbb{Q}$ via $a * b = a + b - ab$. Find an identity e; that is, find $e \in \mathbb{Q}$ such that $a * e = e * a = a$ for all $a \in \mathbb{Q}$. Then decide which elements of $\mathbb{Q}$ have inverses. That is, determine for which $b \in \mathbb{Q}$ there exists a $c \in \mathbb{Q}$ such that $b * c = c * b = e$.

2.5 Modular Arithmetic

When we perform modular arithmetic, we choose an integer $n \geq 2$ and then for any integer a, we concern ourselves only with the remainder when a is divided by n. As the only possible remainders are $0, 1, 2, \ldots, n - 1$, these are the only numbers to worry about.

Definition 2.5. Let $n \geq 2$ be an integer. If $a, b \in \mathbb{Z}$, then we say that a is **congruent** to b **modulo** n, and write $a \equiv b \pmod{n}$, if $n|(a - b)$; that is, if a and b have the same remainder when divided by n.

Example 2.11. As $8|(53 - 21)$, we have $53 \equiv 21 \pmod{8}$. Putting this another way, 53 and 21 both have remainder 5 when divided by 8. We reduce to the remainder and write $53 \equiv 5 \pmod{8}$ and $21 \equiv 5 \pmod{8}$.

We add and multiply modulo n in the usual way, simply reducing to the remainder.

Example 2.12. We observe that

$$5 + 8 \equiv 1 \pmod{12} \text{ and } 5 \cdot 8 \equiv 4 \pmod{12}.$$

Of course, we should be a bit careful here. For instance, since $5 \equiv 17 \pmod{12}$, we had better make sure that $5 + 8 \equiv 17 + 8 \pmod{12}$. This is certainly the case, but it will help if we express things in terms of equivalence classes.

Theorem 2.9. *Let $n \geq 2$ be an integer. Then $a \equiv b$ (mod n) is an equivalence relation on $\mathbb{Z}$. The equivalence class of a consists of all integers having the same remainder as a when divided by n.*

Proof. Reflexivity: We have $n|0 = a - a$, so $a \equiv a$ (mod n). Symmetry: Suppose that $a \equiv b$ (mod n). Then $n|(a - b)$, and hence $n| - (a - b) = b - a$. Thus, $b \equiv a$ (mod n). Transitivity: Suppose that $a \equiv b$ (mod n) and $b \equiv c$ (mod n). Then $n|(a - b)$ and $n|(b - c)$. Hence, $n|((a - b) + (b - c)) = a - c$. That is, $a \equiv c$ (mod n). The statement about the equivalence classes follows from the definition. $\square$

Definition 2.6. Let $n \geq 2$ be an integer. The set of **integers modulo** n, denoted $\mathbb{Z}_n$, is the set of all equivalence classes of $\mathbb{Z}$ with respect to the equivalence relation $a \equiv b$ (mod n). We call these the **congruence classes** modulo n. Specifically, $\mathbb{Z}_n = \{[0], [1], [2], \ldots, [n - 1]\}$.

Example 2.13. The elements of $\mathbb{Z}_4$ are $[0], [1], [2]$ and $[3]$, where

$$[0] = \{\ldots, -, 8, -4, 0, 4, 8, \ldots\}$$
$$[1] = \{\ldots, -7, -3, 1, 5, 9, \ldots\}$$
$$[2] = \{\ldots, -6, -2, 2, 6, 10, \ldots\}$$
$$[3] = \{\ldots, -5, -1, 3, 7, 11, \ldots\}.$$

As usual, in dealing with equivalence classes, the choice of the representative of the class is not unique. For instance, in the above example, we could just as easily have written $[-5]$ or $[7]$ instead of $[3]$. It is, however, customary to reduce final answers in $\mathbb{Z}_n$ to the form $[a]$, where $0 \leq a < n$.

We can now define addition and multiplication on $\mathbb{Z}_n$. These work in the obvious way. Specifically,

$$[a] + [b] = [a + b]$$
$$[a][b] = [ab].$$

Example 2.14. In $\mathbb{Z}_7$, we have $[5] + [2] = [7] = [0]$ and $[5][3] = [15] = [1]$.

Theorem 2.10. *For any integer $n \geq 2$, addition and multiplication on $\mathbb{Z}_n$ are well-defined.*

Proof. Suppose that $a_1 \equiv a_2$ (mod n) and $b_1 \equiv b_2$ (mod n). Then

$$(a_1 + b_1) - (a_2 + b_2) = (a_1 - a_2) + (b_1 - b_2).$$

Since $n|(a_1 - a_2)$ and $n|(b_1 - b_2)$, we see that $n|((a_1 + b_1) - (a_2 + b_2))$. That is, $[a_1 + b_1] = [a_2 + b_2]$, so addition is well-defined. Also,

$$a_1 b_1 - a_2 b_2 = (a_1 b_1 - a_1 b_2) + (a_1 b_2 - a_2 b_2) = a_1(b_1 - b_2) + (a_1 - a_2)b_2.$$

Since $n|(a_1 - a_2)$ and $n|(b_1 - b_2)$, we find that $n|(a_1 b_1 - a_2 b_2)$. That is, $[a_1 b_1] = [a_2 b_2]$, and multiplication is well-defined.

$\square$

Let us discuss a few properties of addition and multiplication in $\mathbb{Z}_n$. These should be compared with the properties of $\mathbb{Z}$ mentioned in Section 2.4.

Theorem 2.11. *Let $n \geq 2$ be an integer, and take any $[a], [b], [c] \in \mathbb{Z}_n$. Then*

1. $[a] + [b] \in \mathbb{Z}_n$ *(closure under addition);*
2. $[a] + ([b] + [c]) = ([a] + [b]) + [c]$ *(associativity);*
3. $[a] + [b] = [b] + [a]$ *(commutativity);*
4. $[a] + [0] = [a]$ *(additive identity); and*
5. $[a] + [-a] = [0]$ *(additive inverse).*

Proof. (1) is clear from the definition. The other parts all work because they work in $\mathbb{Z}$. For instance, $[a] + [b] = [a + b] = [b + a] = [b] + [a]$, proving (3). The remaining parts are left as Exercise 2.35. □

And now, some properties of multiplication.

Theorem 2.12. *Let $n \geq 2$ be an integer, and $[a], [b], [c] \in \mathbb{Z}_n$. Then*

1. $[a][b] \in \mathbb{Z}_n$ *(closure under multiplication);*
2. $[a]([b][c]) = ([a][b])[c]$ *(associativity);*
3. $[a][b] = [b][a]$ *(commutativity);*
4. $[a]([b] + [c]) = [a][b] + [a][c]$ *(distributive law); and*
5. $[a][1] = [a]$ *(multiplicative identity).*

Proof. (1) follows from the definition, and the other parts are true because they are true in $\mathbb{Z}$. For instance,

$$[a]([b] + [c]) = [a][b + c] = [a(b + c)]$$
$$= [ab + ac] = [ab] + [ac] = [a][b] + [a][c],$$

proving (4). The rest is left as Exercise 2.36. □

As in $\mathbb{Z}$, we do not necessarily have multiplicative inverses. For instance, in $\mathbb{Z}_{14}$, we find that $[5][3] = [1]$, but there is no integer a such that $[6][a] = [1]$. However, $\mathbb{Z}_5$ behaves more like $\mathbb{Q}$; indeed, if $[a] \neq [0]$, then there exists a $[b] \in \mathbb{Z}_5$ such that $[a][b] = [1]$. More on this later!

It is worth mentioning that $\mathbb{Z}_n$ does not behave exactly like $\mathbb{Z}$. For instance, in $\mathbb{Z}$, we are used to the fact that if $ab = 0$, then $a = 0$ or $b = 0$. But in $\mathbb{Z}_{12}$, we have $[4][9] = [0]$. We are also accustomed to cancellation in $\mathbb{Z}$; that is, if $ab = ac$, and $a \neq 0$, then $b = c$. Not necessarily true in $\mathbb{Z}_n$! For example, in $\mathbb{Z}_{12}$, we have $[2][3] = [2][9]$, but $[2] \neq [0]$ and $[3] \neq [9]$. So we must be careful with our assumptions.

And now, having acquainted ourselves with $\mathbb{Z}_n$, we are going to make a change in notation. It is rather cumbersome to have to write $[a]$ or $[a] + [b]$ all the time. Therefore, when working in $\mathbb{Z}_n$, we will normally simply write a or $a + b$, as long as the context is clear. We will include the equivalence class brackets if they are needed for greater clarity.

Example 2.15. When working in $\mathbb{Z}_{10}$, we simply write $3 + 8 = 1$ and $3 \cdot 8 = 4$.

We need to prove one last property of modular arithmetic, which dates back to ancient China.

Theorem 2.13 (Chinese Remainder Theorem). *Let $n_1, \ldots, n_k$ be positive integers, all larger than 1, such that $(n_i, n_j) = 1$ whenever $i \neq j$. If $a_1, \ldots, a_k \in \mathbb{Z}$, then there exists an integer b such that $b \equiv a_i \pmod{n_i}$ for all i. Furthermore, if $c \equiv a_i \pmod{n_i}$ for all i, then $b \equiv c \pmod{n_1 n_2 \cdots n_k}$.*

Proof. For each i, let d_i be the product of all of the n_j except for n_i; that is, $d_i = \frac{n_1 n_2 \cdots n_k}{n_i}$. Since $(n_i, n_j) = 1$ when $i \neq j$, Corollary 2.5 shows us that $(n_i, d_i) = 1$. By Corollary 2.1, there exist integers u_i and v_i such that $n_i u_i + d_i v_i = 1$. Thus, $d_i v_i \equiv 1 \pmod{n_i}$. Let

$$b = d_1 v_1 a_1 + d_2 v_2 a_2 + \cdots + d_k v_k a_k.$$

Then since $n_i \mid d_j$ if $i \neq j$, we have

$$b \equiv d_i v_i a_i \equiv a_i \pmod{n_i},$$

for all i, as required.

Finally, if $c \equiv a_i \pmod{n_i}$ for all i as well, then $b \equiv c \pmod{n_i}$ for all i; that is, $n_i \mid (b - c)$ for all i. By Corollary 2.3, $n_1 n_2 \cdots n_k \mid (b - c)$. □

Example 2.16. Let us solve the congruences $b \equiv 3 \pmod 5$, $b \equiv 4 \pmod{11}$ and $b \equiv 6 \pmod{14}$. We have $d_1 = 154$, $d_2 = 70$ and $d_3 = 55$. Solving $5u_1 + 154v_1 = 1$ using the Euclidean algorithm, we get $u_1 = 31$, $v_1 = -1$. When we solve $11u_2 + 70v_2 = 1$, we get $u_2 = -19$, $v_2 = 3$. Finally, a solution to $14u_3 + 55v_3 = 1$ is $u_3 = 4$, $v_3 = -1$. Therefore, $b = 154(-1)(3) + 70(3)(4) + 55(-1)(6) = 48$. Thus, the solution is $b \equiv 48 \pmod{770}$.

Exercises

2.33. Perform each calculation in $\mathbb{Z}_7$. The final answer should be a nonnegative integer no larger than 6.

1. $2 - 3 \cdot 4$
2. $(4 \cdot 5)^{25}$

2.34. Perform each calculation in $\mathbb{Z}_{15}$. The final answer should be a nonnegative integer no larger than 14.

1. $5 \cdot 11 - 3 \cdot 4$
2. 2^{82}

2.35. Complete the proof of Theorem 2.11.

2.36. Complete the proof of Theorem 2.12.

2.37. For each nonzero element $a \in \mathbb{Z}_{20}$, decide if there is a nonzero $b \in \mathbb{Z}_{20}$ such that $ab = 0$ in $\mathbb{Z}_{20}$. If so, provide such an element b.

2.38. For each element $a \in \mathbb{Z}_{20}$, decide if there exists a $b \in \mathbb{Z}_{20}$ such that $ab = 1$ in $\mathbb{Z}_{20}$. If so, provide such an element b.

2.39. Show that if p is prime, then there are at most two elements $a \in \mathbb{Z}_p$ such that $a^2 = 1$ in $\mathbb{Z}_p$. Find an example of a composite p where there are more than two solutions.

2.40. Let a and b be integers. Show that if $a \equiv b \pmod{p}$ for every prime p, then $a = b$.

2.41. Find $a \in \mathbb{Z}$ such that $a \equiv 2 \pmod 3$, $a \equiv 4 \pmod 7$ and $a \equiv 3 \pmod{10}$ simultaneously.

2.42. Find $a \in \mathbb{Z}$ such that $a \equiv 3 \pmod 8$, $a \equiv 4 \pmod{11}$ and $a \equiv 7 \pmod{15}$ simultaneously.

Part II
Groups

Chapter 3
Introduction to Groups

We now begin our study of abstract algebra in earnest! A group is one of the simplest algebraic structures; we take a set, assign an operation to it, impose four basic rules, and see what we can deduce. And yet, the possibilities are endless. Groups show up everywhere, and not just in mathematics. Indeed, it would be difficult to study physics or chemistry without an understanding of group theory. The solution to the famous Rubik's cube is also a problem in groups.

In this chapter, we will define the notion of a group, and give a number of examples. We will also prove several basic properties of groups and subgroups.

3.1 An Important Example

In the next section, we will give the definition of a group. For now, we will look at a motivating example.

Let A be the set $\{1, 2, 3\}$. We would like to consider all of the permutations of A; that is, all the ways of rearranging the numbers 1, 2 and 3. For example, we have the permutation σ, where $\sigma(1) = 2, \sigma(2) = 1$ and $\sigma(3) = 3$. We can easily see that there are going to be exactly 6 such permutations, as there are 3 choices for $\sigma(1)$, then 2 remaining choices for $\sigma(2)$, and once those are known, $\sigma(3)$ is determined.

A bit of notation would be helpful. Let us denote a permutation σ by writing two rows. The elements of A go in the first row, and the numbers to which each of them is sent in the second; that is, we take

$$\sigma = \begin{pmatrix} 1 & 2 & 3 \\ a & b & c \end{pmatrix}$$

to mean that $\sigma(1) = a, \sigma(2) = b$ and $\sigma(3) = c$. Then the permutation we mentioned above would be denoted

© Springer International Publishing AG, part of Springer Nature 2018
G. T. Lee, *Abstract Algebra*, Springer Undergraduate Mathematics Series,
https://doi.org/10.1007/978-3-319-77649-1_3

$$\begin{pmatrix} 1\ 2\ 3 \\ 2\ 1\ 3 \end{pmatrix}.$$

In fact, the complete list of permutations is

$$\begin{pmatrix} 1\ 2\ 3 \\ 1\ 2\ 3 \end{pmatrix}, \ \begin{pmatrix} 1\ 2\ 3 \\ 1\ 3\ 2 \end{pmatrix}, \ \begin{pmatrix} 1\ 2\ 3 \\ 2\ 1\ 3 \end{pmatrix}, \ \begin{pmatrix} 1\ 2\ 3 \\ 2\ 3\ 1 \end{pmatrix}, \ \begin{pmatrix} 1\ 2\ 3 \\ 3\ 1\ 2 \end{pmatrix} \ \text{and} \ \begin{pmatrix} 1\ 2\ 3 \\ 3\ 2\ 1 \end{pmatrix}.$$

Let us now discuss the composition of two permutations. For instance, if

$$\sigma = \begin{pmatrix} 1\ 2\ 3 \\ 2\ 1\ 3 \end{pmatrix} \ \text{and} \ \tau = \begin{pmatrix} 1\ 2\ 3 \\ 3\ 1\ 2 \end{pmatrix},$$

then we see that $(\sigma \circ \tau)(1) = \sigma(\tau(1)) = \sigma(3) = 3, (\sigma \circ \tau)(2) = \sigma(\tau(2)) = \sigma(1) = 2$ and $(\sigma \circ \tau)(3) = \sigma(\tau(3)) = \sigma(2) = 1$. Thus,

$$\sigma \circ \tau = \begin{pmatrix} 1\ 2\ 3 \\ 3\ 2\ 1 \end{pmatrix}.$$

(It is worth noting here that we apply τ first, then σ.)

We can now consider some properties that these permutations enjoy with respect to this composition operation. As we discuss them, please compare with the properties of $\mathbb{Z}$ or $\mathbb{Z}_n$, under addition, with which we are already familiar.

First of all, we have **closure**. That is, if we take two permutations of A and compose them, we obtain another permutation of A. In fact, we proved this in Theorem 1.2, where we saw that the composition of two bijections is a bijection.

Next, we have **associativity**; that is, for any permutations ρ, σ and τ, we have $\rho \circ (\sigma \circ \tau) = (\rho \circ \sigma) \circ \tau$. We have seen this before as well; by Theorem 1.2, the composition of functions is always associative.

Also, we have an **identity**. In particular, if σ is any permutation of A, then

$$\sigma \circ \begin{pmatrix} 1\ 2\ 3 \\ 1\ 2\ 3 \end{pmatrix} = \begin{pmatrix} 1\ 2\ 3 \\ 1\ 2\ 3 \end{pmatrix} \circ \sigma = \sigma.$$

Composing with the permutation that moves nothing cannot change a function.

Finally, we have **inverses**; that is, for each permutation σ, there is another permutation τ such that

$$\sigma \circ \tau = \tau \circ \sigma = \begin{pmatrix} 1\ 2\ 3 \\ 1\ 2\ 3 \end{pmatrix},$$

the identity. This is easy enough to calculate directly; for instance,

$$\begin{pmatrix} 1\ 2\ 3 \\ 2\ 3\ 1 \end{pmatrix} \circ \begin{pmatrix} 1\ 2\ 3 \\ 3\ 1\ 2 \end{pmatrix} = \begin{pmatrix} 1\ 2\ 3 \\ 3\ 1\ 2 \end{pmatrix} \circ \begin{pmatrix} 1\ 2\ 3 \\ 2\ 3\ 1 \end{pmatrix} = \begin{pmatrix} 1\ 2\ 3 \\ 1\ 2\ 3 \end{pmatrix}.$$

However, the existence of such an inverse is guaranteed by Theorem 1.3.

Given our discussion in Sections 2.4 and 2.5, we can agree that all of these properties are shared by $\mathbb{Z}$ and $\mathbb{Z}_n$ under addition. However, we also noted that the addition operation is **commutative**. Not so here! For instance,

$$\begin{pmatrix} 1\ 2\ 3 \\ 2\ 3\ 1 \end{pmatrix} \circ \begin{pmatrix} 1\ 2\ 3 \\ 1\ 3\ 2 \end{pmatrix} = \begin{pmatrix} 1\ 2\ 3 \\ 2\ 1\ 3 \end{pmatrix},$$

whereas

$$\begin{pmatrix} 1\ 2\ 3 \\ 1\ 3\ 2 \end{pmatrix} \circ \begin{pmatrix} 1\ 2\ 3 \\ 2\ 3\ 1 \end{pmatrix} = \begin{pmatrix} 1\ 2\ 3 \\ 3\ 2\ 1 \end{pmatrix}.$$

Thus, in general, $\sigma \circ \tau \neq \tau \circ \sigma$.

These permutations under the composition operation give us a nice example of a group, as we shall see momentarily. There is, of course, nothing very magical about the set $A = \{1, 2, 3\}$ here. Indeed, we could just as easily have used $\{1, 2, 3, \ldots, n\}$, for any positive integer n. The set of all permutations of this set, under the composition operation, is called the **symmetric group** on n letters, and is denoted S_n.

Exercises

3.1. In S_4, let $\sigma = \begin{pmatrix} 1\ 2\ 3\ 4 \\ 3\ 1\ 4\ 2 \end{pmatrix}$ and $\tau = \begin{pmatrix} 1\ 2\ 3\ 4 \\ 3\ 4\ 1\ 2 \end{pmatrix}$. Calculate the following.

1. $\sigma\tau$
2. $\tau\sigma$
3. the inverse of σ

3.2. In S_5, let $\sigma = \begin{pmatrix} 1\ 2\ 3\ 4\ 5 \\ 5\ 3\ 2\ 1\ 4 \end{pmatrix}$ and $\tau = \begin{pmatrix} 1\ 2\ 3\ 4\ 5 \\ 2\ 4\ 1\ 3\ 5 \end{pmatrix}$. Calculate the following.

1. $\sigma\tau\sigma$
2. $\sigma\sigma\tau$
3. the inverse of σ

3.3. How many permutations are there in S_n? In S_5, how many permutations α satisfy $\alpha(2) = 2$?

3.4. Let H be the set of all permutations $\alpha \in S_5$ satisfying $\alpha(2) = 2$. Which of the properties we have discussed (closure, associativity, identity, inverses) does H enjoy under composition of functions?

3.5. Consider the set of all functions from $\{1, 2, 3, 4, 5\}$ to $\{1, 2, 3, 4, 5\}$. Which of the properties (closure, associativity, identity, inverses) does this set enjoy under composition of functions?

3.6. Let G be the set of all permutations of $\mathbb{N}$. Which of the properties (closure, associativity, identity, inverses) does G enjoy under composition of functions?

3.2 Groups

We can now give the general definition of a group.

Definition 3.1. A **group** is a set G, together with a binary operation $*$, satisfying the following conditions:

1. $a * b \in G$ for all $a, b \in G$ (closure);
2. $(a * b) * c = a * (b * c)$ for all $a, b, c \in G$ (associativity);
3. there exists an $e \in G$ such that $a * e = e * a = a$ for all $a \in G$ (existence of identity); and
4. for each $a \in G$, there exists a $b \in G$ such that $a * b = b * a = e$ (existence of inverses).

We will refer to G as a group **under** $*$.

The element e is called the **identity** of the group. If $a \in G$, and $a * b = b * a = e$, then b is called the **inverse** of a, and we write $b = a^{-1}$.

As we discussed in the previous section, the group operation does not have to be commutative. We have a special term for groups that do have this property, named after mathematician Niels H. Abel.

Definition 3.2. A group G is said to be **abelian** if $a * b = b * a$ for all $a, b \in G$.

We devote the remainder of this section to examples of groups.

Example 3.1. As we saw in Sections 2.4 and 2.5, $\mathbb{Z}$ and $\mathbb{Z}_n$ (for any integer $n \geq 2$) are abelian groups under addition. Indeed, 0 is the identity, and the inverse of a is $-a$. In fact, the same can be said for $\mathbb{Q}$, $\mathbb{R}$ and $\mathbb{C}$ under addition.

When a group G has only finitely many elements, we can represent it with a **group table**. We write the elements of G down the first column and along the first row of the table. Then the entry in the row headed by a and the column headed by b is $a * b$. For instance, the group table for $\mathbb{Z}_5$ is given in Table 3.1.

Table 3.1 Group table for the additive group $\mathbb{Z}_5$

	0	1	2	3	4
0	0	1	2	3	4
1	1	2	3	4	0
2	2	3	4	0	1
3	3	4	0	1	2
4	4	0	1	2	3

Example 3.2. Let G be the set of nonzero rational numbers. Then G is an abelian group under multiplication. Indeed, we see that the product of two nonzero rationals is a nonzero rational, hence closure is satisfied. Also, multiplication of rationals is both

associative and commutative. Clearly, $a \cdot 1 = a$, for all $a \in G$, so 1 is the identity. Finally, if $a = m/n \in G$, with m and n nonzero integers, then $a^{-1} = n/m \in G$, since $(m/n)(n/m) = 1$.

This last example merits a second look. In particular, it is worth noting that we cannot do the same thing with the set of nonzero integers. To be sure, the product of two nonzero integers is a nonzero integer, and the multiplication is associative. Also, 1 is the identity. But 2 has no inverse; that is, there is no integer a such that $2a = 1$. In fact, the only integers that would have inverses in this set are 1 and -1. The set $\{1, -1\}$ is easily seen to be a group under multiplication.

Let us see how the integers modulo n compare.

Example 3.3. Let $n \geq 2$ be a positive integer. Let $U(n)$ denote the set of all elements $a \in \mathbb{Z}_n$ such that $(a, n) = 1$. (For instance, $U(10) = \{1, 3, 7, 9\}$.) Let us ensure that this makes sense. That is, if $a \equiv b \pmod{n}$, and $(a, n) = 1$, then it had also better be the case that $(b, n) = 1$. But $a = b + nk$, for some integer k. Then, if c divides both b and n, then c divides a as well. We claim that $U(n)$ is an abelian group under the multiplication operation in $\mathbb{Z}_n$. First, closure. By Corollary 2.5, if $(a, n) = 1$ and $(b, n) = 1$, then $(ab, n) = 1$. We also know that multiplication in $\mathbb{Z}_n$ is associative and commutative, and 1 (which obviously lies in $U(n)$) is the identity. What about inverses? If $a \in U(n)$, then since $(a, n) = 1$, there exist integers u and v such that $au + nv = 1$. That is, in $\mathbb{Z}_n$, $au = ua = 1$. The group table of $U(10)$ is given in Table 3.2.

Table 3.2 Group table for the multiplicative group $U(10)$

	1	3	7	9
1	1	3	7	9
3	3	9	1	7
7	7	1	9	3
9	9	7	3	1

Note that we will use the notation $U(n)$ from the above example throughout the book.

Example 3.4. Let n be a positive integer, and let G be the set of all complex numbers w satisfying $w^n = 1$. Then we claim that G is an abelian group under multiplication. Of course, we know that the multiplication is both associative and commutative, and $1 \in G$ will serve as a multiplicative identity. If $v, w \in G$, then $(vw)^n = v^n w^n = 1$, so $vw \in G$, and we have closure. Also, if $w \in G$, then we know that $1/w \in \mathbb{C}$. But $(1/w)^n = 1/(w^n) = 1$, and therefore $1/w \in G$.

In particular, if $n = 4$ in the above example, then we get the group $\{1, -1, i, -i\}$. Also, if $n = 1$, then we just get the group consisting of the identity. This is known as the **trivial group**.

Of course, not all groups are abelian. Two useful examples follow.

Example 3.5. As we illustrated in the previous section, S_n is a group under composition. If $n \geq 3$, then the group is nonabelian.

Example 3.6. The set of all invertible 2×2 matrices with entries in $\mathbb{R}$ is called the **general linear group** over $\mathbb{R}$, and denoted $GL_2(\mathbb{R})$. It is a group under matrix multiplication. The identity matrix I_2 is the identity of $GL_2(\mathbb{R})$. Also, if $A, B \in GL_2(\mathbb{R})$, then

$$AB(B^{-1}A^{-1}) = A(BB^{-1})A^{-1} = AI_2A^{-1} = AA^{-1} = I_2$$

and, similarly, $(B^{-1}A^{-1})AB = I_2$. Thus, AB is invertible as well, so $GL_2(\mathbb{R})$ is closed under multiplication. Also, matrix multiplication is associative. By definition of $GL_2(\mathbb{R})$, every element A has an inverse, and since $(A^{-1})^{-1} = A$, we know that $A^{-1} \in GL_2(\mathbb{R})$. Thus, $GL_2(\mathbb{R})$ is indeed a group. However, the group is nonabelian. For instance,

$$\begin{pmatrix} 1 & 1 \\ 0 & 1 \end{pmatrix} \begin{pmatrix} 1 & 0 \\ 1 & 1 \end{pmatrix} = \begin{pmatrix} 2 & 1 \\ 1 & 1 \end{pmatrix},$$

whereas

$$\begin{pmatrix} 1 & 0 \\ 1 & 1 \end{pmatrix} \begin{pmatrix} 1 & 1 \\ 0 & 1 \end{pmatrix} = \begin{pmatrix} 1 & 1 \\ 1 & 2 \end{pmatrix}.$$

By changing the entries in the matrices, we can obtain other general linear groups, such as $GL_2(\mathbb{Q})$. We can also use invertible $n \times n$ matrices and obtain $GL_n(\mathbb{R})$.

Let us also present a useful way of obtaining new groups from old ones.

Definition 3.3. Let G be a group with operation $*$ and H a group with operation $\bullet$. On the Cartesian product $G \times H$, define an operation $\diamond$ via

$$(g_1, h_1) \diamond (g_2, h_2) = (g_1 * g_2, h_1 \bullet h_2),$$

for all $g_i \in G$, $h_i \in H$. Under this operation, we call $G \times H$ the **direct product** of G and H.

Theorem 3.1. *The direct product of two groups is a group.*

Proof. Let us adopt the same notation as in the definition. First, we must check that the direct product is closed. But if $g_1, g_2 \in G, h_1, h_2 \in H$, then $(g_1, h_1) \diamond (g_2, h_2) = (g_1 * g_2, h_1 \bullet h_2) \in G \times H$, since $g_1 * g_2 \in G, h_1 * h_2 \in H$. The associativity of $\diamond$ follows from the associativity of $*$ and $\bullet$. Indeed, if $g_1, g_2, g_3 \in G, h_1, h_2, h_3 \in H$, then

$$\begin{aligned}((g_1, h_1) \diamond (g_2, h_2)) \diamond (g_3, h_3) &= (g_1 * g_2, h_1 \bullet h_2) \diamond (g_3, h_3) \\ &= ((g_1 * g_2) * g_3, (h_1 \bullet h_2) \bullet h_3) \\ &= (g_1 * (g_2 * g_3), h_1 \bullet (h_2 \bullet h_3)) \\ &= (g_1, h_1) \diamond ((g_2, h_2) \diamond (g_3, h_3)).\end{aligned}$$

Let e_G and e_H be the identities of G and H respectively. Then for any $g \in G$, $h \in H$, we have

$$(g, h) \diamond (e_G, e_H) = (g * e_G, h \bullet e_H) = (g, h)$$

and, similarly, $(e_G, e_H) \diamond (g, h) = (g, h)$. Thus, (e_G, e_H) is the identity for $G \times H$. Furthermore, $(g, h) \diamond (g^{-1}, h^{-1}) = (g * g^{-1}, h \bullet h^{-1}) = (e_G, e_H)$ and, similarly, $(g^{-1}, h^{-1}) \diamond (g, h) = (e_G, e_H)$. Thus, $(g, h)^{-1} = (g^{-1}, h^{-1})$. The proof is complete. □

Example 3.7. Suppose that $G = \mathbb{Z}_5$ and $H = S_3$. Then in $G \times H$,

$$\left(4, \begin{pmatrix} 1\ 2\ 3 \\ 2\ 3\ 1 \end{pmatrix}\right) \diamond \left(3, \begin{pmatrix} 1\ 2\ 3 \\ 3\ 2\ 1 \end{pmatrix}\right) = \left(4 + 3, \begin{pmatrix} 1\ 2\ 3 \\ 2\ 3\ 1 \end{pmatrix} \circ \begin{pmatrix} 1\ 2\ 3 \\ 3\ 2\ 1 \end{pmatrix}\right) = \left(2, \begin{pmatrix} 1\ 2\ 3 \\ 1\ 3\ 2 \end{pmatrix}\right).$$

Before concluding this section, it seems to be worth mentioning that part of the definition of a group is redundant. We specify that $*$ is a binary operation on G, and then require closure. But closure is part of the definition of a binary operation! Nevertheless, it is a good idea to emphasize this point, as closure must be checked whenever a new group is defined, and it is easy to forget about it if it is buried inside another definition.

Exercises

3.7. Give group tables for the following additive groups.

1. $\mathbb{Z}_3$
2. $\mathbb{Z}_3 \times \mathbb{Z}_2$

3.8. Give group tables for the following groups.

1. $U(12)$
2. S_3

3.9. Show that $G \times H$ is abelian if and only if G and H are abelian.

3.10. Let G be a group containing at most three elements. Show that G is abelian.

3.11. Explain why neither of the following is a group.

1. the set of positive rational numbers under division
2. the set of rational numbers $q \geq 1$ under multiplication

3.12. Is either of the following a group under addition?

1. the set of even integers
2. the set of odd integers

3.13. Let $G = \{a + bi \in \mathbb{C} : a^2 + b^2 = 1\}$. Is G a group under multiplication?

3.14. Let G be the following subset of $\mathbb{Z}_{15}$, namely $\{3, 6, 9, 12\}$. Show that G is a group under multiplication in $\mathbb{Z}_{15}$. Find the identity, and the inverse of each group element.

3.15. Let p be a prime and $G = \{a/p^n : a \in \mathbb{Z}, n \in \mathbb{N}\}$. Is G a group under addition?

3.16. Let G be the set of all matrices of the form $\begin{pmatrix} 1 & a & b \\ 0 & 1 & c \\ 0 & 0 & 1 \end{pmatrix}$, with $a, b, c \in \mathbb{Z}$. Show that G is a group under matrix multiplication. Is it abelian?

3.3   A Few Basic Properties

Let us begin with a small notational change. Usually, when we are working inside a group, we suppress the symbol for the group operation. That is, we write ab instead of $a * b$. The major exception is where the operation is addition, in which case this **multiplicative notation** would be confusing. In that case, we will use **additive notation** and continue to write $a + b$ instead of ab, 0 instead of e and $-a$ instead of a^{-1}.

In the preceding section, we glossed over the uniqueness of the group identity and inverses of group elements. These are important points, if we are to speak of "the" identity, or write a^{-1} and have it mean something. Let us take care of this problem.

Theorem 3.2. *Let G be a group. Then*

1. *the identity of G is unique; and*
2. *if $a \in G$, then a^{-1} is unique.*

Proof. (1) Suppose that e and f are both identities in G. Then as f is an identity, $ef = e$. But as e is an identity, we also have $ef = f$. Therefore, $e = f$.

(2) Suppose that b and c are both inverses of a. Then as b is an inverse of a, $(ba)c = ec = c$. However, as c is an inverse of a, we have $b(ac) = be = b$. Given that our group operation is associative, $b = b(ac) = (ba)c = c$. ☐

We know that in any group, $(ab)c = a(bc)$. Thus, we can write abc without worrying about ambiguity. But we would like to be able to write $abcd$, for instance. To that end, we have the following result.

Theorem 3.3. *Let G be any group, and $a_1, a_2, \ldots, a_n \in G$. Then regardless of how the product $a_1 a_2 \cdots a_n$ is bracketed, the result equals $(\cdots (((a_1 a_2)a_3)a_4) \cdots a_{n-1})a_n$.*

Proof. Our proof is by strong induction upon n. If n is 1 or 2, no bracketing is needed, so there is nothing to do. When $n = 3$, this is the associativity from the definition of a group. Therefore, let $n \geq 4$, and suppose that the theorem is true for any product of fewer than n group elements. Take any bracketing of $w = a_1 \cdots a_n$, and look at the last operation to be performed. Then $w = xy$, where x is the product $a_1 \cdots a_m$ and y is the product $a_{m+1} \cdots a_n$, each with some bracketing. By our inductive hypothesis, $x = (\cdots ((a_1 a_2) a_3) \cdots a_{m-1}) a_m$ and $y = (\cdots ((a_{m+1} a_{m+2}) a_{m+3}) \cdots a_{n-1}) a_n$. If $m = n - 1$, then writing xy in this way, we have our desired conclusion. If not, then by associativity,

$$xy = ((\cdots ((a_1 a_2) a_3) \cdots a_m)(\cdots ((a_{m+1} a_{m+2}) a_{m+3}) \cdots a_{n-1})) a_n.$$

Now applying our inductive hypothesis to the product of the first $n - 1$ terms, we obtain the desired bracketing. □

Therefore, we do not have to use brackets when we write a product of group elements. However, we must always remember that unless our group is abelian, we cannot rearrange terms at will. For instance, $(ab)(cd) = (a(bc))d$, and we can write both as $abcd$, but we cannot write $abcd = cdba$.

Let us also prove a couple of useful facts about inverses.

Theorem 3.4. *Let G be a group, with $a, b \in G$. Then*

1. $(a^{-1})^{-1} = a$; and
2. $(ab)^{-1} = b^{-1} a^{-1}$.

Proof. (1) Since $aa^{-1} = a^{-1} a = e$, we see from the definition of inverses that the inverse of a^{-1} is a.

(2) Notice that $(ab)(b^{-1} a^{-1}) = a(bb^{-1})a^{-1} = aea^{-1} = aa^{-1} = e$ and, similarly, $(b^{-1} a^{-1})(ab) = e$. Therefore, $b^{-1} a^{-1}$ is the inverse of ab. □

Do not make the mistake of thinking that the inverse of ab is $a^{-1} b^{-1}$!

In ordinary arithmetic with real numbers, we know that if $ab = ac$, and $a \neq 0$, then $b = c$. We have something similar for groups.

Theorem 3.5. (Cancellation Law). *Let G be a group and $a, b, c \in G$. If either $ab = ac$ or $ba = ca$, then $b = c$.*

Proof. If $ab = ac$, then $a^{-1} ab = a^{-1} ac$. As $a^{-1} a = e$, we have $eb = ec$, and therefore $b = c$. The proof is similar if $ba = ca$. □

When our group has finitely many elements, the cancellation law has important implications for the group table. Suppose that, in the row headed by a, the group element b occurs twice. Then there exist group elements c and d such that $ac = b = ad$. But then we know that $c = d$. Therefore, a group element can occur only once in each row. In the same way, there will be no repetitions in any column.

Corollary 3.1. *Let G be a group and $a, b \in G$. Then there exists exactly one $c \in G$ such that $ac = b$, and exactly one $d \in G$ such that $da = b$.*

Proof. We showed the uniqueness of c and d above. To show the existence of c and d, let $c = a^{-1} b$ and $d = ba^{-1}$. Then $ac = aa^{-1} b = eb = b$, and $da = ba^{-1} a = be = b$. □

Example 3.8. Suppose G is a group with four elements, a, b, c and d. If we are given the partial group table shown in Table 3.3, we can fill in the missing elements. Indeed, examining the first row, we see that ad cannot be b or d. The last column tells us that it also cannot be a, so $ad = c$. As there must be an a in the first row, $ab = a$. Filling in the rest of the table is left as an exercise.

Table 3.3 Incomplete group table

	a	b	c	d
a	d		b	
b		b		
c				a
d		a		

Exercises

3.17. Simplify each of the following expressions as far as possible in an arbitrary group G, leaving no brackets.

1. $(acb)(cbab)^{-1}$
2. $(a^{-1}bca)^{-1}$

3.18. Repeat the preceding exercise, assuming that G is abelian.

3.19. Fill in the rest of Table 3.3.

3.20. Let $G = \{v, w, x, y, z\}$ be a group with five elements. Further suppose that $vw = y$, $vy = v$, $wx = z$, $xv = w$ and $zw = v$. Fill in the group table for G.

3.21. Show that the following are equivalent for a group G:

1. for every $a, b, c \in G$ satisfying $ab = ca$, we have $b = c$; and
2. G is abelian.

3.22. Suppose that in the definition of a group, we replace the third part with the following weaker condition:
(3') There exists an $e \in G$ such that for every $a \in G$, $ae = a$.
(That is, we do not insist that $ea = a$.) Show that we still get a group.

3.4 Powers and Orders

In group theory, the word *order* is used in two different, but related, ways. One is easy.

Definition 3.4. If G is a group, then its **order**, $|G|$, is the number of elements in the set G. We say that G is a **finite group** if its order is finite; otherwise, it is an **infinite group**.

Example 3.9. If $G = \mathbb{Z}_5$, then $|G| = 5$, and therefore G is a finite group. On the other hand, $\mathbb{Z}$ is an infinite group.

To understand the other use of the word, we need to know about **powers** of group elements. Let G be any group, and $a \in G$. Then for any positive integer n, we let

$$a^n = \underbrace{aa \cdots a}_{n \text{ times}}.$$

(Alternatively, we could define the powers recursively. That is, let $a^1 = a$, and for each positive integer n, let $a^{n+1} = a^n a$.) Also, let $a^0 = e$ and, for each positive integer n, let $a^{-n} = (a^n)^{-1}$.

Example 3.10. In $U(20)$, we calculate $7^3 = 7 \cdot 7 \cdot 7 = 9 \cdot 7 = 3$. If we wanted to know 7^{-3}, then we would calculate $(7^3)^{-1} = 3^{-1} = 7$, since $3 \cdot 7 = 1$.

Powers behave in a rather nice manner, as the following theorem tells us.

Theorem 3.6. *Let G be a group, with $a \in G$, and let m and n be any integers. Then*

1. $a^m a^n = a^{m+n}$;
2. $(a^m)^n = a^{mn}$; *and, in particular,*
3. $a^{-n} = (a^{-1})^n$.

Proof. Exercise 3.26. □

We know that if the group operation is addition, then we will use additive notation, rather than multiplicative. In this case, our exponentiation notation would be confusing, so we will write things in a more familiar manner. Instead of a^n, we will write na (that is, add a to itself n times).

Example 3.11. In $\mathbb{Z}_{12}$, since our operation is addition, instead of writing 5^4, we would write $4 \cdot 5 = 5 + 5 + 5 + 5 = 8$.

Sometimes, a group will consist only of powers of a specific group element.

Definition 3.5. A group G is said to be **cyclic** if there exists an element a such that every element of G is a power of a. In particular, we say that G is **generated** by a, and write $G = \langle a \rangle$.

Example 3.12. The additive group $\mathbb{Z}$ is cyclic; indeed, $\mathbb{Z} = \langle 1 \rangle$. (Remember, in an additive group, the powers are integer multiples, so if $a \in \mathbb{Z}$, then $a = a \cdot 1$.) In fact, $\mathbb{Z} = \langle -1 \rangle$ as well, so the generator of the cyclic group is not unique. In the same way, $\mathbb{Z}_n$ is cyclic.

Example 3.13. Consider the multiplicative group of complex fourth roots of unity discussed in Example 3.4, namely $G = \{1, -1, i, -i\}$. Then G is cyclic. Indeed, $G = \langle i \rangle$ since the powers of i are $i, -1, -i$ and 1.

Not every group is cyclic. For one thing, we have the following fact.

Theorem 3.7. *Every cyclic group is abelian.*

Proof. Let $G = \langle a \rangle$. If $b, c \in G$, then $b = a^m$ and $c = a^n$, for some $m, n \in \mathbb{Z}$. Then $bc = a^m a^n = a^{m+n}$, but $cb = a^n a^m = a^{m+n}$ as well. $\square$

However, abelian groups need not be cyclic.

Example 3.14. The group $U(10)$ is cyclic, but $U(8)$ is not. To see this, observe that $U(10) = \{1, 3, 7, 9\}$. But the powers of 3 are 3, 9, 7 and 1, so $U(10) = \langle 3 \rangle$. On the other hand, $U(8) = \{1, 3, 5, 7\}$. But the powers of 1 are all 1, the powers of 3 are 1 and 3, the powers of 5 are 1 and 5, and the powers of 7 are 1 and 7. Therefore, no element generates $U(8)$.

Now, let us discuss the order of a group element.

Definition 3.6. Let G be a group and $a \in G$. The **order** of a, denoted $|a|$, is the smallest positive integer n such that $a^n = e$, assuming that such an n exists, in which case a has **finite order**. If no such n exists, then a has **infinite order**.

Example 3.15. The identity of a group is the only element having order 1.

Example 3.16. In S_3, the element $\sigma = \begin{pmatrix} 1 & 2 & 3 \\ 2 & 3 & 1 \end{pmatrix}$ has order 3; indeed, $\sigma^2 = \begin{pmatrix} 1 & 2 & 3 \\ 3 & 1 & 2 \end{pmatrix}$, whereas σ^3 is the identity.

Example 3.17. In $\mathbb{Z}$, every element other than 0 has infinite order. For instance, no matter how many times we add 8 to itself, we will never get 0.

Example 3.18. In $\mathbb{Z}_6$, we have $|0| = 1, |1| = |5| = 6, |2| = |4| = 3$ and $|3| = 2$. For instance, $1 \cdot 4 = 4 \neq 0$ and $2 \cdot 4 = 2 \neq 0$, but $3 \cdot 4 = 0$, so $|4| = 3$.

The order of an element tells us a great deal about its powers.

Theorem 3.8. *Let G be a group and $a \in G$. Suppose $i, j \in \mathbb{Z}$. Then*

1. *if a has infinite order, then $a^i = a^j$ if and only if $i = j$; and*
2. *if $|a| = n < \infty$, then $a^i = a^j$ if and only if $i \equiv j \pmod{n}$.*

Proof. (1) Suppose that $a^i = a^j$, but $i \neq j$. Without loss of generality, say $i > j$. Then $a^i (a^j)^{-1} = a^j (a^j)^{-1} = e$. That is, $a^{i-j} = e$. But $i - j$ is a positive integer, and this contradicts the assumption that a has infinite order.

(2) Suppose that $a^i = a^j$. Once again, $a^{i-j} = e$. Using the division algorithm, write $i - j = nq + r$, with $q, r \in \mathbb{Z}$ and $0 \leq r < n$. Then

$$e = a^{i-j} = a^{nq+r} = (a^n)^q a^r.$$

But $a^n = e$. Thus, $a^r = e$. But n is the smallest positive integer having this property, and $r < n$. Therefore, $r = 0$. That is, $n \mid (i - j)$, as required.

Conversely, suppose that $i \equiv j \pmod{n}$. Then let us write $i - j = nk$, for some $k \in \mathbb{Z}$. But we now have

$$a^{i-j} = a^{nk} = (a^n)^k = e^k = e.$$

Thus, $a^{i-j} a^j = e a^j$, and hence $a^i = a^j$. ☐

Example 3.19. In $U(10)$, we see that $|3| = 4$. Thus, $3^i = 3^j$ if and only if $4 \mid (i - j)$. That is, $3^6 = 3^{14}$, but $3^2 \neq 3^{11}$.

Example 3.20. In $\mathbb{Z}$, all the integer multiples of 5 are distinct, because 5 has infinite order.

Corollary 3.2. *Let G be a group, and let $a \in G$ have order $n < \infty$. Then, for any integer i,*

1. $a^i = e$ *if and only if $n \mid i$; and*
2. $|a^i| = n/(i, n)$.

Proof. (1) By the preceding theorem, $a^i = e = a^0$ if and only if $i \equiv 0 \pmod{n}$.
(2) Suppose that, for some positive integer j, we have $(a^i)^j = e$. We see from (1) that since $a^{ij} = e$, we have $n \mid ij$. Write $ij = nk$, with $k \in \mathbb{Z}$. Letting $d = (n, i)$, we have $j(i/d) = k(n/d)$. Now, $(n/d, i/d) = 1$. Thus, by Corollary 2.2, since $n/d \mid j(i/d)$, we must have $n/d \mid j$. Therefore, $|a^i| \geq n/d$. But $(a^i)^{n/d} = a^{in/d} = a^{n(i/d)}$. As i/d is an integer, this is $(a^n)^{i/d} = e^{i/d} = e$. Thus, $|a^i| = n/d$, as required. ☐

Example 3.21. Again considering 3 in $U(10)$, we note that $3^i = 1$ if and only if i is a multiple of 4. Also, $|3^{14}| = 4/(4, 14) = 4/2 = 2$.

If G is a group, and $a, b \in G$, then we say that b is a **conjugate** of a if there exists a $c \in G$ such that $b = c^{-1}ac$.

Theorem 3.9. *In any group, conjugate elements have the same order.*

Proof. Suppose that $b = c^{-1}ac$ and that $a^n = e$, for some positive integer n. Then

$$b^n = (c^{-1}ac)^n = c^{-1}acc^{-1}acc^{-1} \cdots cc^{-1}ac$$
$$= c^{-1}aeae \cdots ac = c^{-1}a^n c = c^{-1}ec = e.$$

That is, $|b| \leq |a|$. But since $b = c^{-1}ac$, we have $a = (c^{-1})^{-1}bc^{-1}$. Thus, by the same argument, $|a| \leq |b|$. Therefore, $|a| = |b|$. ☐

Exercises

3.23. Find the order of each group, and the order of every element of each group.

1. $\mathbb{Z}_{12}$
2. $\mathbb{Z}_2 \times \mathbb{Z}_4$

3.24. Find the order of every element of each group. Is the group cyclic? If so, list all generators.

1. $U(14)$
2. S_3

3.25. Let $G = \langle a \rangle$ be a cyclic group of order 20. Find the orders of a^3, a^{12} and a^{15}.

3.26. Prove Theorem 3.6.

3.27. Let $a \in G$ and $b \in H$. Suppose that $|a| = 12$ and $|b| = 18$. Find the order of (a, b) in $G \times H$.

3.28. Let a and b be elements of odd order in a group. Show that a^2 and b^2 commute if and only if a and b commute. Also show that this does not have to hold if a and b have even order.

3.29. Let a and b be elements of a group. Show that the following pairs of elements have the same order:

1. a and a^{-1}; and
2. ab and ba.

3.30. Let $G = \{a_1, \ldots, a_k\}$ be a finite abelian group. Show that $a_1 a_2 \cdots a_k$ has order 1 or 2.

3.31. Show that it is possible for an abelian group to have exactly three elements of order 2, but not exactly four elements of order 2.

3.32. Suppose that G is a group in which every element has order 1 or 2. Show that G must be abelian.

3.5 Subgroups

One of the most important ways of obtaining new groups is to consider subgroups of a particular group.

Definition 3.7. Let G be a group with operation $*$. Then a subset H of G is called a **subgroup** of G if H is a group under the same operation $*$. In this case, we write $H \leq G$.

Example 3.22. Every group is a subgroup of itself, and $\{e\}$ is a subgroup of every group.

When we refer to a **proper** subgroup of G, we mean any subgroup other than G itself.

Example 3.23. We can see that $\mathbb{Z}$ is a subgroup of $\mathbb{Q}$, and both are subgroups of $\mathbb{R}$.

We do not have to check the entire definition of a group to see if a subset is a subgroup. For instance, we know that the group operation is associative on the entire group, so it is surely associative on every subset. The following theorem will save us some time.

Theorem 3.10. *Let G be a group and H a subset of G. Then H is a subgroup of G if and only if*

1. *$e \in H$ (the subset contains the identity);*
2. *$ab \in H$ for all $a, b \in H$ (the subset is closed); and*
3. *$a^{-1} \in H$ for all $a \in H$ (the subset contains all inverses).*

Proof. Let H be a subgroup of G. Then H has an identity, f. Thus, $ff = f$. But also, $ef = f$. By cancellation, $f = e$, giving (1). Then, by definition of a group, (2) and (3) must hold.

Conversely, suppose that (1)–(3) hold. We must check that H is a group. But by (2), H is closed. As the group operation is associative on G, it is associative on H. By (1) and (3), we have an identity and inverses as well. Therefore, H is a subgroup of G. $\qquad\square$

A remark is in order here. To wit, we could replace condition (1) in the above theorem with the weaker condition

(1') H is not the empty set.

Indeed, if $a \in H$, then we see from (3) that $a^{-1} \in H$, and then (2) tells us that $e = aa^{-1} \in H$. So why not express it that way? Because sometimes, the subset we are checking is not a subgroup. And we can tell immediately that that is the case if the subset does not contain e.

Example 3.24. The set of all even integers, $2\mathbb{Z}$, is a subgroup of $\mathbb{Z}$. Indeed, we certainly have $0 = 2 \cdot 0 \in 2\mathbb{Z}$. If $2m, 2n \in 2\mathbb{Z}$, then $2m + 2n = 2(m + n) \in 2\mathbb{Z}$, so we have closure. Finally, if $2m \in 2\mathbb{Z}$, then its inverse is $-(2m) = 2(-m) \in 2\mathbb{Z}$, and we have inverses. Of course, there is nothing magical about the number 2 here. If a is an integer, than $a\mathbb{Z}$ is a subgroup of $\mathbb{Z}$.

In fact, this last example is a specific case of a more general phenomenon. We have already encountered cyclic groups.

Definition 3.8. Let G be a group and $a \in G$. Then the **cyclic subgroup generated by** a is the set of all powers of a in G, and we write

$$\langle a \rangle = \{a^n : n \in \mathbb{Z}\}.$$

Of course, the group G is cyclic if and only if there exists an $a \in G$ such that $G = \langle a \rangle$.

Theorem 3.11. *If G is a group and $a \in G$, then $\langle a \rangle$ is a subgroup of G.*

Proof. Certainly $e = a^0 \in \langle a \rangle$. Take any $a^m, a^n \in \langle a \rangle$. Then $a^m a^n = a^{m+n} \in \langle a \rangle$. Finally, if $a^m \in \langle a \rangle$, then $(a^m)^{-1} = a^{-m} \in \langle a \rangle$. Now apply Theorem 3.10. □

Example 3.25. In $U(10)$, the powers of 3 are 1, 3, 9 and 7, so $\langle 3 \rangle = \{1, 3, 7, 9\} = U(10)$. Similarly, $\langle 7 \rangle = U(10)$. But the only powers of 9 are 1 and 9, so $\langle 9 \rangle = \{1, 9\}$. Also, $\langle 1 \rangle = \{1\}$.

Example 3.26. In $\mathbb{Z}_{12}$, the multiples of 8 are $1 \cdot 8 = 8, 2 \cdot 8 = 4$ and $3 \cdot 8 = 0$. Thus, we have $\langle 8 \rangle = \{0, 4, 8\}$.

Of course, we do not insist upon commutativity in groups, but it can be useful to know which elements commute with everything.

Definition 3.9. If G is a group, then the **centre** of G, denoted $Z(G)$, is the set of elements of G that commute with everything in G. That is, $Z(G) = \{z \in G : az = za$ for all $a \in G\}$.

Example 3.27. If G is abelian, then $Z(G) = G$.

Example 3.28. The centre of S_3 is the trivial subgroup, $\{e\}$. Verifying this is a matter of considering each element of S_3 other than the identity, and finding another element that does not commute with it.

Example 3.29. The centre of $GL_2(\mathbb{R})$ is the set of all matrices $\begin{pmatrix} a & 0 \\ 0 & a \end{pmatrix}$, where $0 \neq a \in \mathbb{R}$. We leave the proof as Exercise 3.36.

Theorem 3.12. *If G is a group, then $Z(G)$ is a subgroup of G.*

Proof. Certainly $ea = a = ae$ for all $a \in G$, so $e \in Z(G)$. If $y, z \in Z(G)$ and $a \in G$, then $yza = yaz = ayz$; thus, $yz \in Z(G)$. Also, if $z \in Z(G)$ and $a \in G$, then $a^{-1}z = za^{-1}$. Inverting both sides, we get $z^{-1}a = az^{-1}$. Thus, $z^{-1} \in Z(G)$. The proof is complete. □

Some shortcuts are possible when it comes to testing whether a subset is a subgroup.

Theorem 3.13. *Let G be a group and H a subset of G. Then H is a subgroup if and only if*

1. $e \in H$; *and*
2. $ab^{-1} \in H$ *whenever* $a, b \in H$.

Proof. Suppose that H is a subgroup. By Theorem 3.10, we know that $e \in H$ and if $a, b \in H$, then $b^{-1} \in H$, and therefore $ab^{-1} \in H$.

Conversely, suppose that H satisfies (1) and (2). Take any $a, b \in H$. Then since $e \in H$, we have $ea^{-1} = a^{-1} \in H$ and, similarly, $b^{-1} \in H$. Therefore, $a(b^{-1})^{-1} = ab \in H$. In view of Theorem 3.10, H is a subgroup. $\qquad\square$

Once again, instead of checking that $e \in H$, it is enough to verify that H is not empty. We can even make things simpler if H is a finite set.

Theorem 3.14. *Let G be a group and H a finite subset of G. Then $H \leq G$ if and only if*

1. $e \in H$; *and*
2. $ab \in H$ *whenever* $a, b \in H$.

Proof. If H is a subgroup of G, then Theorem 3.10 tells us that (1) and (2) hold. Conversely, suppose that (1) and (2) are true. By Theorem 3.10, we only need to show that if $a \in H$ then $a^{-1} \in H$. In view of (2), we have $aa = a^2 \in H$, and hence $a^2a = a^3 \in H$, and so on; thus, $a^n \in H$ for all positive integers n. But there are infinitely many such powers, and H is finite. Thus, there exist positive integers m and n, with $m > n$, such that $a^m = a^n$. Then $a^{m-n} = e$. If $m - n = 1$, then $a = e$, in which case $a^{-1} = e \in H$. So, suppose that $m - n > 1$. Then $aa^{m-n-1} = a^{m-n-1}a = a^{m-n} = e$. That is, $a^{m-n-1} = a^{-1}$. But $m - n - 1$ is a positive integer, and therefore $a^{m-n-1} \in H$, as required. $\qquad\square$

We must be careful only to use the above theorem when H is finite. To see why, let G be the additive group of integers, and let H be the set of nonnegative integers. Then H contains 0 and is closed under addition, but H is not a subgroup of G, since 1 has no additive inverse.

Example 3.30. Let $G = \mathbb{Z}_8 \times \mathbb{Z}_8$, and let $H = \{(a, b) \in G : 4a = 0\}$. We claim that H is a subgroup of G. Clearly, $(0, 0) \in H$. Also, if $(a, b), (c, d) \in H$, then $(a, b) + (c, d) = (a + c, b + d)$, where $4(a + c) = 4a + 4c = 0 + 0 = 0$. Therefore, H is closed, and hence a subgroup.

We conclude the section with an extended, and important, example. Suppose we have a floor consisting of featureless square ceramic tiles. Let us pry up one of the tiles, and then consider all of the ways in which we can move the tile around in three-dimensional space, and then replace it so that it looks exactly as it did when we began. For convenience, let us label the vertices of the square 1, 2, 3 and 4. Then we can see that each vertex moves to the position of some vertex. Also, two vertices will not move to the same place. Once we have positioned the vertices, we

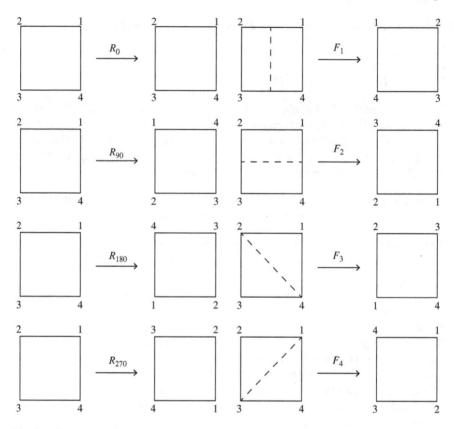

Fig. 3.1 The symmetries in the dihedral group D_8

are done. Therefore, these **symmetries** of a square can be regarded as permutations of the set $\{1, 2, 3, 4\}$; that is, as elements of S_4. Of course, the identity of S_4 is such a symmetry, and if we compose two of these symmetries, then we get another. Thus, by Theorem 3.14, they form a subgroup of S_4, known as the **dihedral group** of order 8, and denoted D_8.

What are the elements of D_8? They are illustrated in Figure 3.1. There are four **rotations**, R_0, R_{90}, R_{180} and R_{270}, where R_α is a counterclockwise rotation by α degrees. We also have four **flips**, F_1 through F_4, about the lines shown in the diagram.

And that is all! Indeed, vertex 1 can go to any of the 4 vertices, but then vertex 2 must be adjacent to it, not diagonally opposite. Once vertices 1 and 2 are positioned, the others fall into place. Therefore, $|D_8| = 8$. The group table of D_8 is shown in Table 3.4.

Remember that when we write $R_{90} F_1 = F_3$, we mean perform F_1 first, then R_{90}. We note that $R_{90} F_1 \neq F_1 R_{90}$, and therefore D_8 is a nonabelian group of order 8. In fact, a quick glance through the table tells us that the centre of D_8 is $\{R_0, R_{180}\}$.

Table 3.4 Group table for the dihedral group D_8

	R_0	R_{90}	R_{180}	R_{270}	F_1	F_2	F_3	F_4
R_0	R_0	R_{90}	R_{180}	R_{270}	F_1	F_2	F_3	F_4
R_{90}	R_{90}	R_{180}	R_{270}	R_0	F_3	F_4	F_2	F_1
R_{180}	R_{180}	R_{270}	R_0	R_{90}	F_2	F_1	F_4	F_3
R_{270}	R_{270}	R_0	R_{90}	R_{180}	F_4	F_3	F_1	F_2
F_1	F_1	F_4	F_2	F_3	R_0	R_{180}	R_{270}	R_{90}
F_2	F_2	F_3	F_1	F_4	R_{180}	R_0	R_{90}	R_{270}
F_3	F_3	F_1	F_4	F_2	R_{90}	R_{270}	R_0	R_{180}
F_4	F_4	F_2	F_3	F_1	R_{270}	R_{90}	R_{180}	R_0

We do not have to begin with a square. Indeed, let us consider any regular n-gon, with $n \geq 3$. Then the symmetries of this n-gon form a subgroup of S_n. By precisely the same arguments as above, it will consist of n rotations and n flips. (There are n possible locations for a given vertex, and once it is fixed, 2 choices for an adjacent vertex. After fixing those vertices, there are no choices remaining.) We call this group of symmetries the dihedral group of order $2n$, and denote it by D_{2n}. In particular, if $n = 3$, we note that D_6 consists of all of S_3, but for larger n, D_{2n} is a proper subgroup of S_n. In any case, we now have an example of a nonabelian group of every even order except 2 and 4.

Exercises

3.33. In each case, is H a subgroup of G?

1. $G = GL_2(\mathbb{R})$, H is the set of matrices with determinant 1
2. $G = D_{10}$, H is the set of flips
3. $G = \mathbb{Q}$, $H = \{a/b : a, b \in \mathbb{Z}, 2 \nmid b\}$

3.34. In each case, is H a subgroup of G?

1. $G = D_{10}$, H is the set of rotations
2. $G = \mathbb{Q}$, H is the set of nonnegative rational numbers
3. G is the multiplicative group of nonzero rational numbers, H is the set of positive rational numbers

3.35. For each positive integer $n \geq 3$, determine the centre of D_{2n}.

3.36. Show that the centre of $GL_2(\mathbb{R})$ consists of the matrices $\begin{pmatrix} a & 0 \\ 0 & a \end{pmatrix}$, for all $0 \neq a \in \mathbb{R}$.

3.37. Show that the intersection of two subgroups of G is also a subgroup. Then extend this to show that if N_i is a subgroup of G for every i in some set T, then $\bigcap_{i \in T} N_i$ is a subgroup of G.

3.38. Let H and K be subgroups of G. Show that $H \cup K$ is a subgroup of G if and only if either $H \subseteq K$ or $K \subseteq H$.

3.39. Find every cyclic subgroup of each of the following groups.

1. $\mathbb{Z}_{20}$
2. $U(16)$

3.40. Let G be an abelian group and $n \in \mathbb{N}$. Let $H = \{a \in G : a^n = e\}$ and $K = \{a^n : a \in G\}$. Show that H and K are subgroups of G.

3.41. In any dihedral group, show that a rotation followed by a rotation, or a flip followed by a flip, is a rotation, whereas a rotation followed by a flip or a flip followed by a rotation is a flip.

3.42. Let G be the set of all sequences of integers $(a_1, a_2, a_3, \ldots)$.

1. Show that G is a group under $(a_1, a_2, \ldots) + (b_1, b_2, \ldots) = (a_1 + b_1, a_2 + b_2, \ldots)$.
2. Let H be the set of all elements $(a_1, a_2, \ldots)$ of G such that only finitely many a_i are different from 0 (and $(0, 0, 0, \ldots) \in H$). Show that H is a subgroup of G.

3.6 Cyclic Groups

Cyclic groups have a very straightforward structure. Let us prove a few basic facts. First, we can illustrate the link between the order of an element and the order of a group.

Theorem 3.15. *Let $G = \langle a \rangle$ be cyclic. If a has infinite order, then all powers of a are distinct. If $|a| = n < \infty$, then the distinct elements of G are $e, a, a^2, \ldots, a^{n-1}$. In particular, $|a| = |\langle a \rangle|$.*

Proof. If a has infinite order, then we can use Theorem 3.8. Suppose $|a| = n < \infty$. If $m \in \mathbb{Z}$, then write $m = nq + r$, where $q, r \in \mathbb{Z}$ and $0 \le r < n$. Then as $m \equiv r$ (mod n), Theorem 3.8 tells us that $a^m = a^r$. In particular, every element of G is equal to some a^i ($0 \le i < n$). Now, suppose that $a^i = a^j$, with $0 \le i < j < n$. Then by Theorem 3.8, $i \equiv j$ (mod n). But given the range of values for i and j, this is impossible. □

The subgroups of cyclic groups are also easy to determine.

Theorem 3.16. *Every subgroup of a cyclic group is cyclic.*

Proof. Let $G = \langle a \rangle$, and let $H \le G$. If $H = \{e\}$, then $H = \langle e \rangle$, and we are done, so assume that H is not the trivial subgroup. Then H contains a^m, for some nonzero integer m. If $m < 0$, then H also contains $(a^m)^{-1} = a^{-m}$, so H contains a positive power of a. Let n be the smallest positive integer such that $a^n \in H$. We claim that $H = \langle a^n \rangle$. Surely H contains every power of a^n, so $\langle a^n \rangle \le H$. But suppose $a^k \in H$. Then write $k = nq + r$, with $q, r \in \mathbb{Z}$ and $0 \le r < n$. Now, H contains a^k and $(a^n)^{-q}$, and therefore $a^k(a^n)^{-q} = a^{k-nq} = a^r$. But n is the smallest positive integer such that $a^n \in H$. As $r < n$, we can only have $r = 0$. Thus, $a^k = (a^n)^q \in \langle a^n \rangle$. That is, $H \le \langle a^n \rangle$, proving the claim. We are done. □

Actually, we can say more.

Corollary 3.3. *Let $G = \langle a \rangle$, where $|a| = n < \infty$. Then the order of every subgroup of G is a divisor of n. Furthermore, if m is a positive divisor of n, then G has exactly one subgroup of order m, namely $\langle a^{n/m} \rangle$.*

Proof. By the preceding theorem, every subgroup is of the form $\langle a^k \rangle$, for some $k \in \mathbb{Z}$. But Corollary 3.2 tells us that the order of every power of a is a divisor of n.

Let m be a positive divisor of n. Again using Corollary 3.2, we see that $|a^{n/m}| = n/(n, n/m) = n/(n/m) = m$. Thus, $\langle a^{n/m} \rangle$ is indeed a subgroup of order m. Let us check that it is unique. Suppose that $\langle a^k \rangle$ is a subgroup of order m. Then $|a^k| = m$, and so $(a^k)^m = e$. That is, $a^{km} = e$, hence $n|km$. But then $n/m|k$. Writing $k = (n/m)i$, with $i \in \mathbb{Z}$, we have $a^k = (a^{n/m})^i \in \langle a^{n/m} \rangle$. Thus, $\langle a^k \rangle \leq \langle a^{n/m} \rangle$. But these two subgroups have the same order. Therefore, they are equal. $\square$

Example 3.31. Let $G = \langle a \rangle$, where $|a| = 20$. Then G has exactly one subgroup of order 5, namely $\langle a^4 \rangle = \{e, a^4, a^8, a^{12}, a^{16}\}$.

Thus, a cyclic group can only have one subgroup of any given order. This is a special property of cyclic groups; indeed, D_8 and $U(8)$ are easily seen to have several different cyclic subgroups of order 2.

We can also discuss the number of elements of a particular order in a cyclic group. Some notation will be helpful. The following function is named after Leonhard Euler.

Definition 3.10. The **Euler phi-function** is a function $\varphi : \mathbb{N} \to \mathbb{N}$, where $\varphi(n)$ is the number of positive integers less than or equal to n that are relatively prime to n.

Example 3.32. Of the integers from 1 to 10, only 1, 3, 7 and 9 are relatively prime to 10, so $\varphi(10) = 4$. The first few values of φ are given in Table 3.5.

From the definition of the group $U(n)$, we immediately obtain the following.

Theorem 3.17. *For any positive integer n, $|U(n)| = \varphi(n)$.*

But we can also use the Euler function to count the elements of a particular order in a finite cyclic group.

Theorem 3.18. *Let $G = \langle a \rangle$ be a cyclic group of order n. Let m be a positive divisor of n. Then the number of elements of order m in G is $\varphi(m)$.*

Proof. If b is an element of order m in G, then $\langle b \rangle$ must be the unique cyclic subgroup of order m. That is, all of the elements of order m in G are in the cyclic subgroup of order m. Thus, we may as well assume that G is cyclic of order m. We must

Table 3.5 Values of the Euler phi-function

n	1	2	3	4	5	6	7	8	9	10	11	12	13	14	15	16
$\varphi(n)$	1	1	2	2	4	2	6	4	6	4	10	4	12	6	8	8

therefore decide which elements of this group have order m. But by Corollary 3.2, the order of a^k is m if and only if $(k, m) = 1$. By definition, the number of such k, with $1 \le k \le m$, is $\varphi(m)$. (By Theorem 3.15, the elements of $\langle a \rangle$ are precisely a^i, with $0 \le i < m$, but as $a^0 = a^m = e$, this is the same as considering a^i with $1 \le i \le m$.)

$\square$

Example 3.33. Let $G = \langle a \rangle$ be cyclic of order 50. Then we know that there are $\varphi(10) = 4$ elements of order 10 in G. They lie in the subgroup of order 10, namely $\langle a^{50/10} \rangle = \langle a^5 \rangle$. Indeed, the precise elements will be $(a^5)^k$, where $(k, 10) = 1$. This means that $k \in \{1, 3, 7, 9\}$, so the elements of order 10 are a^5, a^{15}, a^{35} and a^{45}. It is worth noting that the number of elements of order 10 in a cyclic group of order one million is also $\varphi(10) = 4$.

For relatively small numbers, $\varphi(n)$ is easy to determine, but for large n, it would be tedious to go through all the numbers from 1 to n in order to see if they are relatively prime to n. Happily, there is a shortcut. The first part of the following theorem is Exercise 3.45. It will make more sense if we postpone the proof of the second part until Section 4.4.

Theorem 3.19. *Let p be a prime number, and let m and n be positive integers. Then*

1. $\varphi(p^n) = p^n - p^{n-1}$*; and*
2. *if $(m, n) = 1$, then $\varphi(mn) = \varphi(m)\varphi(n)$.*

Thus, we can determine $\varphi(n)$ by writing n as a product of powers of primes and then using the above theorem.

Example 3.34. We have $\varphi(81) = 81 - 27 = 54$ and $\varphi(540) = \varphi(4)\varphi(27)\varphi(5) = (4 - 2)(27 - 9)(5 - 1) = 144$.

Exercises

3.43. 1. Let $G = \langle a \rangle$ be a cyclic group of order 12. List every subgroup of G.
2. List every subgroup of $\mathbb{Z}_{12}$.

3.44. 1. Let $G = \langle a \rangle$ be a cyclic group of order 120. List all of the elements of order 12 in G.
2. How many elements of order 12 are there in a cyclic group of order 1200?

3.45. Let p be a prime and n a positive integer. Show that $\varphi(p^n) = p^n - p^{n-1}$.

3.46. Find all positive integers n such that $|U(n)| = 24$.

3.47. Let G be a nonabelian group. If H and K are cyclic subgroups of G, does it follow that $H \cap K$ is also a cyclic subgroup? Prove that it does, or provide a counterexample.

3.48. Let $G = \langle a \rangle$ be infinite cyclic. If m and n are positive integers, find a generator for $\langle a^m \rangle \cap \langle a^n \rangle$.

3.49. Let n be a positive integer and let T be the set of positive integers that divide n. Show that $\sum_{k \in T} \varphi(k) = n$.

3.50. For precisely which positive integers n is $U(2^n)$ cyclic?

3.51. Let G be any group and n a positive integer.

1. If H and K are subgroups of order n in G, and $H \neq K$, show that $H \cap K$ does not contain any elements of order n.
2. Show that the number of elements of order n in G is either a multiple of $\varphi(n)$ or infinite.

3.52. Show that a nontrivial group G has no nontrivial proper subgroups if and only if G is cyclic of prime order. (Do not assume, to begin with, that G is finite.)

3.7 Cosets and Lagrange's Theorem

One important fact we learned in the preceding section is that if G is a finite cyclic group, then the order of every subgroup of G divides the order of G. As it turns out, this is true for all finite groups, but a different proof will be required. To this end, we need some new terminology.

Definition 3.11. Let G be a group and H a subgroup. If $a, b \in G$, we say that a is **congruent to b modulo H**, and we write $a \equiv b \pmod{H}$, if $a^{-1}b \in H$ (or, in the case of an additive group, if $-a + b \in H$).

Example 3.35. Let $G = \mathbb{Z}$ and $H = 5\mathbb{Z}$. Then as $-1 + 16 = 15 \in H$, we see that $1 \equiv 16 \pmod{H}$. In this particular case, the notion is identical to congruence modulo 5.

Example 3.36. Let $G = U(20) = \{1, 3, 7, 9, 11, 13, 17, 19\}$, and let $H = \langle 3 \rangle$; namely, $H = \{1, 3, 7, 9\}$. Then we note that $13^{-1} \cdot 19 = 17 \cdot 19 = 3 \in H$. Thus, $13 \equiv 19 \pmod{H}$.

Lemma 3.1. *Let G be a group and H a subgroup. Then congruence modulo H is an equivalence relation on G.*

Proof. Reflexivity: If $a \in G$, then $a^{-1}a = e \in H$, and therefore $a \equiv a \pmod{H}$. Symmetry: If $a, b \in G$ and $a \equiv b \pmod{H}$, then $a^{-1}b \in H$, and therefore $(a^{-1}b)^{-1} = b^{-1}a$ lies in H as well. But this means that $b \equiv a \pmod{H}$. Transitivity: Suppose that $a, b, c \in G$, where $a \equiv b \pmod{H}$ and $b \equiv c \pmod{H}$. Then $a^{-1}b, b^{-1}c \in H$. But in this case, H contains their product, $a^{-1}bb^{-1}c = a^{-1}c$. Thus, $a \equiv c \pmod{H}$. We are done. $\qquad \square$

What are the equivalence classes?

Lemma 3.2. *Let G be a group and H a subgroup. If $a \in G$, then its equivalence class with respect to congruence modulo H is the set $\{ah : h \in H\}$.*

Proof. If $a \equiv b \pmod{H}$, then $a^{-1}b \in H$, so $a^{-1}b = h$, for some $h \in H$. Thus, $b = ah$, which is in our set. Conversely, if $b = ah$, for some $h \in H$, then $a^{-1}b = h \in H$, and therefore $a \equiv b \pmod{H}$. □

We need a name for this set.

Definition 3.12. Let G be a group, $H \leq G$ and $a \in G$. Then the **left coset** of a with respect to H is the set $\{ah : h \in H\}$, which is denoted aH. (Note: If the group operation is addition, then we will write $a + H = \{a + h : h \in H\}$.)

Example 3.37. If $G = U(20)$, let $H = \langle 9 \rangle = \{1, 9\}$. Then $3H = \{3 \cdot 1, 3 \cdot 9\} = \{3, 7\}$. Also, $7H = \{7 \cdot 1, 7 \cdot 9\} = \{3, 7\}$, so $3H = 7H$. Furthermore, $1H = 9H = H$, $11H = 19H = \{11, 19\}$ and $13H = 17H = \{13, 17\}$. Note that these left cosets partition G.

Example 3.38. Let $G = \mathbb{Z}$ and $H = 3\mathbb{Z}$. Then there are three left cosets: $0 + H = H$, $1 + H = \{\cdots, -5, -2, 1, 4, 7, \ldots\}$ and $2 + H = \{\cdots, -4, -1, 2, 5, 8, \ldots\}$. Note that $2 + H = 5 + H = -13 + H$, and so on. Again, the left cosets partition G.

In general, we know that equivalence classes always partition a set. Therefore, we can record the following result.

Theorem 3.20. *Let G be a group and H a subgroup. Then the left cosets of H in G partition G. In particular,*

1. *each $a \in G$ is in exactly one left coset, namely aH; and*
2. *if $a, b \in G$, then either $aH = bH$ or $aH \cap bH = \varnothing$.*

Two points should be kept in mind here. First, left cosets are not subgroups! Remember, the left cosets partition G, and therefore the identity can only be in one of them, namely, $eH = H$. The rest cannot possibly be subgroups. Second, as we have already seen, when we write aH, the element a is not unique. Indeed, since the left cosets are equivalence classes, we have $aH = bH$ if and only if $a^{-1}b \in H$.

We can now prove our first big result on finite groups, due to Joseph-Louis Lagrange.

Theorem 3.21. (Lagrange's Theorem). *Let G be a finite group and H a subgroup. Then $|H|$ divides $|G|$.*

Proof. We have already seen that G is partitioned into left cosets; in particular, $|G|$ is the sum of the sizes of these left cosets. But for any $a \in G$, $aH = \{ah : h \in H\}$. Now, if $ah_1 = ah_2$, with $h_1, h_2 \in H$, then by the cancellation law, $h_1 = h_2$. Therefore, aH consists of precisely $|H|$ distinct elements. It now follows that the order of G is $|H|$ multiplied by the number of left cosets. In particular, $|H|$ divides $|G|$. □

Definition 3.13. Let G be a group and $H \leq G$. Then the **index** of H in G, denoted $[G : H]$, is the number of left cosets of H in G.

Corollary 3.4. *If G is a finite group and H is a subgroup, then $[G : H] = |G|/|H|$.*

Proof. This is immediate from the proof of the above theorem.   □

Example 3.39. Let $G = D_8$ and $H = \langle R_{90} \rangle = \{R_0, R_{90}, R_{180}, R_{270}\}$. Then $[G : H] = |G|/|H| = 8/4 = 2$. Thus, there are two left cosets. One is $R_0 H = H$. The other must be $F_1 H = \{F_1, F_2, F_3, F_4\}$. If $K = \langle F_1 \rangle = \{R_0, F_1\}$, then it must have $8/2 = 4$ left cosets. One is $R_0 K = K$. To find another, just choose an element of G that we have not yet found, say R_{90}. Then we get $R_{90} K = \{R_{90}, F_3\}$. We haven't yet used F_2, so take $F_2 K = \{F_2, R_{180}\}$. Finally, we can take $R_{270} K = \{R_{270}, F_4\}$.

Example 3.40. Note that the subgroups of an infinite group can be of finite or infinite index. For instance, we saw above that $0 + 3\mathbb{Z}$, $1 + 3\mathbb{Z}$ and $2 + 3\mathbb{Z}$ are the distinct left cosets of $3\mathbb{Z}$ in $\mathbb{Z}$. Thus, $[\mathbb{Z} : 3\mathbb{Z}] = 3$. On the other hand, $\mathbb{Z}$ has infinite index in $\mathbb{Q}$. To see this, observe that for all positive integers n, the left cosets $1/n + \mathbb{Z}$ are distinct. And there are still more!

Lagrange's theorem has a beautiful consequence.

Corollary 3.5. *Let G be a finite group, and $a \in G$. Then the order of a divides the order of G.*

Proof. The order of a is the order of the cyclic subgroup generated by a, and that must divide the order of G.   □

Example 3.41. Note that $|D_8| = 8$, the identity has order 1, R_{180} and the flips all have order 2 and $|R_{90}| = |R_{270}| = 4$. All of the orders are divisors of 8.

Of course, it does not follow that because a number n divides the order of a group, then the group has an element of that order. Indeed, if that were always true, then a group of order n would have to have an element of order n, and therefore every finite group would be cyclic, which is not the case.

One important thing that we can do is to try to classify all the groups of some particular order. We can now make a step in that direction.

Corollary 3.6. *Every group of prime order is cyclic.*

Proof. Take $e \neq a \in G$, where $|G|$ is a prime. As $|a|$ divides $|G|$, and $|a| \neq 1$, we must have $|a| = |G|$. But then $|G| = |\langle a \rangle|$, and therefore $G = \langle a \rangle$.   □

Not surprisingly, there is also such a thing as a right coset. Indeed, if we had defined $a \equiv b \pmod{H}$ to mean that $ab^{-1} \in H$, then we would have found that this is still an equivalence relation, and the equivalence classes would have been as follows.

Definition 3.14. Let G be a group and $H \leq G$. Then for any $a \in G$, the **right coset** of a with respect to H is $Ha = \{ha : h \in H\}$. (If G is an additive group, then we write $H + a = \{h + a : h \in H\}$.)

If G is abelian, then there is no distinction between left and right cosets. In nonabelian groups, right cosets also partition G, but possibly in a different way.

Example 3.42. Take G, H and K as in Example 3.39. Then we can see that one right coset of H in G is $HR_0 = H = R_0H$ and the other must be $HF_1 = \{F_1, F_2, F_3, F_4\} = F_1H$. Here, the left and right cosets agree. But it is not the same for K. For instance, $R_{90}K = \{R_{90}, F_3\}$, but $KR_{90} = \{R_{90}, F_4\}$.

Would it have made a difference if we had defined the index of H in G using right cosets instead of left? Fortunately, no. This is clear if G is finite, as Lagrange's theorem works equally well using right cosets. But what if G is an infinite group having a subgroup H of index $n < \infty$? Then notice that $aH = bH$ if and only if $a^{-1}b \in H$, but also $Ha^{-1} = Hb^{-1}$ if and only if $a^{-1}b \in H$. Thus, if the distinct left cosets of H in G are $a_1H, a_2H, \ldots, a_nH$, then the distinct right cosets are $Ha_1^{-1}, Ha_2^{-1}, \ldots, Ha_n^{-1}$.

Exercises

3.53. For each group G and subgroup H, find all the left cosets and right cosets of H in G.

1. $G = \mathbb{Z}$, $H = 4\mathbb{Z}$
2. $G = D_8$, $H = \langle F_2 \rangle$

3.54. For each group G and subgroup H, find all the left cosets and right cosets of H in G.

1. $G = U(13)$, $H = \langle 8 \rangle$
2. $G = S_3$, $H = \left\langle \begin{pmatrix} 1 & 2 & 3 \\ 2 & 1 & 3 \end{pmatrix} \right\rangle$

3.55. Let G be a group whose order is the product of two (not necessarily distinct) primes. Show that every proper subgroup of G is cyclic.

3.56. Let G be a group of order p^n, for some prime p and positive integer n. Show that G has an element of order p.

3.57. Let G be a group having a subgroup H of order 28 and a subgroup K of order 65. Show that $H \cap K = \{e\}$.

3.58. Let G be a finite group having an element of order k, for each $1 \le k \le 10$. What is the smallest possible order of G? Show that a group of that order exists having this property.

3.59. Let $G = \{a_1, \ldots, a_k\}$ be an abelian group of odd order k. Show that $a_1 a_2 \cdots a_k = e$.

3.60. Show that every group of order 55 contains an element of order 5 and an element of order 11.

3.61. Let G be a group with subgroups H and K. If $[G : K] = n$, show that $[H : H \cap K] \le n$.

3.62. Let G be a group with subgroups H and K such that $K \le H$. Suppose that $[G : H] = m$ and $[H : K] = n$. Show that $[G : K] = mn$. (Do not assume that G is finite.)

Chapter 4
Factor Groups and Homomorphisms

In the previous chapter, we tended to consider just one group at a time. But we need to find ways of relating groups to each other. For instance, we would like to know if two groups are, in every meaningful sense, the same. This would be the case if we took a group and created a new one by simply changing the labels on the group elements, but left the structure otherwise intact. Surely, we would not wish to think of these as different sorts of groups.[1] This is where the notion of a group homomorphism and, in particular, an isomorphism, will come into the picture.

But first, we will discuss factor groups. These constitute an important way of creating new groups from old ones. As we shall see, there is a natural connection between factor groups and homomorphisms. In order to define a factor group, we require a special sort of subgroup, called a normal subgroup. Let us begin there.

4.1 Normal Subgroups

Let H be a subgroup of G. We would like to form a group whose elements are the left cosets aH. Unfortunately, as we shall see in the next section, not just any subgroup will suffice; we need an extra condition. This is where normal subgroups come in.

Recall that if $H \leq G$, then the left cosets of H do not necessarily coincide with the right cosets. We need to consider subgroups for which they do coincide.

Definition 4.1. Let G be a group and N a subgroup. We say that N is a **normal subgroup** of G if $aN = Na$ for all $a \in G$.

Example 4.1. For every group G, G is a normal subgroup of itself, as $aG = Ga = G$ for all a. Also, $\{e\}$ is normal. Indeed, $a\{e\} = \{e\}a = \{a\}$ for all a.

---

[1] Upon reading this sentence aloud, the author failed to stop himself from writing "And don't call me Shirley." We miss you, Leslie Nielsen!

© Springer International Publishing AG, part of Springer Nature 2018
G. T. Lee, *Abstract Algebra*, Springer Undergraduate Mathematics Series,
https://doi.org/10.1007/978-3-319-77649-1_4

Example 4.2. The centre of every group is a normal subgroup. Indeed, writing $Z = Z(G)$, we have $aZ = \{az : z \in Z\} = \{za : z \in Z\} = Za$. In fact, every subgroup of $Z(G)$ is normal in G, for precisely the same reason. In particular, every subgroup of an abelian group is normal.

Be warned: this last example can be a bit misleading. Remember, when we say that $aN = Na$, we do not necessarily mean that $an = na$ for all $n \in N$. Indeed, we could have $an = n_1 a$, for some different $n_1 \in N$. The following example may be helpful.

Example 4.3. Refer to Example 3.42. We saw that in D_8, the subgroup $\langle R_{90} \rangle$ is normal. That is, $a\langle R_{90} \rangle = \langle R_{90} \rangle a$ for all $a \in D_8$. This does not mean that $aR_{90} = R_{90}a$, however. Indeed, $F_1 R_{90} = R_{270} F_1$. But as $R_{270} \in \langle R_{90} \rangle$, this is fine. We also saw in that example that $\langle F_1 \rangle$ is not a normal subgroup of D_8, as $R_{90}\langle F_1 \rangle \neq \langle F_1 \rangle R_{90}$.

There is one special case in which we do not need to worry about normality.

Theorem 4.1. *If G is a group, then any subgroup of index 2 is normal in G.*

Proof. Let H be a subgroup of index 2. Then one of the left cosets is H, and the other must consist of everything outside of H. In particular, $aH = H$ if $a \in H$ and aH is the other left coset if $a \notin H$. But exactly the same thing can be said for right cosets! So the left and right cosets agree. □

Example 4.4. It is worth noting that if N is a normal subgroup of G and H is a normal subgroup of N, it does not necessarily follow that H is normal in G. For instance, $N = \{R_0, R_{180}, F_1, F_2\}$ is a normal subgroup of D_8. (To check that it is a subgroup, use Theorem 3.14. To check that it is normal, use Theorem 4.1.) Also, $H = \langle F_1 \rangle$ is normal in N. (Again, it has index 2.) But as we saw in Example 4.3, H is not normal in D_8.

Let us define a new subgroup.

Definition 4.2. Let H be a subgroup of G. Then for any $a \in G$, we write $a^{-1}Ha = \{a^{-1}ha : h \in H\}$.

Theorem 4.2. *If H is a subgroup of G and $a \in G$, then $a^{-1}Ha$ is a subgroup of G. Furthermore, $|a^{-1}Ha| = |H|$.*

Proof. We have $e \in H$, and therefore $e = a^{-1}ea \in a^{-1}Ha$. If $a^{-1}h_1a, a^{-1}h_2a \in a^{-1}Ha$, then

$$(a^{-1}h_1a)(a^{-1}h_2a) = a^{-1}h_1(aa^{-1})h_2a = a^{-1}h_1eh_2a = a^{-1}h_1h_2a \in a^{-1}Ha,$$

since $h_1h_2 \in H$. Finally, if $a^{-1}ha \in a^{-1}Ha$, then $(a^{-1}ha)^{-1} = a^{-1}h^{-1}a \in a^{-1}Ha$, since $h^{-1} \in H$. Thus, $a^{-1}Ha$ is a subgroup of G. Also, given the definition of $a^{-1}Ha$, it is clear that we can only get one element for each element of H. But if $a^{-1}h_1a = a^{-1}h_2a$, then by cancellation, $h_1 = h_2$. Thus, $|a^{-1}Ha| = |H|$. □

We can use this to give several different ways of saying that a subgroup is normal.

Theorem 4.3. *Let G be a group and H a subgroup. Then the following are equivalent:*

1. *H is normal in G;*
2. $a^{-1}ha \in H$ *for all* $h \in H$ *and all* $a \in G$;
3. $a^{-1}Ha \subseteq H$ *for all* $a \in G$; *and*
4. $a^{-1}Ha = H$ *for all* $a \in G$.

Proof. It is clear that (4) implies (3) and (3) implies (2). Let us show that (2) implies (1). Suppose that (2) holds. Take any $a \in G$. Then for any $h \in H$, we have $a^{-1}ha = h_1$, for some $h_1 \in H$. Thus, $ha = ah_1 \in aH$. That is, $Ha \subseteq aH$. Also, $(a^{-1})^{-1}ha^{-1} = h_2$, for some $h_2 \in H$. That is, $aha^{-1} = h_2$, and therefore $ah = h_2a \in Ha$. Thus, $aH \subseteq Ha$, so $aH = Ha$ and (1) is proved.

Finally, let us show that (1) implies (4). Let H be a normal subgroup of G. Take any $a \in G$. Then $Ha = aH$. Thus, for any $h \in H$, we have $ha \in aH$, and therefore $ha = ah_1$, for some $h_1 \in H$. That is, $a^{-1}ha = h_1 \in H$. Therefore, $a^{-1}Ha \subseteq H$. But using a^{-1} in place of a, we also get $aHa^{-1} \subseteq H$. Hence, if $h \in H$, then $aha^{-1} = h_2$, for some $h_2 \in H$. But now $h = a^{-1}h_2a \in a^{-1}Ha$. That is, $H \subseteq a^{-1}Ha$, and we are done. $\square$

Example 4.5. Let $SL_n(\mathbb{R})$ denote the set of all matrices in $GL_n(\mathbb{R})$ having determinant 1. We call this the **special linear group**. In view of Exercise 3.33, we know that $SL_n(\mathbb{R})$ is a subgroup of $GL_n(\mathbb{R})$. In fact, it is a normal subgroup. Indeed, if $A \in SL_n(\mathbb{R})$ and $B \in GL_n(\mathbb{R})$, then

$$\det(B^{-1}AB) = \det(B^{-1})\det(A)\det(B) = \det(B^{-1})\det(B)\det(A),$$

since the determinants are just real numbers. But this is $\det(B^{-1}B)\det(A) = 1$, since $B^{-1}B$ is the identity matrix and $\det(A) = 1$. Therefore, $B^{-1}AB \in SL_n(\mathbb{R})$, and by Theorem 4.3, $SL_n(\mathbb{R})$ is indeed normal.

Another useful construction is the following.

Definition 4.3. If H and K are subgroups of G, then we write $HK = \{hk : h \in H, k \in K\}$. (If the group operation is addition, write $H+K = \{h+k : h \in H, k \in K\}$.)

Note that HK is a subset of G, not necessarily a subgroup! It is easy to come up with examples where HK is not a subgroup, but the following theorem will lead us to some that cannot possibly work.

Theorem 4.4. *If H and K are finite subgroups of a group G, then*

$$|HK| = \frac{|H||K|}{|H \cap K|}.$$

Proof. Considering all possible $h \in H$ and $k \in K$, it is clear that we can produce at most $|H||K|$ elements hk, but we must determine how many times each unique group element appears in such a list. Note that if $h_1 k_1 = h_2 k_2$, with $h_i \in H$ and $k_i \in K$, then $h_2^{-1} h_1 = k_2 k_1^{-1} \in H \cap K$. Thus, $h_1 = h_2 g$ and $k_2 = g k_1$, for some $g \in H \cap K$. Conversely, if $h_1 = h_2 g$ and $k_2 = g k_1$, with $g \in H \cap K$, then $h_1 k_1 = h_2 g k_1 = h_2 g g^{-1} k_2 = h_2 k_2$. In other words, each hk will occur once for every element of $H \cap K$. The result follows. □

Example 4.6. Let $G = S_3$ and let H and K be any two different subgroups of order 2. Then $H \cap K$ can only contain the identity, and therefore $|HK| = 4$. But by Lagrange's theorem, a group of order 6 cannot have a subgroup of order 4. Therefore, HK is not a subgroup.

But HK will be a subgroup if either H or K is normal.

Theorem 4.5. *Let H and K be subgroups of G. Then*

1. *if either H or K is normal in G, then HK is a subgroup of G; and*
2. *if both H and K are normal in G, then HK is normal as well.*

Proof. (1) Observe that $e = ee \in HK$. Suppose that H is normal. Let us show closure. If $h_i \in H$ and $k_i \in K$, then

$$(h_1 k_1)(h_2 k_2) = h_1 (k_1 h_2 k_1^{-1}) k_1 k_2.$$

Since H is normal, $k_1 h_2 k_1^{-1} \in H$, and therefore $h_1 k_1 h_2 k_1^{-1} \in H$, $k_1 k_2 \in K$, as required. Also,

$$(h_1 k_1)^{-1} = k_1^{-1} h_1^{-1} = (k_1^{-1} h_1^{-1} k_1) k_1^{-1}.$$

Again, since H is normal, $k_1^{-1} h_1^{-1} k_1 \in H$, so $(h_1 k_1)^{-1} \in HK$. If K is normal, the proof is similar and left as an exercise.

(2) Take $h \in H$, $k \in K$ and $a \in G$. Then

$$a^{-1} hka = (a^{-1} ha)(a^{-1} ka).$$

But $a^{-1} ha \in H$ and $a^{-1} ka \in K$. Thus, $a^{-1} hka \in HK$. □

Exercises

4.1. Is each of the following sets a normal subgroup of $GL_2(\mathbb{R})$?

1. $H = \{A \in GL_2(\mathbb{R}) : \det(A) \in \mathbb{Q}\}$
2. the set of diagonal matrices $\begin{pmatrix} a & 0 \\ 0 & b \end{pmatrix}$ in $GL_2(\mathbb{R})$

4.2. Find every normal subgroup of S_3.

4.3. If N is a normal subgroup of G, and $|N| = 2$, show that $N \leq Z(G)$.

4.4. Let N be a normal subgroup of G. Let H be the set of all elements h of G such that $hn = nh$ for all $n \in N$. Show that H is a normal subgroup of G.

4.5. Show that the intersection of two normal subgroups of G is also a normal subgroup. Then extend this to show that if N_i is a normal subgroup of G for every i in some set T, then $\bigcap_{i \in T} N_i$ is a normal subgroup of G.

4.6. Let $N_1 \leq N_2 \leq N_3 \leq \cdots$ be normal subgroups of G. Show that $\bigcup_{i=1}^{\infty} N_i$ is a normal subgroup of G.

4.7. Let G be a group having exactly one subgroup H of order n. Show that H is normal in G.

4.8. Let $G = H \times K$. If N is a normal subgroup of H and L is a normal subgroup of K, show that $N \times L$ is a normal subgroup of G. Is every normal subgroup of G of this form?

4.9. Suppose that H is a subgroup of G and $a^{-1}b^{-1}ab \in H$, for all $a, b \in G$. Show that H is normal.

4.10. Let H and K be subgroups of G. Show that HK is a subgroup if and only if $HK = KH$.

4.2 Factor Groups

We are now in a position to construct a new sort of group.

Definition 4.4. Let G be a group and N a normal subgroup. Then the **factor group** (or **quotient group**) G/N is the set of all left cosets aN, with $a \in G$, under the operation $(aN)(bN) = abN$.

The fact that the factor group is indeed a group needs proving. Then we can look at some examples.

Theorem 4.6. *If G is any group and N is a normal subgroup, then G/N is a group of order* $[G : N]$.

Proof. The main point is to verify that the operation is well-defined. The rest will follow easily from the fact that G is a group. In other words, suppose that $a_1 N = a_2 N$ and $b_1 N = b_2 N$. We must show that $a_1 b_1 N = a_2 b_2 N$. Otherwise, this operation is nonsensical. But as $a_1 N = a_2 N$, we have $a_1^{-1} a_2 = n_1$, for some $n_1 \in N$. Similarly, $b_1^{-1} b_2 = n_2 \in N$. Then

$$(a_1 b_1)^{-1} a_2 b_2 = b_1^{-1} a_1^{-1} a_2 b_2 = b_1^{-1} n_1 b_2 = (b_1^{-1} n_1 b_1)(b_1^{-1} b_2) = b_1^{-1} n_1 b_1 n_2.$$

Now, as N is normal, $b_1^{-1} n_1 b_1 \in N$. Thus, $(a_1 b_1)^{-1} a_2 b_2 \in N$, which means that $a_1 b_1 N = a_2 b_2 N$, as required.

Let us check the group properties. As for closure, if aN and bN are left cosets, then so is abN. Also, for any $a, b, c \in G$, we have

$$aN(bNcN) = aNbcN = a(bc)N = (ab)cN = (aNbN)cN,$$

so associativity is proved. If $a \in G$, then $aNeN = aN = eNaN$; thus, eN is the identity of G/N. Finally, $aNa^{-1}N = eN = a^{-1}NaN$; that is, $a^{-1}N$ is the inverse of aN. Therefore, G/N is a group. The group consists of the left cosets, so the order is the number of left cosets, which is $[G : N]$. The proof is complete.   □

Notice that the proposed group operation would not even be well-defined if N were not a normal subgroup.

Example 4.7. Let $G = U(15) = \{1, 2, 4, 7, 8, 11, 13, 14\}$, and let $N = \langle 14 \rangle = \{1, 14\}$. There is no need to worry about normality, since G is abelian. The left cosets are $1N = \{1, 14\}$, $2N = \{2, 13\}$, $4N = \{4, 11\}$ and $7N = \{7, 8\}$. Thus, $G/N = \{1N, 2N, 4N, 7N\}$. We note that $(4N)(7N) = 13N = 2N$ and $(2N)(4N) = 8N = 7N$. The rest of the group table is given in Table 4.1. We can also use this table to find inverses; for instance, $2N7N = 1N$. Since $1N$ is the identity, $(2N)^{-1} = 7N$.

Table 4.1 Group table for $U(15)/\langle 14 \rangle$

	$1N$	$2N$	$4N$	$7N$
$1N$	$1N$	$2N$	$4N$	$7N$
$2N$	$2N$	$4N$	$7N$	$1N$
$4N$	$4N$	$7N$	$1N$	$2N$
$7N$	$7N$	$1N$	$2N$	$4N$

Example 4.8. Let $G = \mathbb{Z}$ and $N = 5\mathbb{Z}$. Again, N is certainly a normal subgroup of G. Also, $G/N = \{0 + N, 1 + N, 2 + N, 3 + N, 4 + N\}$. Addition in the factor group behaves like modular arithmetic; indeed, $(1 + N) + (2 + N) = 3 + N$ and $(3 + N) + (4 + N) = 7 + N = 2 + N$. The full group table is given in Table 4.2. Note that $\mathbb{Z}/5\mathbb{Z}$ has precisely the same group table as $\mathbb{Z}_5$ (see Table 3.1).

Table 4.2 Group table for $\mathbb{Z}/5\mathbb{Z}$

	$0+N$	$1+N$	$2+N$	$3+N$	$4+N$
$0+N$	$0+N$	$1+N$	$2+N$	$3+N$	$4+N$
$1+N$	$1+N$	$2+N$	$3+N$	$4+N$	$0+N$
$2+N$	$2+N$	$3+N$	$4+N$	$0+N$	$1+N$
$3+N$	$3+N$	$4+N$	$0+N$	$1+N$	$2+N$
$4+N$	$4+N$	$0+N$	$1+N$	$2+N$	$3+N$

Example 4.9. Let $G = D_8$ and $N = \langle R_{90} \rangle$. As N has index 2, it is necessarily a normal subgroup, by Theorem 4.1. In fact, there are only two left cosets, $R_0 N$, which consists of all of the rotations, and $F_1 N$, which consists of all of the flips. The group table is given in Table 4.3.

Table 4.3 Group table for $D_8 / \langle R_{90} \rangle$

	$R_0 N$	$F_1 N$
$R_0 N$	$R_0 N$	$F_1 N$
$F_1 N$	$F_1 N$	$R_0 N$

Observe that powers of group elements in a quotient group work as we would expect. Indeed, $(aN)^m = a^m N$, for any integer m. In particular, $(aN)^{-1} = a^{-1} N$. Let us prove a few other useful facts.

Theorem 4.7. *Let G be a group, with $a \in G$ and N a normal subgroup of G. Then*

1. if G is abelian, then so is G/N;
2. if G is cyclic, then so is G/N; and
3. if $|a| = m < \infty$, then $|aN|$ divides m.

Proof. (1) If $b, c \in G$, then $(bN)(cN) = bcN = cbN = (cN)(bN)$.

(2) If $G = \langle b \rangle$, then for any $cN \in G/N$, let us say that $c = b^k$. Then $cN = b^k N = (bN)^k$. Thus, $G/N = \langle bN \rangle$.

(3) Note that $(aN)^m = a^m N = eN$. Thus, by Corollary 3.2, the order of aN divides m. $\square$

A small word of caution is in order. Do not assume that the order of a equals that of aN. All we know is that $|aN|$ divides $|a|$. Also, if a has infinite order, then we know nothing about $|aN|$; it could be finite or infinite.

The following theorem tells us how to determine the subgroups of a factor group. The proof, however, is left as Exercise 4.18.

Theorem 4.8. *Let G be a group and N a normal subgroup. Then the subgroups of G/N are precisely of the form H/N, where H is a subgroup of G containing N. Furthermore, H/N is normal in G/N if and only if H is normal in G.*

Here is one more rather neat fact about factor groups.

Theorem 4.9. *Let G be any group. If $G/Z(G)$ is cyclic, then G is abelian.*

Proof. Let $Z = Z(G)$, and suppose that $G/Z = \langle aZ \rangle$. Take any $b, c \in G$. Then $bZ = a^m Z$, for some integer m, and $cZ = a^n Z$, for some integer n. Thus, $b = a^m y$ and $c = a^n z$, for some $y, z \in Z$. But noting that powers of a commute with each other and elements of Z commute with everything, we have $bc = a^m y a^n z = a^n z a^m y = cb$. Thus, G is abelian. $\square$

Corollary 4.1. *The centre of a group cannot have prime index in that group.*

Proof. If $G/Z(G)$ has prime order, then by Corollary 3.6, $G/Z(G)$ is cyclic. But then the preceding theorem tells us that G is abelian; therefore, $Z(G) = G$ has index 1, which is not prime.   □

Note that it is entirely possible for G to be nonabelian but $G/Z(G)$ to be abelian. See Exercise 4.13.

Exercises

4.11. Let G be a group having a normal subgroup N. Suppose that in G/N, the order of aN is 5. If $|N| = 14$, what are the possible orders of a? Show that each order you find can actually occur in some group.

4.12. Write the group table for

1. $D_8/\langle R_{180}\rangle$
2. $U(40)/\langle 3\rangle$.

4.13. Find a nonabelian group G such that

1. $G/Z(G)$ is abelian
2. G is infinite, but $G/Z(G)$ is finite.

4.14. Show that an element of the factor group $\mathbb{R}/\mathbb{Z}$ has finite order if and only if it is in $\mathbb{Q}/\mathbb{Z}$.

4.15. Let G be a finite group having a normal subgroup N. If G/N has an element of order 42, show that G has an element of order 42. Does the same hold for infinite groups?

4.16. Let N be a normal subgroup of G. Show that G/N is abelian if and only if $a^{-1}b^{-1}ab \in N$ for all $a, b \in G$.

4.17. Suppose that G has normal subgroups K and N such that G/K and G/N are abelian. If $K \cap N = \{e\}$, show that G is abelian.

4.18. Let G have a normal subgroup N. Show that the subgroups of G/N are precisely of the form H/N, where H is a subgroup of G with $N \subseteq H$. Furthermore, show that H is normal in G if and only if H/N is normal in G/N.

4.19. Let G be an abelian group. Show that the elements of finite order in G form a normal subgroup N, and that the only element of finite order in G/N is the identity.

4.20. Let G be a nonabelian group. Show that there exists a subgroup H of G such that $Z(G) \subsetneq H \subsetneq G$.

4.3 Homomorphisms

We would like to talk about functions from one group to another. But an arbitrary function is not necessarily very useful. We need it to respect the group operation. This is the first step towards our goal (realized in the next section) of describing a way of determining if two groups have the same structure.

Definition 4.5. Let G and H be groups. Then a **group homomorphism** (or, simply, **homomorphism**) from G to H is a function $\alpha : G \rightarrow H$ such that

$$\alpha(g_1 g_2) = \alpha(g_1)\alpha(g_2)$$

for all $g_1, g_2 \in G$.

Note that in the above definition, the product $g_1 g_2$ is the product in G, whereas the product $\alpha(g_1)\alpha(g_2)$ takes place in H. These group operations need not be the same.

Definition 4.6. If $\alpha : G \rightarrow H$ is a homomorphism, then the **kernel** of α is the set

$$\ker(\alpha) = \{g \in G : \alpha(g) = e\}.$$

Example 4.10. If $n \geq 2$ is a positive integer, then $\alpha : \mathbb{Z} \rightarrow \mathbb{Z}_n$ given by $\alpha(a) = [a]$ (where we insert the equivalence class brackets for clarity) is a homomorphism. Indeed, $\alpha(a + b) = [a + b] = [a] + [b] = \alpha(a) + \alpha(b)$, for all $a, b \in \mathbb{Z}$. Here, $\ker(\alpha) = \{a \in \mathbb{Z} : [a] = [0]\} = n\mathbb{Z}$.

Example 4.11. Let G be the additive group of integers, and let H be the multiplicative group of nonzero rational numbers. Then the function $\alpha : G \rightarrow H$ given by $\alpha(a) = 2^a$ is a homomorphism. To check this, we had first better verify that α does indeed map G into H. But if a is an integer, then 2^a is a nonzero rational number. Also, $\alpha(a + b) = 2^{a+b} = 2^a 2^b = \alpha(a)\alpha(b)$, as required. We see that $\ker(\alpha) = \{a \in \mathbb{Z} : 2^a = 1\} = \{0\}$.

Example 4.12. Let G be any group, and consider the map $\alpha : G \times G \rightarrow G$ given by $\alpha((g_1, g_2)) = g_2$. Then α is a homomorphism. Indeed, if $g_i \in G$, then

$$\alpha((g_1, g_2)(g_3, g_4)) = \alpha((g_1 g_3, g_2 g_4)) = g_2 g_4,$$

and this is also equal to $\alpha((g_1, g_2))\alpha((g_3, g_4))$. Furthermore, the kernel is $\{(g, e) : g \in G\} = G \times \{e\}$.

Example 4.13. If G and H are any groups, then $\alpha : G \rightarrow H$ given by $\alpha(g) = e$ for all $g \in G$ is a homomorphism. Indeed, $\alpha(g_1 g_2) = e$, and $\alpha(g_1)\alpha(g_2) = e^2 = e$. The kernel of α is all of G.

We can give a few basic properties of homomorphisms.

Theorem 4.10. *Let* $\alpha : G \to H$ *be a homomorphism, and take any* $g \in G$. *Then*

1. $\alpha(e) = e$;
2. $\alpha(g^n) = (\alpha(g))^n$, *for any integer n; and*
3. *if* $|g| = m < \infty$, *then the order of* $\alpha(g)$ *divides m.*

Proof. (1) Note that

$$\alpha(e) = \alpha(ee) = \alpha(e)\alpha(e).$$

Cancelling, we find that $\alpha(e)$ is the identity of H.
  (2) If $n > 0$, then note that

$$\alpha(g^n) = \alpha(\underbrace{gg \cdots g}_{n\text{ times}}) = \underbrace{\alpha(g)\alpha(g) \cdots \alpha(g)}_{n\text{ times}} = (\alpha(g))^n.$$

If $n = 0$, then use part (1). If $n = -1$, then note that

$$\alpha(g)\alpha(g^{-1}) = \alpha(gg^{-1}) = \alpha(e) = e.$$

Similarly, $\alpha(g^{-1})\alpha(g) = e$. Therefore, $\alpha(g^{-1}) = (\alpha(g))^{-1}$. Combining what we already know, the case where $n < -1$ follows immediately.
  (3) We have $(\alpha(g))^m = \alpha(g^m) = \alpha(e) = e$. Thus, by Corollary 3.2, the order of $\alpha(g)$ divides m. □

The kernel of a homomorphism is rather important, as the following result suggests.

Theorem 4.11. *Let* $\alpha : G \to H$ *be a homomorphism. Then*

1. $\ker(\alpha)$ *is a normal subgroup of G; and*
2. α *is one-to-one if and only if* $\ker(\alpha) = \{e\}$.

Proof. Let $K = \ker(\alpha)$. Let us show that K is a subgroup of G. By Theorem 4.10, $\alpha(e) = e$, so $e \in K$. Suppose $k_1, k_2 \in K$. Then $\alpha(k_1 k_2) = \alpha(k_1)\alpha(k_2) = ee = e$; hence, $k_1 k_2 \in K$. Also, $\alpha(k_1^{-1}) = (\alpha(k_1))^{-1} = e^{-1} = e$. Thus, $k_1^{-1} \in K$, and K is a subgroup. If $k \in K$ and $g \in G$, then

$$\alpha(g^{-1}kg) = \alpha(g^{-1})\alpha(k)\alpha(g) = \alpha(g^{-1})e\alpha(g) = (\alpha(g))^{-1}\alpha(g) = e.$$

Therefore, $g^{-1}kg \in K$, and K is normal.
  Now, suppose that α is one-to-one. Since $\alpha(e) = e$, we know that if $\alpha(g) = e$, then $g = e$. Therefore, the kernel is simply $\{e\}$. Conversely, suppose that $\ker(\alpha) = \{e\}$. If $\alpha(g_1) = \alpha(g_2)$, then $\alpha(g_1)(\alpha(g_2))^{-1} = e$. But this means that $\alpha(g_1 g_2^{-1}) = e$, and therefore $g_1 g_2^{-1} \in K = \{e\}$. That is, $g_1 = g_2$, and α is one-to-one. □

Two other sorts of subgroups are also useful.

Definition 4.7. Let $\alpha : G \to H$ be a homomorphism. If L is any subgroup of G, then the **image** of L is $\alpha(L) = \{\alpha(l) : l \in L\}$. If M is any subgroup of H, then the **preimage** (or **inverse image**) of M is the set $\alpha^{-1}(M) = \{g \in G : \alpha(g) \in M\}$.

Note that the use of the notation $\alpha^{-1}(M)$ does not imply that the function α is invertible. It may or may not be.

Example 4.14. Consider Example 4.11. If $L = 3\mathbb{Z}$, then $\alpha(L) = \{2^{3a} : a \in \mathbb{Z}\} = \{8^a : a \in \mathbb{Z}\}$. If $M = \{\pm 4^a : a \in \mathbb{Z}\}$, then $\alpha^{-1}(M) = 2\mathbb{Z}$.

Example 4.15. Let $G = \mathbb{Z}$, and consider $\alpha : G \times G \to G$, as in Example 4.12. Let $L = 3\mathbb{Z} \times 5\mathbb{Z}$. Then $\alpha(L) = 5\mathbb{Z}$. If $M = 6\mathbb{Z}$, then $\alpha^{-1}(M) = \mathbb{Z} \times 6\mathbb{Z}$.

We conclude with a few properties of images and preimages.

Theorem 4.12. *Let $\alpha : G \to H$ be a homomorphism. Then*

1. *if L is a subgroup of G, then $\alpha(L)$ is a subgroup of H;*
2. *if L is normal in G, then $\alpha(L)$ is normal in $\alpha(G)$;*
3. *if L is cyclic, then $\alpha(L)$ is cyclic;*
4. *if L is abelian, then $\alpha(L)$ is abelian;*
5. *α is onto if and only if $\alpha(G) = H$;*
6. *if $M \leq H$, then $\alpha^{-1}(M) \leq G$; and*
7. *if M is a normal subgroup of H, then $\alpha^{-1}(M)$ is normal in G.*

Proof. (1) We have $e \in L$, so $e = \alpha(e) \in \alpha(L)$. If $\alpha(l_1), \alpha(l_2) \in \alpha(L)$, then $\alpha(l_1)\alpha(l_2) = \alpha(l_1 l_2) \in \alpha(L)$, since $l_1 l_2 \in L$. Also, $(\alpha(l_1))^{-1} = \alpha(l_1^{-1}) \in \alpha(L)$, since $l_1^{-1} \in L$.

(2) If $l \in L$, $g \in G$, then $(\alpha(g))^{-1}\alpha(l)\alpha(g) = \alpha(g^{-1}lg) \in \alpha(L)$, since $g^{-1}lg \in L$.

(3) If $L = \langle k \rangle$, then for any $\alpha(l) \in \alpha(L)$, we have $l = k^m$, for some integer m. Then $\alpha(l) = \alpha(k^m) = (\alpha(k))^m$. Thus, $\alpha(L) = \langle \alpha(k) \rangle$.

(4) If $l_1, l_2 \in L$, then $\alpha(l_1)\alpha(l_2) = \alpha(l_1 l_2) = \alpha(l_2 l_1) = \alpha(l_2)\alpha(l_1)$.

(5) This is the definition of "onto".

(6) Notice that $\alpha(e) = e \in M$; hence, $e \in \alpha^{-1}(M)$. Also, if $g_1, g_2 \in \alpha^{-1}(M)$, then $\alpha(g_1 g_2) = \alpha(g_1)\alpha(g_2) \in M$, since $\alpha(g_1), \alpha(g_2) \in M$. Thus, $g_1 g_2 \in \alpha^{-1}(M)$. Furthermore, $\alpha(g_1^{-1}) = (\alpha(g_1))^{-1} \in M$, since $\alpha(g_1) \in M$. Thus, $g_1^{-1} \in \alpha^{-1}(M)$.

(7) Take $a \in \alpha^{-1}(M)$, $g \in G$. Then $\alpha(g^{-1}ag) = (\alpha(g))^{-1}\alpha(a)\alpha(g) \in M$, since $\alpha(a) \in M$ and M is normal. Thus, $g^{-1}ag \in \alpha^{-1}(M)$. $\qquad\square$

Exercises

4.21. Are α and β, described below, homomorphisms? If so, are they one-to-one and onto?

1. G is the group of positive real numbers under multiplication, H is $\mathbb{R}$ (under addition), $\alpha : G \to H$ via $\alpha(a) = \log_{10} a$
2. $\beta : \mathbb{Z} \to \mathbb{Z}$, $\beta(a) = a + 1$

4.22. Let $\alpha : \mathbb{Z}_9 \times \mathbb{Z}_{27} \to \mathbb{Z}_{27}$ be given by $\alpha((a, b)) = 3b$, for all $a \in \mathbb{Z}_9$, $b \in \mathbb{Z}_{27}$. Show that α is a homomorphism. Also, find $\ker(\alpha)$, and decide if α is onto.

4.23. Define $\alpha : U(16) \times U(16) \to U(16)$ via $\alpha((a, b)) = ab^{-1}$. Show that α is a homomorphism, and find $\alpha^{-1}(\langle 7 \rangle)$.

4.24. Describe every homomorphism $\alpha : \mathbb{Z}_{10} \to \mathbb{Z}_{15}$.

4.25. Let G be a finite group and $\alpha : G \to H$ an onto homomorphism.

1. If G has an element of order n, must H have one?
2. If H has an element of order n, must G have one?

4.26. Let $\alpha : G \to H$ be a homomorphism, and suppose that $\alpha(g) = h$. For any $a \in G$, show that $\alpha(a) = h$ if and only if $a = gk$ for some $k \in \ker(\alpha)$.

4.27. Define $\alpha : G \times G \to G$ via $\alpha((g, h)) = gh$. If α is a homomorphism, show that G is abelian.

4.28. Show that a group G is cyclic if and only if there exists an onto homomorphism from $\mathbb{Z}$ to G.

4.29. Let N be a normal subgroup of G. Show that there exist a group H and a homomorphism $\alpha : G \to H$ with kernel N.

4.30. Let G be the multiplicative group of nonzero complex numbers and H the multiplicative group of nonzero real numbers. Does there exist a one-to-one homomorphism from G to H?

4.4   Isomorphisms

One of our goals is to establish if two groups are, in effect, the same. To this end, we need to strengthen the notion of a homomorphism.

Definition 4.8. Let G and H be groups. Then a **group isomorphism** (or, simply, **isomorphism**) from G to H is a homomorphism from G to H that is bijective. When such an isomorphism exists, we say that G and H are **isomorphic** groups.

Isomorphic groups have precisely the same structure. The isomorphism simply provides new labels for the group elements.

Theorem 4.13. *On any collection of groups, isomorphism is an equivalence relation.*

Proof. Reflexivity: Use the function $\alpha : G \to G$ given by $\alpha(g) = g$ for all g. It is easily seen to be an isomorphism. Symmetry: Suppose that $\alpha : G \to H$ is an isomorphism. By Theorem 1.3, there exists a function $\beta : H \to G$ given by $\beta(h) = g$, where $\alpha(g) = h$, and this β is also bijective. We must check that it is a homomorphism. Take any $h_1, h_2 \in H$, and suppose that $\beta(h_i) = g_i$. Then $\alpha(g_1 g_2) = \alpha(g_1)\alpha(g_2) = h_1 h_2$; thus, $\beta(h_1 h_2) = g_1 g_2 = \beta(h_1)\beta(h_2)$, as required. Transitivity: Suppose that $\alpha : G \to H$ and $\beta : H \to K$ are isomorphisms. Let $\gamma = \beta \circ \alpha$. By Theorem 1.2, γ is bijective. We must check that it is a homomorphism. Take any $g_1, g_2 \in G$. Then

$$\gamma(g_1 g_2) = \beta(\alpha(g_1 g_2)) = \beta(\alpha(g_1)\alpha(g_2)) = \beta(\alpha(g_1))\beta(\alpha(g_2)) = \gamma(g_1)\gamma(g_2).$$

We are done. $\qquad\qquad\qquad\qquad\qquad\qquad\qquad\qquad\qquad\qquad\qquad\qquad$ $\square$

Therefore, it makes sense to say that G and H are isomorphic; we do not have to specify that G is isomorphic to H. In order to verify that a particular function is an isomorphism, we have to check three things: it must respect the group operation, it must be one-to-one and it must be onto. We can use Theorem 4.11 for the second of these; to show that it is one-to-one, it is enough to show that the kernel is trivial.

Example 4.16. Let us show that $\mathbb{Z}_3 \times \mathbb{Z}_5$ and $\mathbb{Z}_{15}$ are isomorphic groups. We define $\alpha : \mathbb{Z}_{15} \to \mathbb{Z}_3 \times \mathbb{Z}_5$ via $\alpha(a) = (a, a)$. First, is this well-defined? If $a = b$ in $\mathbb{Z}_{15}$, then $15|(a-b)$, so $3|(a-b)$ and $5|(a-b)$, and therefore $(a, a) = (b, b)$ in $\mathbb{Z}_3 \times \mathbb{Z}_5$. Check that it is a homomorphism. If $a, b \in \mathbb{Z}_{15}$, then $\alpha(a + b) = (a + b, a + b) = (a, a) + (b, b) = \alpha(a) + \alpha(b)$. Next, let us show that it is one-to-one. If $a \in \ker(\alpha)$, then $(a, a) = (0, 0)$. That is, $3|a$ and $5|a$. Therefore, $15|a$, so $a = 0$ in $\mathbb{Z}_{15}$, and α is one-to-one. In this case, we do not need to check surjectivity, because the 15 elements of $\mathbb{Z}_{15}$ map to 15 different elements of $\mathbb{Z}_3 \times \mathbb{Z}_5$. But $\mathbb{Z}_3 \times \mathbb{Z}_5$ only has 15 elements! Hence, the function must be onto.

Example 4.17. Lest we get too comfortable, $\mathbb{Z}_{24}$ is not isomorphic to $\mathbb{Z}_4 \times \mathbb{Z}_6$. Why not? Notice that 1 has order 24 in $\mathbb{Z}_{24}$. If these groups had precisely the same structure, then $\mathbb{Z}_4 \times \mathbb{Z}_6$ would have to have an element of order 24 as well. But it is easy to see that $12(a, b) = (0, 0)$ for every $(a, b) \in \mathbb{Z}_4 \times \mathbb{Z}_6$, so every element has order dividing 12.

Example 4.18. As we noted following Example 3.4, the set $G = \{1, -1, i, -i\}$ (where i is the complex number) is a group under multiplication. We claim that it is isomorphic to the additive group $\mathbb{Z}_4$. To see this, we define $\alpha : \mathbb{Z}_4 \to G$ via $\alpha(0) = 1, \alpha(1) = i, \alpha(2) = -1$ and $\alpha(3) = -i$. This function is clearly bijective, and we can check that it respects the group operations by comparing the group tables. The tables for $\mathbb{Z}_4$ and G are shown in Tables 4.4 and 4.5. Note that if we replace 0 and $\alpha(0)$ with A, 1 and $\alpha(1)$ with B, and so on, we see that both groups have Table 4.6. Thus, α is just a relabelling of the group elements.

In fact, we can classify all cyclic groups up to isomorphism.

Table 4.4 Group table for the additive group $\mathbb{Z}_4$

$$
\begin{array}{c|cccc}
 & 0 & 1 & 2 & 3 \\
\hline
0 & 0 & 1 & 2 & 3 \\
1 & 1 & 2 & 3 & 0 \\
2 & 2 & 3 & 0 & 1 \\
3 & 3 & 0 & 1 & 2 \\
\end{array}
$$

Table 4.5 Group table for the multiplicative group $\{1, -1, i, -i\}$

$$
\begin{array}{c|cccc}
 & 1 & i & -1 & -i \\
\hline
1 & 1 & i & -1 & -i \\
i & i & -1 & -i & 1 \\
-1 & -1 & -i & 1 & i \\
-i & -i & 1 & i & -1 \\
\end{array}
$$

Table 4.6 Group table for both Tables 4.4 and 4.5 after relabelling

$$
\begin{array}{c|cccc}
 & A & B & C & D \\
\hline
A & A & B & C & D \\
B & B & C & D & A \\
C & C & D & A & B \\
D & D & A & B & C \\
\end{array}
$$

Theorem 4.14. *Let $G = \langle a \rangle$ be a cyclic group. If a has infinite order, then G is isomorphic to $\mathbb{Z}$. If a has order $n < \infty$, then G is isomorphic to $\mathbb{Z}_n$.*

Proof. Let G be infinite cyclic. Define $\alpha : \mathbb{Z} \to G$ via $\alpha(i) = a^i$. We claim that α is an isomorphism. If $i, j \in \mathbb{Z}$, then $\alpha(i + j) = a^{i+j} = a^i a^j = \alpha(i)\alpha(j)$, as required. If $i \in \ker(\alpha)$, then $a^i = e = a^0$. By Theorem 3.8, $i = 0$. Thus, α is one-to-one. Furthermore, if $i \in \mathbb{Z}$, then $a^i \in \alpha(\mathbb{Z})$, as $\alpha(i) = a^i$. Thus, α is onto as well, and therefore an isomorphism.

Now suppose that $|a| = n < \infty$. Define $\alpha : \mathbb{Z}_n \to G$ via $\alpha(i) = a^i$. Here, we must check that α is well-defined. But if $i = j$ in $\mathbb{Z}_n$, then $n | (i - j)$. Thus, by Theorem 3.8, $a^i = a^j$. The fact that α is an onto homomorphism follows as above. If $i \in \ker(\alpha)$, then $a^i = e$, and by Corollary 3.2, n divides i. Thus, in $\mathbb{Z}_n$, $i = 0$. $\square$

Corollary 4.2. *If a group G has prime order p, then G is isomorphic to $\mathbb{Z}_p$.*

Proof. Combine Corollary 3.6 and Theorem 4.14. $\square$

So, groups of prime order have as nice a structure as we could ask. With a little more work, we can also classify the groups with order twice a prime.

Lemma 4.1. *Let G be a group having distinct commuting elements a and b of order 2. Then G has a subgroup isomorphic to $\mathbb{Z}_2 \times \mathbb{Z}_2$.*

Proof. Given the conditions upon a and b, we can see that $H = \{e, a, b, ab\}$ is a subgroup. (It contains the identity, and closure is easily checked.) Also, H contains four distinct elements. (Clearly, e, a and b are distinct. If $ab = e = bb$, then $a = b$. If $ab = a = ae$, then $b = e$. If $ab = b = eb$, then $a = e$. These are all impossible.) We claim that it is isomorphic to $\mathbb{Z}_2 \times \mathbb{Z}_2$. Let $\alpha : H \to \mathbb{Z}_2 \times \mathbb{Z}_2$ be given by $\alpha(e) = (0, 0)$, $\alpha(a) = (1, 0)$, $\alpha(b) = (0, 1)$ and $\alpha(ab) = (1, 1)$. This function is clearly bijective, and running through the possible pairs of group elements, we see that it is a homomorphism. $\square$

Corollary 4.3. *Every group G of order 4 is isomorphic to either $\mathbb{Z}_4$ or $\mathbb{Z}_2 \times \mathbb{Z}_2$.*

Proof. In a group of order 4, every nonidentity element has order 2 or 4. If there is an element of order 4, G is cyclic and, by Theorem 4.14, isomorphic to $\mathbb{Z}_4$. Otherwise, every nonidentity element has order 2. By Exercise 3.32, G is abelian, and the preceding lemma tells us that G has a subgroup isomorphic to $\mathbb{Z}_2 \times \mathbb{Z}_2$. Given the order of the group, we are done. $\square$

When p is a prime larger than 2, we are already aware of two possible groups of order $2p$; the cyclic group and the dihedral one. In fact, those are all of the options.

Theorem 4.15. *Let $|G| = 2p$, where p is an odd prime. Then G is isomorphic to either $\mathbb{Z}_{2p}$ or D_{2p}.*

Proof. The possible orders for nonidentity elements of G are 2, p and $2p$. If G has an element of order $2p$ then it is cyclic and, by Theorem 4.14, isomorphic to $\mathbb{Z}_{2p}$. So, assume that every nonidentity element has order 2 or p.

If every nonidentity element has order 2, then once again, G is abelian and, by Lemma 4.1, G has a subgroup of order 4, contradicting Lagrange's theorem. Therefore, let $a \in G$ have order p. Take any $b \notin \langle a \rangle$. Suppose that $|b| = p$. Then noting that $\langle a \rangle \cap \langle b \rangle$ is a subgroup of both $\langle a \rangle$ and $\langle b \rangle$ (see Exercise 3.37), Lagrange's theorem tells us that it can only have order 1 or p. As $b \notin \langle a \rangle$, it must be 1. Thus, by Theorem 4.4, $|\langle a \rangle \langle b \rangle| = |\langle a \rangle||\langle b \rangle|/|\langle a \rangle \cap \langle b \rangle| = p^2/1 = p^2$. But this exceeds the order of G. Therefore, $|b| = 2$.

Now, as $\langle a \rangle$ has index 2, Theorem 4.1 tells us that it is normal. Thus, $b^{-1}ab \in \langle a \rangle$, say $b^{-1}ab = a^i$. But then

$$a = b^{-2}ab^2 = b^{-1}(b^{-1}ab)b = b^{-1}a^i b = (b^{-1}ab)^i = (a^i)^i = a^{i^2}.$$

As a has order p, we have $i^2 \equiv 1 \pmod{p}$. That is, $p|(i^2 - 1) = (i - 1)(i + 1)$. As p is prime, $i \equiv \pm 1 \pmod{p}$. Thus, $b^{-1}ab = a$ or a^{-1}.

Suppose that $b^{-1}ab = a$. Then a and b commute. But consider the order of ab. If $(ab)^n = e$, then $a^n = b^{-n} \in \langle a \rangle \cap \langle b \rangle = \{e\}$, as a and b have different prime orders. Thus, $p|n$ and $2|n$, so $2p|n$. That is, ab has order $2p$, which we have excluded. Therefore, $b^{-1}ab = a^{-1}$.

We now know everything about the group. As $\langle a \rangle$ has index 2 and $b \notin \langle a \rangle$, the elements of G are precisely a^i and ba^i, $0 \le i < p$. Furthermore, we know how to find the product of any two elements. Indeed, $a^i a^j = a^{i+j}$ (reducing the exponent modulo p if necessary), $ba^i a^j = ba^{i+j}$,

$$a^i ba^j = b(b^{-1} a^i b)a^j = b(b^{-1} ab)^i a^j = b(a^{-1})^i a^j = ba^{j-i}$$

and $ba^i ba^j = b(ba^{j-i}) = a^{j-i}$. Thus, we can fill in the entire group table for G, and we have precisely the same group structure as in the dihedral group! Indeed, letting F be any flip in D_{2p}, we define $\alpha : G \to D_{2p}$ via $\alpha(a^i) = R_{360i/p}$ and $\alpha(ba^i) = FR_{360i/p}$. Then α is certainly bijective and it is a homomorphism as well. $\square$

We can also mop up a proof we postponed.

Theorem 4.16. *If m and n are relatively prime, then $U(mn)$ is isomorphic to $U(m) \times U(n)$.*

Proof. Define $\alpha : U(mn) \to U(m) \times U(n)$ via $\alpha(a) = (a, a)$. If $(a, mn) = 1$, then $(a, m) = (a, n) = 1$, so we have $(a, a) \in U(m) \times U(n)$ whenever $a \in U(mn)$. Let us verify that α is well-defined. But if $a = b$ in $\mathbb{Z}_{mn}$, then $mn|(a - b)$, so $m|(a - b)$ and $n|(a - b)$, and therefore $(a, a) = (b, b)$ in $U(m) \times U(n)$. It is also a homomorphism; indeed, if $a, b \in U(mn)$, then $\alpha(ab) = (ab, ab) = (a, a)(b, b) = \alpha(a)\alpha(b)$. Let us check that α is one-to-one. But if $a \in \ker(\alpha)$, then $(a, a) = (1, 1)$ in $U(m) \times U(n)$; that is, $m|(a - 1)$ and $n|(a - 1)$. As m and n are relatively prime, $mn|(a - 1)$. That is, $a = 1$ in $U(mn)$; hence, α is one-to-one. Finally, we must show that α is onto. Take any $(c, d) \in U(m) \times U(n)$. By the Chinese Remainder Theorem, there exists an a such that $a \equiv c \pmod{m}$ and $a \equiv d \pmod{n}$. Furthermore, to show that a is in $U(mn)$, it suffices to show that it is relatively prime to both m and n. Without loss of generality, suppose that $(a, m) = k > 1$. Then as $k|a$ and $k|m$, we see that $k|c$ as well. But then $(c, m) \ne 1$, which is impossible. Therefore, $a \in U(mn)$ and $\alpha(a) = (c, d)$. Thus, α is indeed an isomorphism. $\square$

This gives us the second part of Theorem 3.19.

Corollary 4.4. *If m and n are relatively prime, then $\varphi(mn) = \varphi(m)\varphi(n)$.*

Proof. The order of $U(k)$ is $\varphi(k)$. As isomorphic groups have the same order, the preceding theorem completes the proof. $\square$

We wish to add one more point to Corollary 4.3 and Theorem 4.15. If we are to classify groups of a particular order up to isomorphism, we had better ensure that the groups we have listed are not isomorphic to each other. Proving that two groups are not isomorphic generally involves finding a property that one has but the other lacks. For instance, $\mathbb{Z}_{2p}$ is cyclic for all primes p, but neither $\mathbb{Z}_2 \times \mathbb{Z}_2$ nor D_{2p} is cyclic (indeed, D_{2p} is not even abelian). Some properties that can be useful follow.

Theorem 4.17. *Let G and H be isomorphic groups. Then*

1. *G is abelian if and only if H is abelian;*
2. *G is cyclic if and only if H is cyclic;*
3. $|G| = |H|$;
4. *for any positive integer n, G and H have the same number of elements of order n (which could be an infinite number);*
5. *for any positive integer n, G and H have the same number of subgroups of order n (which could be an infinite number); and*
6. *for any positive integer n, G and H have the same number of normal subgroups of order n (which could be an infinite number).*

Proof. Let $\alpha : G \to H$ be an isomorphism. (1) As $\alpha(G) = H$, we see from Theorem 4.12 that if G is abelian, so is H. But the same can be said for $\alpha^{-1} : H \to G$.

(2) Same idea.

(3) An isomorphism is a bijection.

(4) Take $g \in G$ of order n. By Theorem 4.10, $|\alpha(g)|$ divides $|g|$. But by the same argument, $|g| = |\alpha^{-1}(\alpha(g))|$ divides $|\alpha(g)|$. Thus, $|g| = |\alpha(g)|$. That is, the elements of order n in G are in one-to-one correspondence with the elements of order n in H.

(5) Let L be a subgroup of G of order n. Then $\alpha(L)$ is a subgroup of H, and it is isomorphic to L; hence, it has the same order. If M is some other subgroup of G, then since α is one-to-one, $\alpha(M)$ is a different group. Thus, H has at least as many subgroups of order n as G does. But applying α^{-1}, we find that G has at least as many subgroups of order n as H does.

(6) Let L be a normal subgroup of order n in G. Then by Theorem 4.12, $\alpha(L)$ is a normal subgroup of $\alpha(G) = H$. Now proceed as in (5). $\qquad\square$

Example 4.19. As $U(10) = \langle 3 \rangle$ is cyclic of order 4, we know that $U(10)$ is isomorphic to $\mathbb{Z}_4$. Now, $U(8)$ is an abelian group of order 4, but it is not cyclic, so it is not isomorphic to $U(10)$. By Corollary 4.3, $U(8)$ is isomorphic to $\mathbb{Z}_2 \times \mathbb{Z}_2$.

Example 4.20. Consider the groups $U(20)$ and $U(8) \times U(3)$. Each is an abelian group of order 8 and neither is cyclic; however, they are not isomorphic. To see this, note that $U(20)$ has exactly three elements of order 2; namely, 9, 11 and 19. However, $U(8) \times U(3)$ has too many elements of order 2; in fact, all seven nonidentity elements have that order.

Exercises

4.31. For each of the following pairs of groups, explain why they are not isomorphic.

1. $\mathbb{Z}_4 \times \mathbb{Z}_4$ and $\mathbb{Z}_4 \times \mathbb{Z}_2 \times \mathbb{Z}_2$
2. $GL_2(\mathbb{R})$ and $\mathbb{R}$
3. $\mathbb{Z}$ and $\mathbb{Z} \times \mathbb{Z}$

4.32. For each of the following pairs of groups, explain why they are or are not isomorphic.

1. $\mathbb{Z}_9 \times \mathbb{Z}_3$ and $\mathbb{Z}_3 \times \mathbb{Z}_3 \times \mathbb{Z}_3$
2. $\mathbb{Z}_{21}$ and $\mathbb{Z}_3 \times \mathbb{Z}_7$
3. $U(22)$ and $\mathbb{Z}_{10}$
4. D_{20} and $\mathbb{Z}_2 \times \mathbb{Z}_{10}$

4.33. Let G be the set of all matrices of the form $\begin{pmatrix} 1 & 0 \\ a & 1 \end{pmatrix}$, for all integers a. Show that G is a subgroup of $GL_2(\mathbb{R})$. To what familiar group is it isomorphic?

4.34. Show that $\mathbb{Z}$ is not isomorphic to $\mathbb{Q}$.

4.35. Let $H \leq G$ and $a \in G$. Show that H and $a^{-1}Ha$ are isomorphic.

4.36. Show that $G \times H$ is isomorphic to $H \times G$.

4.37. Show that $\mathbb{Z}$ is isomorphic to a proper subgroup of itself.

4.38. Let G be any group. Let H consist of the same set of elements as G, but with a new operation given by $a * b = ba$, for all a and b. Show that H is a group, and that it is isomorphic to G.

4.39. Consider the group G from Exercise 3.42. Show that it is isomorphic to a proper subgroup of itself.

4.40. Consider the group H from Exercise 3.42. Show that it is isomorphic to the multiplicative group of positive rational numbers.

4.5 The Isomorphism Theorems for Groups

In this section, we will discuss three theorems that can aid us in showing that certain groups are isomorphic. The first of these theorems is the most important, and is used to prove the other two.

Theorem 4.18 (First Isomorphism Theorem for Groups). *Let $\alpha : G \to H$ be a homomorphism. Then $G/\ker(\alpha)$ is isomorphic to $\alpha(G)$.*

Proof. Let $K = \ker(\alpha)$. We know that K is a normal subgroup of G. Define $\beta : G/K \to \alpha(G)$ via $\beta(aK) = \alpha(a)$. We claim that β is an isomorphism.

First, we must show that β is well-defined. Suppose that $aK = bK$. Then $a^{-1}b \in K$, and therefore $\alpha(a^{-1}b) = e$. That is, $(\alpha(a))^{-1}\alpha(b) = e$, so $\alpha(a) = \alpha(b)$. Thus, β is well-defined.

Also, β is a homomorphism. Indeed,

$$\beta(aKbK) = \beta(abK) = \alpha(ab) = \alpha(a)\alpha(b) = \beta(aK)\beta(bK).$$

Next, let us check that β is one-to-one. Suppose that $aK \in \ker(\beta)$. Then $\alpha(a) = e$, which means that $a \in K$, so $aK = eK$. That is, $\ker(\beta) = \{eK\}$, and β is one-to-one.

Finally, we must verify that β is onto. Take $\alpha(a) \in \alpha(G)$. Then $\beta(aK) = \alpha(a)$. We are done. $\square$

The First Isomorphism Theorem is a crucial tool in proving that groups are iso-morphic. It is also an enormous time-saver! Whenever we are asked to show that something along the lines of G/N is isomorphic to H, all we need to do is find a homomorphism from G onto H with kernel N. We do not need to define a function on cosets and check that it is well-defined.

Example 4.21. For any integer $n \geq 2$, $\mathbb{Z}/n\mathbb{Z}$ is isomorphic to $\mathbb{Z}_n$. Indeed, define $\alpha : \mathbb{Z} \to \mathbb{Z}_n$ via $\alpha(a) = [a]$ (where we insert the equivalence class brackets for clarity). This is a homomorphism, as $\alpha(a+b) = [a+b] = [a]+[b] = \alpha(a)+\alpha(b)$, for all $a, b \in \mathbb{Z}$. Also $\ker(\alpha) = \{a \in \mathbb{Z} : a \equiv 0 \pmod{n}\} = n\mathbb{Z}$. Finally, if $[a] \in \mathbb{Z}_n$, then $\alpha(a) = [a]$, so α is onto. The First Isomorphism Theorem completes the proof.

Example 4.22. We claim that $GL_2(\mathbb{R})/SL_2(\mathbb{R})$ is isomorphic to the multiplicative group of nonzero real numbers, which we denote by H. Indeed, define $\alpha : GL_2(\mathbb{R}) \to H$ via $\alpha(A) = \det(A)$. As an invertible matrix has a nonzero determinant, the image of $GL_2(\mathbb{R})$ is indeed contained in H. Also, if $A, B \in GL_2(\mathbb{R})$, then $\alpha(AB) = \det(AB)$ and since determinants respect products, this is $\det(A)\det(B) = \alpha(A)\alpha(B)$. Thus, α is a homomorphism. By definition, its kernel is $SL_2(\mathbb{R})$. Finally, if $a \in H$, then

$$\alpha\left(\begin{pmatrix} a & 0 \\ 0 & 1 \end{pmatrix}\right) = a,$$

and therefore α is onto. Now we apply the First Isomorphism Theorem.

Example 4.23. Let us show that if G and H are any groups, then $G \times H$ has a factor group isomorphic to G. Define $\alpha : G \times H \to G$ via $\alpha((g, h)) = g$, for all $g \in G$, $h \in H$. Check that α is a homomorphism. If $g_i \in G$, $h_i \in H$, then

$$\alpha((g_1, h_1)(g_2, h_2)) = \alpha((g_1 g_2, h_1 h_2)) = g_1 g_2 = \alpha((g_1, h_1))\alpha((g_2, h_2)).$$

Also, if $g \in G$, then $\alpha((g, e)) = g$, so α is onto. Therefore, $(G \times H)/\ker(\alpha)$ is isomorphic to H. If we wish to specify the group being factored out, note that $\ker(\alpha) = \{(e, h) : h \in H\} = \{e\} \times H$.

Theorem 4.19 (Second Isomorphism Theorem for Groups). *Let G be a group with H and N subgroups, such that N is normal. Then $H/(H \cap N)$ is isomorphic to HN/N.*

Proof. We will show that $H \cap N$ is normal in H by demonstrating that it is the kernel of a homomorphism. Also, by Theorem 4.5, HN is a subgroup of G, since N is normal. Define $\alpha : H \to HN/N$ via $\alpha(h) = hN$. As $H \subseteq HN$, we see that $hN \in HN/N$. Observe that α is a homomorphism. Indeed, if $h_1, h_2 \in H$, then $\alpha(h_1 h_2) = h_1 h_2 N = (h_1 N)(h_2 N) = \alpha(h_1)\alpha(h_2)$. Also, if $hn \in HN$, then $\alpha(h) = hN = hnN$, since $h^{-1}hn = n \in N$. Thus, α is onto. Finally, $\ker(\alpha) = \{h \in H : hN = eN\} = \{h \in H : h \in N\} = H \cap N$. The First Isomorphism Theorem finishes the proof.   □

Theorem 4.20 (Third Isomorphism Theorem for Groups). *Let G be a group and suppose that N and K are normal subgroups, with $K \subseteq N$. Then $(G/K)/(N/K)$ is isomorphic to G/N.*

Proof. Define $\alpha : G/K \to G/N$ via $\alpha(aK) = aN$, for any $a \in G$. First, let us check that this is well-defined. But if $aK = bK$, then $a^{-1}b \in K \subseteq N$, so $aN = bN$. Next, let us show that α is a homomorphism. But

$$\alpha((aK)(bK)) = \alpha(abK) = abN = (aN)(bN) = \alpha(aK)\alpha(bK).$$

Furthermore, if $aN \in G/N$, then $\alpha(aK) = aN$, so α is onto. Finally,

$$\ker(\alpha) = \{aK \in G/K : aN = eN\} = \{aK \in G/K : a \in N\} = N/K.$$

We now apply the First Isomorphism Theorem.   □

Example 4.24. The Third Isomorphism Theorem tells us that $(\mathbb{Z}/12\mathbb{Z})/(4\mathbb{Z}/12\mathbb{Z})$ is isomorphic to $\mathbb{Z}/4\mathbb{Z}$. (Admittedly, we could have worked this out by noting that $\mathbb{Z}$ is cyclic, so its factor group is cyclic, and the factor group of the factor group is cyclic, and that every cyclic group of order 4 is isomorphic to $\mathbb{Z}_4$, which in turn is isomorphic to $\mathbb{Z}/4\mathbb{Z}$. But isn't this faster?)

Exercises

4.41. Let $G = \mathbb{Z} \times \mathbb{Z}$ and $N = \{(a, a) : a \in \mathbb{Z}\}$. Show that G/N is isomorphic to $\mathbb{Z}$.

4.42. For any groups G and H, show that $(G \times H)/(G \times \{e\})$ is isomorphic to H.

4.43. Show that $\mathbb{R}/\mathbb{Z}$ is isomorphic to the multiplicative group $H = \{a + bi \in \mathbb{C} : a^2 + b^2 = 1\}$.

4.44. Let G be an abelian group and n a positive integer. Consider the groups H and K from Exercise 3.40. Show that G/H is isomorphic to K.

4.45. Let G be the group from Exercise 3.16.

1. Find $Z(G)$.
2. Show that $G/Z(G)$ is isomorphic to $\mathbb{Z} \times \mathbb{Z}$.

4.46. Let G be a group having subgroups N and K of index 2, such that $N \neq K$.

1. Show that $[N : N \cap K] = 2$.
2. Show that $G/(N \cap K)$ is isomorphic to $\mathbb{Z}_2 \times \mathbb{Z}_2$.

4.6 Automorphisms

One particular type of isomorphism deserves special mention.

Definition 4.9. Let G be any group. Then an **automorphism** of G is an isomorphism $\alpha : G \to G$. The set of all automorphisms of G is called the **automorphism group** of G, and is denoted $\mathrm{Aut}(G)$.

Example 4.25. Let G be any abelian group. Then the function $\alpha : G \to G$ given by $\alpha(a) = a^{-1}$ for all $a \in G$ is an automorphism. To see that α is a homomorphism, note that $\alpha(ab) = (ab)^{-1} = a^{-1}b^{-1} = \alpha(a)\alpha(b)$. (Note that this would not work if G were nonabelian, as $(ab)^{-1} = b^{-1}a^{-1}$!) If $\alpha(a) = e$, then $a^{-1} = e$, so $a = e$ and α is one-to-one. Also, if $a \in G$, then $\alpha(a^{-1}) = a$. Thus, α is onto as well.

Theorem 4.21. *For any group G, the automorphism group of G is a group under composition of functions.*

Proof. As we noted in Theorem 4.13, the composition of two isomorphisms is an isomorphism; therefore, the same follows for automorphisms, so $\mathrm{Aut}(G)$ is closed. By Theorem 1.2, the composition of functions is always associative. Certainly, the identity function that fixes every element of G is an automorphism, and serves as an identity for $\mathrm{Aut}(G)$. Finally, we saw in Theorem 4.13 that every isomorphism has an inverse isomorphism; thus, each automorphism has an inverse. □

Generally speaking, determining $\mathrm{Aut}(G)$ for an arbitrary group G is a difficult problem. But we can, at least, solve it when G is cyclic.

Theorem 4.22. *Let $G = \langle a \rangle$ be a cyclic group. Then*

1. *if a has infinite order, then $\mathrm{Aut}(G)$ is isomorphic to $\mathbb{Z}_2$; and*
2. *if $|a| = n < \infty$, then $\mathrm{Aut}(G)$ is isomorphic to $U(n)$.*

Proof. Let $\alpha \in \mathrm{Aut}(G)$. If $\alpha(a) = a^i$, then for every $j \in \mathbb{Z}$, we have $\alpha(a^j) = (\alpha(a))^j = (a^i)^j$. In particular, $G = \alpha(G) = \langle a^i \rangle$. Thus, a^i must generate G. Conversely, suppose that $G = \langle a^i \rangle$, and $\alpha(a) = a^i$. Then we can only have $\alpha(a^j) = a^{ij}$ for all integers j. We claim such an α is an automorphism. Indeed,

$$\alpha(a^j a^k) = \alpha(a^{j+k}) = a^{i(j+k)} = a^{ij}a^{ik} = \alpha(a^j)\alpha(a^k),$$

so α is a homomorphism. If $\alpha(a^j) = e$, then $(a^i)^j = e$. If a has infinite order, then $ij = 0$, and therefore $j = 0$. If $|a| = n < \infty$, then $n|ij$. But as $|a| = |a^i| = |G|$, Corollary 3.2 tells us that $(n, i) = 1$. This means that $n|j$, so $a^j = e$. Either way, $\ker(\alpha) = \{e\}$. As a^i is a generator, it follows immediately that α is onto. The claim is proved. Thus, the automorphisms of G are precisely given by $\alpha(a^j) = a^{ij}$, where a^i is a fixed generator of G.

If a has infinite order, then the only generators of $\langle a \rangle$ are a and a^{-1}. Indeed, if a^m were a generator, then we would have to have $a = (a^m)^l$, for some $l \in \mathbb{Z}$. But then $ml = 1$, which means that $m \in \{1, -1\}$. It is clear, on the other hand, that both a and a^{-1} are generators. Thus, $\mathrm{Aut}(G)$ has order 2. By Corollary 4.2, $\mathrm{Aut}(G)$ is isomorphic to $\mathbb{Z}_2$.

Now suppose that $|a| = n < \infty$. Let us define $\gamma : \mathrm{Aut}(G) \to U(n)$ via $\gamma(\alpha) = i$, where $\alpha(a) = a^i$. Again, Corollary 3.2 tells us that since $|a| = |a^i| = |G|$, we have $(i, n) = 1$, so $i \in U(n)$. Now, if $\alpha, \beta \in \mathrm{Aut}(G)$, with $\alpha(a) = a^i$ and $\beta(a) = a^j$, then

$$(\alpha \circ \beta)(a) = \alpha(\beta(a)) = \alpha(a^j) = (\alpha(a))^j = a^{ij}.$$

Thus, $\gamma(\alpha \circ \beta) = ij$ (reducing modulo n if necessary). But $\gamma(\alpha)\gamma(\beta) = ij$ as well, so γ is a homomorphism. If $\gamma(\alpha) = 1$, then $\alpha(a) = a$, and hence α is the identity automorphism. Therefore, $\ker(\gamma)$ is trivial. Finally, if $i \in U(n)$, then as we have observed, a^i is a generator of G, and we obtain $\alpha \in \mathrm{Aut}(G)$ such that $\alpha(a) = a^i$. Therefore, $\gamma(\alpha) = i$, and γ is onto. Hence, γ is the isomorphism we seek.   □

In particular, we see that the automorphism group of a cyclic group is abelian. It would be a mistake to think that the automorphism group of an abelian group is necessarily abelian, as the following example indicates.

Example 4.26. Let $G = \mathbb{Z}_2 \times \mathbb{Z}_2$. Then define $\alpha : G \to G$ via $\alpha((0, 0)) = (0, 0)$, $\alpha((1, 0)) = (0, 1)$, $\alpha((0, 1)) = (1, 0)$ and $\alpha((1, 1)) = (1, 1)$. Also, let $\beta((0, 0)) = (0, 0)$, $\beta((1, 0)) = (1, 0)$, $\beta((0, 1)) = (1, 1)$ and $\beta((1, 1)) = (0, 1)$. Clearly, α and β are both bijective. The group is also small enough that one can check all of the possibilities and find that they are homomorphisms. Therefore, $\alpha, \beta \in \mathrm{Aut}(G)$. But $\alpha(\beta((1, 0))) = \alpha((1, 0)) = (0, 1)$, whereas $\beta(\alpha((1, 0))) = \beta((0, 1)) = (1, 1)$. Thus, $\alpha \circ \beta \neq \beta \circ \alpha$, so $\mathrm{Aut}(G)$ is nonabelian. In Exercise 4.49, we must show that $\mathrm{Aut}(G)$ is isomorphic to D_6.

Let us define a particular type of automorphism.

Definition 4.10. Let G be a group and $a \in G$. Then the **inner automorphism** induced by a is $\theta_a : G \to G$ given by $\theta_a(g) = a^{-1}ga$ for all $g \in G$. The **inner automorphism group** of G is $\mathrm{Inn}(G) = \{\theta_a : a \in G\}$.

Inner automorphisms are only interesting when the group G is nonabelian; for abelian groups, every inner automorphism is the identity function, as $a^{-1}ga = g$. Let us list a few basic properties of inner automorphisms.

Lemma 4.2. *Let G be a group and $a, b \in G$. Then*

1. $\theta_a \in \mathrm{Aut}(G)$;
2. $\theta_a \circ \theta_b = \theta_{ba}$; *and*
3. $(\theta_a)^{-1} = \theta_{a^{-1}}$.

Proof. (1) First, let us show that θ_a is a homomorphism. If $g, h \in G$, then

$$\theta_a(gh) = a^{-1}gha = (a^{-1}ga)(a^{-1}ha) = \theta_a(g)\theta_a(h).$$

If $\theta_a(g) = e$, then $a^{-1}ga = e$, so $g = aea^{-1} = e$; thus, $\ker(\theta_a) = \{e\}$, and θ_a is one-to-one. Finally, if $g \in G$, then $\theta_a(aga^{-1}) = a^{-1}aga^{-1}a = g$; thus, θ_a is onto.

(2) If $g \in G$, then $\theta_a(\theta_b(g)) = \theta_a(b^{-1}gb) = a^{-1}b^{-1}gba = \theta_{ba}(g)$. Thus, $\theta_a \circ \theta_b = \theta_{ba}$.

(3) If $g \in G$, then $\theta_a(\theta_{a^{-1}}(g)) = \theta_a(aga^{-1}) = a^{-1}aga^{-1}a = g$; thus, $\theta_a \circ \theta_{a^{-1}}$ is the identity function. Similarly, so is $\theta_{a^{-1}} \circ \theta_a$. □

Theorem 4.23. *For any group G, Inn(G) is a normal subgroup of* Aut(G).

Proof. By the preceding lemma, $\text{Inn}(G) \subseteq \text{Aut}(G)$. Certainly $\theta_e \in \text{Inn}(G)$ is the identity automorphism. Also, the preceding lemma shows that $\text{Inn}(G)$ is closed under composition and the taking of inverses. Therefore, $\text{Inn}(G) \leq \text{Aut}(G)$. To show normality, take $\alpha \in \text{Aut}(G)$ and $\theta_a \in \text{Inn}(G)$. Then

$$\begin{aligned}
(\alpha^{-1} \circ \theta_a \circ \alpha)(g) &= \alpha^{-1}(\theta_a(\alpha(g))) \\
&= \alpha^{-1}(a^{-1}\alpha(g)a) \\
&= \alpha^{-1}(a^{-1})\alpha^{-1}(\alpha(g))\alpha^{-1}(a) \\
&= (\alpha^{-1}(a))^{-1}g\alpha^{-1}(a) \\
&= \theta_{\alpha^{-1}(a)}(g),
\end{aligned}$$

for all $g \in G$. That is, $\alpha^{-1} \circ \theta_a \circ \alpha = \theta_{\alpha^{-1}(a)} \in \text{Inn}(G)$, and $\text{Inn}(G)$ is normal. □

It is certainly possible for $\text{Aut}(G)$ to be larger than G; indeed, Example 4.26 provides such a group. But there is only one inner automorphism for each group element. However, θ_a does not have to be different from θ_b if $a \neq b$. For instance, if a and b are both central, then θ_a and θ_b are both equal to the identity automorphism. The following theorem tells the tale.

Theorem 4.24. *Let G be a group. Then*

1. *if $a, b \in G$, then $\theta_a = \theta_b$ if and only if $ba^{-1} \in Z(G)$; and*
2. *$G/Z(G)$ is isomorphic to* Inn(G).

Proof. (1) Take any $a, b \in G$. Then $\theta_a = \theta_b$ if and only if $a^{-1}ga = b^{-1}gb$ for all $g \in G$. But this occurs if and only if $ba^{-1}g = gba^{-1}$ for all $g \in G$. In other words, if and only if ba^{-1} is central.

(2) Define $\alpha : G \to \text{Inn}(G)$ via $\alpha(a) = \theta_{a^{-1}}$. Let us show that α is a homomorphism. If $a, b \in G$, then

$$\alpha(ab) = \theta_{(ab)^{-1}} = \theta_{b^{-1}a^{-1}} = \theta_{a^{-1}} \circ \theta_{b^{-1}} = \alpha(a)\alpha(b),$$

making use of Lemma 4.2. Also, if $\theta_a \in \text{Inn}(G)$, then $\alpha(a^{-1}) = \theta_a$, so α is onto. Furthermore, $a \in \ker(\alpha)$ if and only if $\theta_a = \theta_e$. By (1), this happens if and only if $a = ae^{-1} \in Z(G)$. Now apply the First Isomorphism Theorem. □

Example 4.27. As the centre of S_3 is trivial, we see that $\text{Inn}(S_3)$ is isomorphic to S_3.

Example 4.28. As $Z(D_8) = \langle R_{180} \rangle$, the distinct elements of $\text{Inn}(D_8)$ are of the form θ_a, where we take one a for each left coset of $\langle R_{180} \rangle$ in D_8. That is, $\text{Inn}(D_8) = \{\theta_{R_0}, \theta_{R_{90}}, \theta_{F_1}, \theta_{F_3}\}$. In particular, it is a group of order 4, so by Corollary 4.3, it is isomorphic to either $\mathbb{Z}_4$ or $\mathbb{Z}_2 \times \mathbb{Z}_2$. But every flip in D_8 already has order 2, and the square of every rotation is in $Z(D_8)$. Therefore, we see that there is no element of order 4 in $D_8/Z(D_8)$; thus, it must be isomorphic to $\mathbb{Z}_2 \times \mathbb{Z}_2$.

Exercises

4.47. Let G be an abelian group of order n, and let m be a positive integer relatively prime to n. Show that $\alpha : G \rightarrow G$ given by $\alpha(a) = a^m$ is an automorphism of G.

4.48. Let G be a group with automorphism α and H a group with automorphism β. Show that $\gamma : G \times H \rightarrow G \times H$ given by $\gamma((g, h)) = (\alpha(g), \beta(h))$ is an automorphism.

4.49. Show that the automorphism group of $\mathbb{Z}_2 \times \mathbb{Z}_2$ is isomorphic to D_6.

4.50. Let G and H be isomorphic groups. Show that their automorphism groups are also isomorphic.

4.51. Let α be an automorphism of G. Show that $\{a \in G : \alpha(a) = a\}$ is a subgroup of G.

4.52. Let α and β be any two automorphisms of G. Show that $\{a \in G : \alpha(a) = \beta(a)\}$ is a subgroup of G.

4.53. For any group G, an automorphism α of G is said to be a power automorphism if $\alpha(H) \subseteq H$ for every subgroup H of G. If $G = \langle a \rangle \times \langle b \rangle$ is the direct product of two cyclic groups, and α is a power automorphism of G, show that there exists a $k \in \mathbb{Z}$ such that $\alpha(g) = g^k$ for all $g \in G$.

4.54. To what familiar group is the inner automorphism group of D_{12} isomorphic?

4.55. Let α be an automorphism of $\mathbb{Q}$. Show that for every $q \in \mathbb{Q}$, we have $\alpha(q) = q\alpha(1)$.

4.56. Let G be a group such that the automorphism group of G is trivial.

1. Show that G is abelian.
2. Show that $a^2 = e$ for every $a \in G$.

Chapter 5
Direct Products and the Classification of Finite Abelian Groups

We can now determine the structure of finite abelian groups. In particular, every such group is isomorphic to a direct product of cyclic groups, each having prime power order. The proof of this result is our main goal in the present chapter.

5.1 Direct Products

We defined the direct product of two groups in Definition 3.3. There is no particular reason that we need to restrict ourselves to two.

Definition 5.1. Let $G_1, \ldots, G_k$ be any groups. Then the **(external) direct product** $G_1 \times G_2 \times \cdots \times G_k$ is the Cartesian product of the groups G_i under the operation $(a_1, \ldots, a_k)(b_1, \ldots, b_k) = (a_1 b_1, \ldots, a_k b_k)$, for all $a_i, b_i \in G_i$. (We allow $k = 1$ here, in which case $G = G_1$.)

Theorem 5.1. *If $G_1, \ldots, G_k$ are groups, then $G_1 \times \cdots \times G_k$ is a group.*

Proof. The proof is essentially identical to that of Theorem 3.1. $\square$

The reason we used the word "external" in the above definition is that the groups G_i are not subgroups of the direct product; indeed, they are not even subsets. However, G_1 is, for instance, isomorphic in a natural way to $G_1 \times \{e\} \times \cdots \times \{e\}$, which is a subgroup of the direct product. What we would like is a way of showing that a group is isomorphic to the direct product of certain subgroups. To this end, let us consider the following.

Definition 5.2. Let G be a group, and let $N_1, \ldots, N_k$ be subgroups of G. Then we say that G is the **internal direct product** of $N_1, \ldots, N_k$ if

© Springer International Publishing AG, part of Springer Nature 2018
G. T. Lee, *Abstract Algebra*, Springer Undergraduate Mathematics Series,
https://doi.org/10.1007/978-3-319-77649-1_5

1. each N_i is normal;
2. $N_1 N_2 \cdots N_k = G$; and
3. for each i, $1 \leq i < k$, we have $(N_1 N_2 \cdots N_i) \cap N_{i+1} = \{e\}$.

(Again, we allow $k = 1$, in which case $G = N_1$.)

In particular, G is the internal direct product of normal subgroups N_1 and N_2 if and only if $N_1 N_2 = G$ and $N_1 \cap N_2 = \{e\}$.

Example 5.1. Let $G = \mathbb{Z}_{20}$, $N_1 = \langle 4 \rangle$ and $N_2 = \langle 5 \rangle$. As G is abelian, every subgroup is normal. Also, $N_1 = \{0, 4, 8, 12, 16\}$ and $N_2 = \{0, 5, 10, 15\}$. Thus, $N_1 \cap N_2 = \{0\}$. For each $a \in G$, we could find $n_1 \in N_1$ and $n_2 \in N_2$ such that $a = n_1 + n_2$ but, in fact, we can avoid this by noting that $|N_1 + N_2| = |N_1||N_2|/|N_1 \cap N_2| = 5 \cdot 4/1 = 20$ (see Theorem 4.4). Thus, $N_1 + N_2 = G$, and G is the internal direct product of N_1 and N_2.

Note that if there are more than two groups, then we need to check more than just that each $N_i \cap N_j = \{e\}$ for the final part of the definition.

Example 5.2. Let $G = \mathbb{Z}_{30}$, $N_1 = \langle 15 \rangle$, $N_2 = \langle 10 \rangle$ and $N_3 = \langle 6 \rangle$. Again, normality is not an issue. It is easy to see that $N_1 \cap N_2 = \{0\}$. Thus, $|N_1 + N_2| = |N_1||N_2| = 2 \cdot 3 = 6$. As every element of $N_1 + N_2$ is in $\langle 5 \rangle$, we see immediately that $N_1 + N_2 = \langle 5 \rangle$. But now we observe that $(N_1 + N_2) \cap N_3 = \{0\}$. Then the same argument shows that $|N_1 + N_2 + N_3| = 30$, and we know that $N_1 + N_2 + N_3 = G$. Therefore, G is the internal direct product of N_1, N_2 and N_3.

Let us see how internal direct products behave. Here are some highly useful facts.

Lemma 5.1. *Let G be a group with normal subgroups K and N. If $K \cap N = \{e\}$, then $kn = nk$ for all $k \in K$, $n \in N$.*

Proof. Let $h = (nk)^{-1}(kn) = k^{-1}n^{-1}kn$. As K is normal, $n^{-1}kn \in K$, so $h \in K$. As N is normal, $k^{-1}n^{-1}k \in N$, so $h \in N$. Since $K \cap N = \{e\}$, we have $(nk)^{-1}(kn) = e$, and therefore $kn = nk$, as required. $\qquad\square$

Lemma 5.2. *If G is the internal direct product of $N_1, \ldots, N_k$, then every element of G can be written in exactly one way as $n_1 n_2 \cdots n_k$, with each $n_i \in N_i$.*

Proof. Since $G = N_1 \cdots N_k$, we know that every element of G can be written in such a way. We only need to show uniqueness. Our proof is by induction on k. If $k = 1$, there is nothing to do, as $G = N_1$. Assume that $k > 1$ and the result holds for groups written as an internal direct product of a smaller number of subgroups. Suppose that $n_1 \cdots n_{k-1} n_k = h_1 \cdots h_{k-1} h_k$, with $n_i, h_i \in N_i$. Then

$$h_k n_k^{-1} = (h_1 \cdots h_{k-1})^{-1}(n_1 \cdots n_{k-1}) \in N_k \cap (N_1 \cdots N_{k-1}) = \{e\}.$$

Therefore, $n_k = h_k$, and we have $n_1 \cdots n_{k-1} = h_1 \cdots h_{k-1}$ in $N_1 N_2 \cdots N_{k-1}$, which is an internal direct product of $k - 1$ subgroups. By our inductive hypothesis, $n_i = h_i$ for all i. $\qquad\square$

Example 5.3. As we saw in Example 5.2, $\mathbb{Z}_{30}$ is the internal direct product of $\langle 15 \rangle$, $\langle 10 \rangle$ and $\langle 6 \rangle$. Note, for instance, that $23 = 15 + 20 + 18$. By the above lemma, there is no other way to write 23 as a sum of elements in $\langle 15 \rangle$, $\langle 10 \rangle$ and $\langle 6 \rangle$.

And now, the big reason why we are interested in these internal direct products.

Theorem 5.2. *Let G be a group, and suppose that it is the internal direct product of normal subgroups $N_1, \ldots, N_k$. Then G is isomorphic to the external direct product $N_1 \times \cdots \times N_k$.*

Proof. Define $\alpha : N_1 \times \cdots \times N_k \to G$ via $\alpha((n_1, \ldots, n_k)) = n_1 \cdots n_k$. We claim that α is an isomorphism. In view of Lemma 5.2, α is bijective. Thus, it remains to show that it is a homomorphism. Take $n_i, h_i \in N_i$. Then

$$\alpha((n_1, \ldots, n_k)(h_1, \ldots, h_k)) = \alpha((n_1 h_1, \ldots, n_k h_k)) = n_1 h_1 n_2 h_2 n_3 h_3 \cdots n_k h_k.$$

As N_1 and N_2 are normal subgroups, and $N_1 \cap N_2 = \{e\}$, Lemma 5.1 says that $h_1 n_2 = n_2 h_1$. Thus,

$$n_1 h_1 n_2 h_2 n_3 h_3 \cdots n_k h_k = n_1 n_2 h_1 h_2 n_3 h_3 \cdots n_k h_k.$$

By Theorem 4.5, $N_1 N_2$ is a normal subgroup of G, and we know that $(N_1 N_2) \cap N_3 = \{e\}$. Therefore, $h_1 h_2 n_3 = n_3 h_1 h_2$. We now have

$$n_1 h_1 n_2 h_2 n_3 h_3 \cdots n_k h_k = n_1 n_2 n_3 h_1 h_2 h_3 n_4 h_4 \cdots n_k h_k.$$

Repeating this procedure, we find that

$$\alpha((n_1, \ldots, n_k)(h_1, \ldots, h_k)) = n_1 n_2 \cdots n_k h_1 h_2 \cdots h_k = \alpha((n_1, \ldots, n_k))\alpha((h_1, \ldots, h_k)).$$

Thus, α is a homomorphism, and the proof is complete. $\square$

As a result of this theorem, we will engage in a small abuse of notation and write $G = N_1 \times N_2 \times \cdots \times N_k$ when G is the internal direct product of $N_1, \ldots, N_k$, as well as for the external direct product.

Example 5.4. By Example 5.2, $\mathbb{Z}_{30} = \langle 15 \rangle \times \langle 10 \rangle \times \langle 6 \rangle$.

Example 5.5. We claim that $U(8) = \langle 3 \rangle \times \langle 7 \rangle$. As the group is abelian, normality is not an issue. Also, $|3| = |7| = 2$, so the intersection of these cyclic subgroups must be trivial. Furthermore, $1 = 1 \cdot 1, 3 = 3 \cdot 1, 7 = 1 \cdot 7$ and $5 = 3 \cdot 7$, so $U(8) = \langle 3 \rangle \langle 5 \rangle$ (or just use an order argument). Thus, we have an internal direct product.

Exercises

5.1. Write $U(32)$ as the internal direct product of two proper subgroups.

5.2. Let $G = H \times K$. If $h \in H$ has order m and $k \in K$ has order n, what is the order of (h, k)?

5.3. How many elements of order 5 are there in $\mathbb{Z}_5 \times \mathbb{Z}_{25}$? How many elements of order 25?

5.4. How many cyclic subgroups of order 5 are there in $\mathbb{Z}_5 \times \mathbb{Z}_{25}$? How many cyclic subgroups of order 25?

5.5. Show that D_8 is not the internal direct product of two proper subgroups.

5.6. Let $|a| = 4$ and $|b| = 2$. Write $\langle a \rangle \times \langle b \rangle$ as the internal direct product of two proper subgroups in every possible way.

5.7. Show that in Definition 5.2, it is not sufficient to replace the third condition with the stipulation that $N_i \cap N_j = \{e\}$ whenever $i \neq j$. In particular, find a group G with normal subgroups N_1, N_2 and N_3 such that $N_1 N_2 N_3 = G$ and $N_i \cap N_j = \{e\}$ whenever $i \neq j$, but $G \neq N_1 \times N_2 \times N_3$.

5.8. Let $G = \langle a \rangle$ be cyclic of order 84. Show that $G = \langle a^{12} \rangle \times \langle a^{21} \rangle \times \langle a^{28} \rangle$.

5.9. Suppose that $G = N_1 \times N_2$ is an internal direct product. If $\alpha : G \to H$ is an onto homomorphism, does it follow that $H = \alpha(N_1) \times \alpha(N_2)$? Prove that it does, or give an explicit counterexample.

5.10. Let G be a group having finite normal subgroups $N_1, \ldots, N_k$, such that the gcd of $|N_i|$ and $|N_j|$ is 1 whenever $i \neq j$. Show that $N_1 N_2 \cdots N_k = N_1 \times N_2 \times \cdots \times N_k$.

5.2   The Fundamental Theorem of Finite Abelian Groups

Let us now classify the finite abelian groups. We will break our proof down into stages. For the first stage, we need a definition.

Definition 5.3. Let p be a prime number. Furthermore, let G be a group and $a \in G$. We say that a is a p-**element** if the order of a is p^n for some integer $n \geq 0$. If every element of G is a p-element, then G is a p-**group**.

Example 5.6. The dihedral group D_8 is a 2-group, as every element has order 1, 2 or 4. On the other hand, $\mathbb{Z}_{24}$ is not a p-group. Indeed, 12 and 18 are both 2-elements and 8 is a 3-element, so it cannot be a p-group. In fact, 1 is not a p-element, for any prime p.

Lemma 5.3. *Let p be a prime and G an abelian group. Then the p-elements of G form a subgroup.*

Proof. Let H be the set of all p-elements of G. As e has order p^0, we have $e \in H$. Let $a, b \in H$. Then say that $|a| = p^n$ and $|b| = p^m$. Let k be the larger of m and n. Then as G is abelian, $(ab)^{p^k} = a^{p^k} b^{p^k} = e^2 = e$, as $|a|$ and $|b|$ both divide p^k. Thus, $|ab|$ divides p^k, and therefore $ab \in H$. Finally, if $a \in H$, then $|a| = |a^{-1}|$, so $a^{-1} \in H$. Thus, H is indeed a subgroup of G. $\qquad\square$

Note that the preceding lemma does not work for nonabelian groups. Indeed, in S_3, we can see that $\begin{pmatrix} 1\ 2\ 3 \\ 2\ 1\ 3 \end{pmatrix}$ and $\begin{pmatrix} 1\ 2\ 3 \\ 1\ 3\ 2 \end{pmatrix}$ both have order 2, but their product, $\begin{pmatrix} 1\ 2\ 3 \\ 2\ 3\ 1 \end{pmatrix}$, has order 3.

The following result is also very handy.

Lemma 5.4. *Let G be any group and let $e \neq a \in G$ be such that a has finite order. Then $a = a^{n_1} a^{n_2} \cdots a^{n_k}$ for some integers $n_1, \ldots, n_k$, where each a^{n_i} is a p_i-element, for some prime p_i dividing $|a|$.*

Proof. Our proof is by induction on the number of distinct primes, l, dividing $|a|$. If $l = 1$, then a is a p-element, so just let $n_1 = 1$. Suppose that $l > 1$ and that the result is true for smaller values of l. Let p be a prime dividing $|a|$, and say that $|a| = p^m q$, with $(p, q) = 1$. By Corollary 2.1, there exist $u, v \in \mathbb{Z}$ such that $p^m u + qv = 1$. Then

$$a = a^1 = a^{p^m u + qv} = (a^{p^m})^u (a^q)^v.$$

Now, $(a^q)^{p^m} = a^{p^m q} = e$; hence, a^q is a p-element and so is $(a^q)^v$. So, let $p_1 = p$ and $n_1 = qv$. Similarly, the order of $(a^{p^m})^u$ divides q, and q has fewer primes dividing it than $|a|$. Thus, by our inductive hypothesis, $a^{u p^m}$ can be written as a product of powers (which are also powers of a) in the manner stated in the theorem. The proof is complete. $\qquad\square$

We can now simplify our task by breaking a finite abelian group down into a direct product of p-groups.

Lemma 5.5. *Let G be a nontrivial finite abelian group, and let $p_1, \ldots, p_k$ be the distinct primes dividing $|G|$. Then $G = H_1 \times H_2 \times \cdots \times H_k$, where H_i is the subgroup of G consisting of all of the p_i-elements of G.*

Proof. Lemma 5.3 tells us that the H_i are subgroups and, as G is abelian, we do not have to worry about normality. Let us show that $G = H_1 H_2 \cdots H_k$. But taking any $a \in G$, we see from Lemma 5.4 that a can be written as a product of elements from various H_i. (If $a = e$, there is obviously nothing to worry about.) Finally, we must show that for each i, $1 \leq i < k$, we have $(H_1 \cdots H_i) \cap H_{i+1} = \{e\}$. But suppose that $a \in H_{i+1}$ and, simultaneously, $a = a_1 \cdots a_i$, with $a_j \in H_j$. Then letting $|a_j| = p_j^{m_j}$, and $m = p_1^{m_1} \cdots p_i^{m_i}$, we have $a^m = a_1^m \cdots a_i^m$, and since each $|a_j|$ divides m, we conclude that $a^m = e$. Thus, $|a|$ divides m. But also, a is a p_{i+1}-element. As $(m, p_{i+1}) = 1$, the only possible conclusion is that $a = e$, and we have an internal direct product. $\qquad\square$

We can now focus our attention on finite abelian p-groups. The following lemma does the biggest part of the work. It is the most difficult proof we have encountered so far, and will take some time to absorb.

Lemma 5.6. *Let G be a finite abelian p-group, and let $a \in G$ be an element of largest possible order. Then $G = \langle a \rangle \times H$, for some subgroup H of G.*

Proof. Our proof is by strong induction on $|G|$. If $|G| = 1$, then $a = e$ and using $H = \langle e \rangle$ will work. So, assume that $|G| > 1$ and that the lemma holds for groups of smaller order.

Let $|a| = p^n$, with n a positive integer. If $\langle a \rangle = G$, then we can use $H = \langle e \rangle$, so assume that $\langle a \rangle \neq G$. Take $b \in G$ such that $b \notin \langle a \rangle$. As b is a p-element, we know that $b^{p^k} = e \in \langle a \rangle$, for some positive integer k. Let m be the smallest positive integer such that $b^{p^m} \in \langle a \rangle$, and let $c = b^{p^{m-1}}$. Then $c \notin \langle a \rangle$, but $c^p = b^{p^m} \in \langle a \rangle$. In particular, let us say that $c^p = a^i$, with $i \in \mathbb{Z}$.

Now, as G is a p-group, and the largest element order is p^n, we must have $c^{p^n} = e$. Thus, $|c^p|$ divides p^{n-1}. Suppose that $(p, i) = 1$. Then by Corollary 3.2, $|c^p| = |a^i| = p^n$, which is impossible. Thus, p divides i; let us say that $i = pj$. Then let $d = a^{-j}c$. Note that $a^j \in \langle a \rangle$; thus, if $d \in \langle a \rangle$, then $c = a^j d \in \langle a \rangle$, which is a contradiction. Therefore, $d \notin \langle a \rangle$. However, $d^p = a^{-jp}c^p = (a^i)^{-1}c^p = e$; thus, $|d| = p$.

Now, let us consider the group $M = G/\langle d \rangle$. (As G is abelian, we do not have to worry about $\langle d \rangle$ being normal.) We note that M is still abelian (by Theorem 4.7), its order is $[G : \langle d \rangle] = |G|/p$ and it is a p-group with the orders of elements dividing orders of elements of G (by Theorem 4.7). Also, we claim that $|a\langle d \rangle| = p^n$. As its order must divide p^n, suppose that $a^{p^{n-1}} \in \langle d \rangle$. Since $a^{p^{n-1}} \neq e$, we must have $a^{p^{n-1}} = d^s$, with $0 < s < p$. But then $(s, p) = 1$, so by Corollary 2.1, there exist $u, v \in \mathbb{Z}$ such that $su + pv = 1$. Thus, $d = d^{su+pv} = (d^s)^u(d^p)^v = a^{p^{n-1}u}e \in \langle a \rangle$, giving us a contradiction. Therefore, $|a\langle d \rangle| = p^n$, as claimed.

It now follows that $a\langle d \rangle$ is an element of largest order in M. As M is an abelian p-group of smaller order than G, our inductive hypothesis tells us that there is a subgroup K of M such that $M = N \times K$, where N is the subgroup of M generated by $a\langle d \rangle$. By Theorem 4.8, $K = H/\langle d \rangle$, where H is a subgroup of G containing $\langle d \rangle$.

We claim that $G = \langle a \rangle \times H$. Normality is not an issue. Suppose that $a^i \in \langle a \rangle \cap H$. Then $a^i \langle d \rangle \in N \cap K$, and as the product $N \times K$ is direct, this means that $a^i \langle d \rangle = e\langle d \rangle$. But we demonstrated above that the order of $a\langle d \rangle$ is p^n, which means that p^n divides i, and therefore $a^i = e$. Thus, $\langle a \rangle \cap H = \{e\}$.

Now, take any $g \in G$. Then as $M = N \times K$, we have $g\langle d \rangle = xy$, for some $x \in N$, $y \in K$. Let us write $x = a^t \langle d \rangle$ and $y = w\langle d \rangle$, with $t \in \mathbb{Z}$ and $w \in H$. Then $g = a^t wd^l$, for some $l \in \mathbb{Z}$. As $a^t \in \langle a \rangle$ and $wd^l \in H$, we now see that $\langle a \rangle H = G$. Thus, we have the required direct product, and our proof is complete. $\qquad\square$

And now, the payoff for our hard work!

Theorem 5.3 (Fundamental Theorem of Finite Abelian Groups). *Let G be a finite abelian group. Then G is the direct product of subgroups, $H_1 \times \cdots \times H_k$, with* each H_i cyclic of order $p_i^{n_i}$, where the p_i are (not necessarily distinct) primes, and the n_i are nonnegative integers.

Proof. If G is the trivial group, there is nothing to do. Otherwise, by Lemma 5.5, G is the direct product of p-subgroups. Therefore, we may as well assume that G is a finite abelian p-group. Our proof is by strong induction on $|G|$. If $|G| = 1$, again, there is nothing to do, so let G be nontrivial and suppose that our theorem holds for groups of smaller order. Let a be an element of largest possible order in G. Then by Lemma 5.6, $G = \langle a \rangle \times H$, for some subgroup H. But then $|H| = |G|/|a|$, so H has smaller order, and by our inductive hypothesis, H is a direct product of cyclic groups of prime power order. However, $\langle a \rangle$ is also a cyclic group of prime power order, and we are done. $\qquad\square$

We can express this slightly differently.

Corollary 5.1. *Let G be a nontrivial finite abelian group. Then G is isomorphic to $\mathbb{Z}_{p_1^{n_1}} \times \mathbb{Z}_{p_2^{n_2}} \times \cdots \times \mathbb{Z}_{p_k^{n_k}}$, where the p_i are some (not necessarily distinct) primes, and the n_i are positive integers.*

Proof. Combine Theorems 5.2 and 5.3 with Theorem 4.14. $\qquad\square$

Example 5.7. Up to isomorphism, the abelian groups of order 16 are $\mathbb{Z}_{16}$, $\mathbb{Z}_8 \times \mathbb{Z}_2$, $\mathbb{Z}_4 \times \mathbb{Z}_4$, $\mathbb{Z}_4 \times \mathbb{Z}_2 \times \mathbb{Z}_2$ and $\mathbb{Z}_2 \times \mathbb{Z}_2 \times \mathbb{Z}_2 \times \mathbb{Z}_2$.

Example 5.8. Note that $U(32)$ is an abelian group of order $\varphi(32) = 16$, so it must be isomorphic to one of the groups in the preceding example. But which one? Examining the orders of the elements, we find that there is no element of order 16, so it is not $\mathbb{Z}_{16}$. However, $|3| = 8$. As none of the other groups in the preceding example have an element of order 8, $U(32)$ is isomorphic to $\mathbb{Z}_8 \times \mathbb{Z}_2$.

Example 5.9. As $200 = 2^3 5^2$, the finite abelian groups of order 200 are all isomorphic to one of the following, namely $\mathbb{Z}_8 \times \mathbb{Z}_{25}$, $\mathbb{Z}_4 \times \mathbb{Z}_2 \times \mathbb{Z}_{25}$, $\mathbb{Z}_2 \times \mathbb{Z}_2 \times \mathbb{Z}_2 \times \mathbb{Z}_{25}$, $\mathbb{Z}_8 \times \mathbb{Z}_5 \times \mathbb{Z}_5$, $\mathbb{Z}_4 \times \mathbb{Z}_2 \times \mathbb{Z}_5 \times \mathbb{Z}_5$ and $\mathbb{Z}_2 \times \mathbb{Z}_2 \times \mathbb{Z}_2 \times \mathbb{Z}_5 \times \mathbb{Z}_5$.

We might be momentarily concerned about the absence of $\mathbb{Z}_{200}$ in the preceding example. However, it is isomorphic to $\mathbb{Z}_8 \times \mathbb{Z}_{25}$, as the following theorem shows us.

Theorem 5.4. *Let $G = H_1 \times \cdots \times H_k$, where each H_i is cyclic of order n_i. Then G is cyclic if and only if $(n_i, n_j) = 1$ whenever $i \neq j$.*

Proof. Let $H_i = \langle a_i \rangle$. If the n_i are all relatively prime, then we claim that $(a_1, \ldots, a_k)$ has order $n_1 \cdots n_k = |G|$, and therefore G is cyclic. Suppose that $(a_1, \ldots, a_k)^m = (e, \ldots, e)$. Then each $a_i^m = e$, so $n_i | m$. As the n_i are relatively prime, $n_1 \cdots n_k | m$, by Corollary 2.3. Since $|G| = n_1 \cdots n_k$, the largest possible order of an element is $n_1 \cdots n_k$, and the claim is proved.

On the other hand, suppose that the n_i are not relatively prime. Without loss of generality, say that some prime p divides both n_1 and n_2. Then for any $r_i \in \mathbb{Z}$, we have $(a_1^{r_1}, \ldots, a_k^{r_k})^{n_1 \cdots n_k / p} = (e, \ldots, e)$, since each n_i divides $n_1 \cdots n_k / p$. (For $i = 1$, we have $n_1(n_2/p)n_3 \cdots n_k$, and for $i \geq 2$, we have $(n_1/p)n_2n_3 \cdots n_k$.) Thus, every element of G has order dividing $n_1n_2 \cdots n_k/p$, and therefore there is no element of order $|G|$, so G is not cyclic. $\square$

As a result of our classification, we can prove a special case of a famous result due to Augustin-Louis Cauchy.

Theorem 5.5 (Cauchy's Theorem for Abelian Groups). *Let G be a finite abelian group, and suppose that p is a prime dividing $|G|$. Then G has an element of order p.*

Proof. If $|G|$ is divisible by a prime, then G is not the trivial group. Letting G be as in Corollary 5.1, we see that $|G| = p_1^{n_1} p_2^{n_2} \cdots p_k^{n_k}$. If p divides $|G|$, then $p = p_i$, for some i. But then G has a subgroup isomorphic to $\mathbb{Z}_{p^{n_i}}$, for some $n_i > 0$. However, in $\mathbb{Z}_{p^{n_i}}$, the element p^{n_i-1} has order p. The proof is complete. $\square$

Corollary 5.2. *A finite abelian p-group has order p^n, for some $n \geq 0$.*

Proof. Let G be a finite abelian p-group. If the corollary is false, then the order of G is divisible by q, for some prime $q \neq p$. But then G has an element of order q, which is impossible. $\square$

Exercises

5.11. Give a list of abelian groups of each of the following orders, such that every abelian group of that order is isomorphic to one of the groups in the list.

1. 21
2. 81
3. 9800

5.12. Give a list of abelian groups of each of the following orders, such that every abelian group of that order is isomorphic to one of the groups in the list.

1. 144
2. 243
3. 55125

5.13. Write $U(56)$ as an external direct product of cyclic groups of prime power order, as in Corollary 5.1.

5.14. Write $(\mathbb{Z}_{20} \times \mathbb{Z}_6)/\langle(10, 2)\rangle$ as an external direct product of cyclic groups of prime power order, as in Corollary 5.1.

5.15. Let p be a prime. Suppose that G is a nontrivial finite abelian group in which every element has order 1 or p. Show that G is isomorphic to a group of the form $\mathbb{Z}_p \times \mathbb{Z}_p \times \cdots \times \mathbb{Z}_p$.

5.16. Suppose that n is an integer that is a product of distinct primes. If G is a finite abelian group, and $|G|$ is divisible by n, show that G has a cyclic subgroup of order n.

5.17. If $\langle a \rangle$ is a cyclic group of order 35, write a as the product of a 5-element and a 7-element.

5.18. If $\langle a \rangle$ is a cyclic group of order 90, write a as the product of p-elements, for various primes p.

5.19. Prove Theorem 5.5 in a different way, as follows. Let p be a prime dividing $|G|$. Show that G has an element a of some prime order, say q. If $q = p$, we are done. Otherwise, what can be said about $G/\langle a \rangle$? Complete the proof.

5.20. Let G be a finite abelian group and let n be a positive integer dividing $|G|$. Show that G has a subgroup of order n.

5.3 Elementary Divisors and Invariant Factors

For any positive integer n, we now know all possible abelian groups of order n, up to isomorphism. Indeed, we determine the prime factorization of n, and then proceed as in Examples 5.7 and 5.9. But we have not yet made certain that the groups we found are not isomorphic to each other. Let us work on that.

Definition 5.4. Let G be a nontrivial finite abelian group, and say that $G = H_1 \times H_2 \times \cdots \times H_k$, where each H_i is cyclic of order $p_i^{n_i}$, for some prime p_i and positive integer n_i. Then the **elementary divisors** of G are the numbers $p_1^{n_1}, p_2^{n_2}, \ldots, p_k^{n_k}$, where the order in this list is irrelevant, but each number must be listed as many times as it occurs. The trivial group has no elementary divisors.

Example 5.10. The elementary divisors of $\mathbb{Z}_9 \times \mathbb{Z}_9 \times \mathbb{Z}_3 \times \mathbb{Z}_{125}$ are $9, 9, 3, 125$.

Example 5.11. To find the elementary divisors of $\mathbb{Z}_{300} \times \mathbb{Z}_3$, we use Theorem 5.4 to see that the group is isomorphic to $\mathbb{Z}_{25} \times \mathbb{Z}_4 \times \mathbb{Z}_3 \times \mathbb{Z}_3$, so the elementary divisors are $4, 3, 3, 25$.

Definition 5.5. Let G be an abelian group and n a positive integer. Then we write $G^n = \{a^n : a \in G\}$.

Lemma 5.7. *Let G and H be abelian groups and n a positive integer. Then*

1. G^n *is a subgroup of G; and.*
2. *if $\alpha : G \to H$ is an onto homomorphism, then $\alpha(G^n) = H^n$.*

Proof. (1) See Exercise 3.40.

(2) If $g^n \in G^n$, then $\alpha(g^n) = (\alpha(g))^n \in H^n$. Also, if $h^n \in H^n$, then as α is onto, write $h = \alpha(g)$, with $g \in G$. Then $h^n = (\alpha(g))^n = \alpha(g^n) \in \alpha(G^n)$, completing the proof. $\square$

The elementary divisors are very important, as they uniquely determine a finite abelian group, up to isomorphism.

Theorem 5.6. *Let G and H be finite abelian groups. Then G and H are isomorphic if and only if they have the same elementary divisors.*

Proof. If G and H have the same elementary divisors, then each is isomorphic to a direct product of cyclic groups, and the groups appearing in the direct product in G have the same order as those appearing in H, so they are isomorphic. (We must be a bit careful, as the cyclic groups may appear in a different order in the direct product, but $M \times N$ is always isomorphic to $N \times M$, so this is not a problem. See Exercise 4.36.) Note that if neither G nor H has any elementary divisors, then each is the trivial group, so they are isomorphic.

On the other hand, let $\alpha : G \to H$ be an isomorphism. Take any prime p. Now, by Lemma 5.3, the p-elements of G form a subgroup, as do those of H. Furthermore, as isomorphisms preserve the orders of group elements, α provides an isomorphism from one of these p-subgroups to the other. As the elementary divisors come from these p-subgroups, we may as well assume to begin with that G and H are both p-groups. We proceed by strong induction on $|G|$. If $|G| = 1$, then G and H are both the trivial group, so neither has elementary divisors. Therefore, assume that $|G| > 1$ and the result holds for groups of smaller order.

In particular, say $G = G_1 \times \cdots \times G_k$ and $H = H_1 \times \cdots \times H_l$, where $G_i = \langle g_i \rangle$ is cyclic of order p^{n_i}, and $H_i = \langle h_i \rangle$ is cyclic of order p^{m_i}. Rearranging the terms if necessary, we may assume that $n_1 \geq n_2 \geq \cdots \geq n_k > 0$ and $m_1 \geq m_2 \geq \cdots \geq m_l > 0$. By the above lemma, $\alpha(G^p) = H^p$. Thus, $\alpha(G_1^p \times \cdots \times G_k^p) = H_1^p \times \cdots \times H_l^p$. But $G_i^p = \langle g_i^p \rangle$, and since $|g_i| = p^{n_i}$, we have $|g_i^p| = p^{n_i-1}$, by Corollary 3.2. Similarly, $|h_i^p| = p^{m_i-1}$. Thus, G^p is a p-group of strictly smaller order than G, and by our inductive hypothesis, the elementary divisors of G^p and H^p are the same. But the elementary divisors of G^p are $p^{n_1-1}, p^{n_2-1}, \ldots, p^{n_r-1}$, where $n_r > 1$ but $n_u = 1$ whenever $u > r$. (When $n_u = 1$, we have $p^{n_u-1} = 1$, which does not count as an elementary divisor. If $n_1 = 1$, then G^p has no elementary divisors.) Similarly, the elementary divisors of H^p are $p^{m_1-1}, \ldots, p^{m_s-1}$, where $m_s > 1$ but $m_v = 1$ whenever $v > s$. Therefore, $r = s$ and $m_i - 1 = n_i - 1$ whenever $i \leq r$. But then $m_i = n_i$, for all $i \leq r$. Also, $n_i = 1$ for all $i > r$ and $m_i = 1$ for all $i > s$. In order to prove that G and H have the same elementary divisors, it remains only to show that $k = l$. But $|G| = p^{n_1} \cdots p^{n_r} p^{k-r}$ and $|H| = p^{n_1} \cdots p^{n_r} p^{l-r}$. As isomorphic groups have the same order, $p^{k-r} = p^{l-r}$, and therefore $k = l$. If G^p has no elementary divisors, then neither does H^p, and we simply get $p^k = p^l$, hence $k = l$. $\qquad\square$

Example 5.12. The five abelian groups of order 16 listed in Example 5.7 are all non-isomorphic, as they have different elementary divisors. Similarly for the six abelian groups of order 200 given in Example 5.9.

Example 5.13. Let $G = \mathbb{Z}_{200} \times \mathbb{Z}_8 \times \mathbb{Z}_6$, $H = \mathbb{Z}_{120} \times \mathbb{Z}_{10} \times \mathbb{Z}_4 \times \mathbb{Z}_2$ and $K = \mathbb{Z}_{25} \times \mathbb{Z}_{24} \times \mathbb{Z}_8 \times \mathbb{Z}_2$. These are all abelian groups of order 9600. However, using Theorem 5.4, we see that G is isomorphic to $\mathbb{Z}_8 \times \mathbb{Z}_{25} \times \mathbb{Z}_8 \times \mathbb{Z}_3 \times \mathbb{Z}_2$, so its elementary divisors are 8, 8, 2, 3, 25. Similarly, H is isomorphic to $\mathbb{Z}_3 \times \mathbb{Z}_8 \times \mathbb{Z}_5 \times \mathbb{Z}_5 \times \mathbb{Z}_2 \times \mathbb{Z}_4 \times \mathbb{Z}_2$, so its elementary divisors are 8, 4, 2, 2, 3, 5, 5 and K is isomorphic to $\mathbb{Z}_{25} \times \mathbb{Z}_3 \times \mathbb{Z}_8 \times \mathbb{Z}_8 \times \mathbb{Z}_2$, so its elementary divisors are 8, 8, 2, 3, 25. Therefore, G and K are isomorphic, but H is not isomorphic to either of them.

There is another interesting way to express a finite abelian group as a direct product of cyclic groups.

Theorem 5.7 (Invariant Factor Decomposition). *Suppose that G is a nontrivial finite abelian group. Then $G = H_1 \times H_2 \times \cdots \times H_k$, where each H_i is a cyclic subgroup of G of order m_i, with $m_1 > 1$ and $m_i | m_{i+1}$, for $1 \le i < k$.*

Proof. We will explain how to construct the H_i, assuming that G has been expressed as a direct product of cyclic groups of prime power order, as in Corollary 5.1. Let $p_1, \ldots, p_r$ be the primes dividing $|G|$. For each j, find the largest power $p_j^{n_j}$ such that $\mathbb{Z}_{p_j^{n_j}}$ appears in Corollary 5.1. Letting $m_k = p_1^{n_1} p_2^{n_2} \cdots p_r^{n_r}$, Theorem 5.4 says that $H_k = \mathbb{Z}_{p_1^{n_1}} \times \cdots \times \mathbb{Z}_{p_r^{n_r}}$ is isomorphic to $\mathbb{Z}_{m_k}$. Now, delete all of the terms from the direct product in Corollary 5.1 that we have used (deleting only one copy, if multiple copies of the same group appear). For each j, let $p_j^{s_j}$ be the largest power appearing in the remaining terms (where $s_j = 0$ is entirely possible). Let $m_{k-1} = p_1^{s_1} \cdots p_r^{s_r}$. By construction, each $s_j \le n_j$, so $m_{k-1} | m_k$. Again, $H_{k-1} = \mathbb{Z}_{p_1^{s_1}} \times \cdots \times \mathbb{Z}_{p_r^{s_r}}$ is isomorphic to $\mathbb{Z}_{m_{k-1}}$. Delete all of these terms that we have just used, and repeat until we exhaust the entire direct product in Corollary 5.1. $\square$

Definition 5.6. If G is isomorphic to $\mathbb{Z}_{m_1} \times \cdots \times \mathbb{Z}_{m_k}$, where $m_1 > 1$ and $m_i | m_{i+1}$, for $1 \le i < k$, then the numbers $m_1, \ldots, m_k$ are called the **invariant factors** of G.

Example 5.14. Let us use our work in Example 5.9 to find the invariant factors of the abelian groups of order 200. We apply the method from Theorem 5.7. Considering $\mathbb{Z}_4 \times \mathbb{Z}_2 \times \mathbb{Z}_{25}$, we see that the highest power of 2 that appears is 4, and the highest power of 5 is 25. Therefore, $m_k = 4 \cdot 25 = 100$. Deleting $\mathbb{Z}_4$ and $\mathbb{Z}_{25}$, we are left with $\mathbb{Z}_2$, so $m_{k-1} = 2$, and we are finished. Thus, our group is isomorphic to $\mathbb{Z}_2 \times \mathbb{Z}_{100}$, so the invariant factors are 2, 100. When we examine $\mathbb{Z}_2 \times \mathbb{Z}_2 \times \mathbb{Z}_2 \times \mathbb{Z}_5 \times \mathbb{Z}_5$, we see that $m_k = 2 \cdot 5 = 10$. Deleting $\mathbb{Z}_2$ and $\mathbb{Z}_5$, we are left with $\mathbb{Z}_2 \times \mathbb{Z}_2 \times \mathbb{Z}_5$. Thus, $m_{k-1} = 2 \cdot 5 = 10$. Deleting $\mathbb{Z}_2$ and $\mathbb{Z}_5$, we are left only with $\mathbb{Z}_2$. Thus, $m_{k-2} = 2$, and we are finished. Therefore, our group is isomorphic to $\mathbb{Z}_2 \times \mathbb{Z}_{10} \times \mathbb{Z}_{10}$, which gives invariant factors of 2, 10, 10. Considering $\mathbb{Z}_8 \times \mathbb{Z}_{25}$, we simply get $\mathbb{Z}_{200}$, so 200 is the only invariant factor. Looking at $\mathbb{Z}_2 \times \mathbb{Z}_2 \times \mathbb{Z}_2 \times \mathbb{Z}_{25}$, we have $\mathbb{Z}_2 \times \mathbb{Z}_2 \times \mathbb{Z}_{50}$, so the invariant factors are 2, 2, 50. When we examine $\mathbb{Z}_8 \times \mathbb{Z}_5 \times \mathbb{Z}_5$, we obtain $\mathbb{Z}_5 \times \mathbb{Z}_{40}$, so the invariant factors are 5, 40. Finally, if we take $\mathbb{Z}_4 \times \mathbb{Z}_2 \times \mathbb{Z}_5 \times \mathbb{Z}_5$, then we get $\mathbb{Z}_{10} \times \mathbb{Z}_{20}$, so the invariant factors are 10, 20.

In the above example, the nonisomorphic groups produced different lists of invariant factors. As it turns out, this always happens.

Theorem 5.8. *Let G and H be nontrivial finite abelian groups. Then G and H are isomorphic if and only if they have the same invariant factors.*

Proof. Let G be isomorphic to $\mathbb{Z}_{m_1} \times \cdots \times \mathbb{Z}_{m_k}$ with $m_1 > 1$ and $m_i | m_{i+1}, 1 \leq i < k$. Similarly, write H as $\mathbb{Z}_{n_1} \times \cdots \times \mathbb{Z}_{n_l}$, with $n_1 > 1$ and $n_i | n_{i+1}, 1 \leq i < l$. If G and H have the same invariant factors, then they are both isomorphic to the same direct product, and therefore to each other.

On the other hand, suppose that G and H are isomorphic. We will show that they have the same invariant factors. Our proof is by strong induction on $|G|$. If $|G| = 2$, then the only possible invariant factor list is 2 for both G and H, so there is nothing to do. Assume that $|G| > 2$ and that the result is true for groups of smaller order. If we take $(g_1, \ldots, g_k) \in G$, then each g_i has order dividing m_i, and therefore all g_i have order dividing m_k. On the other hand $(0, 0, \ldots, 0, 1)$ has order m_k. Thus, m_k is the largest possible order of an element of G. Similarly, n_l is the largest possible order of any element of H. Therefore, as isomorphisms preserve orders of group elements, $m_k = n_l$. Now, expressing each m_i as a product of prime powers, we note that the elementary divisors of G are those that come from $\mathbb{Z}_{m_1} \times \cdots \times \mathbb{Z}_{m_{k-1}}$ together with those from $\mathbb{Z}_{m_k}$. Similarly, the elementary divisors of H are those coming from $\mathbb{Z}_{n_1} \times \cdots \times \mathbb{Z}_{n_{l-1}}$ together with those from $\mathbb{Z}_{n_l} = \mathbb{Z}_{m_k}$. As G and H are isomorphic, Theorem 5.6 tells us that they have the same elementary divisors. Deleting those from $\mathbb{Z}_{m_k}$, the groups $\mathbb{Z}_{m_1} \times \cdots \times \mathbb{Z}_{m_{k-1}}$ and $\mathbb{Z}_{n_1} \times \cdots \times \mathbb{Z}_{n_{l-1}}$ have the same elementary divisors. Thus, by Theorem 5.6, these groups are isomorphic. As they have smaller order than G, our inductive hypothesis tells us that $k - 1 = l - 1$ and each $m_i = n_i$. Therefore, the invariant factors are identical.

(We have to be a bit careful if either $k = 1$ or $l = 1$, as then we have nothing left when we remove the term $\mathbb{Z}_{m_k}$ or $\mathbb{Z}_{n_l}$. But in this case, comparing orders, we must have $k = l = 1$, and the only invariant factor is m_1 for both groups.) $\qquad \square$

Exercises

5.21. Find the elementary divisors for each of the following groups.

1. $\mathbb{Z}_{42} \times \mathbb{Z}_{4200}$
2. $\mathbb{Z}_6 \times \mathbb{Z}_{18} \times \mathbb{Z}_{54}$

5.22. Find the invariant factors for each of the following groups.

1. $\mathbb{Z}_3 \times \mathbb{Z}_3 \times \mathbb{Z}_9 \times \mathbb{Z}_{25} \times \mathbb{Z}_{11} \times \mathbb{Z}_{121}$
2. $\mathbb{Z}_4 \times \mathbb{Z}_8 \times \mathbb{Z}_8 \times \mathbb{Z}_{16} \times \mathbb{Z}_5 \times \mathbb{Z}_{25} \times \mathbb{Z}_{49}$

5.23. Let p, q and r be distinct primes. Give the list of elementary divisors for every possible abelian group of order $p^3 q^2 r$.

5.24. Let p, q and r be distinct primes. Give the list of invariant factors for every possible abelian group of order $p^3 q^2 r$.

5.25. For which positive integers n are all abelian groups of order n isomorphic?

5.26. Find the smallest positive integer n such that there are exactly four nonisomorphic abelian groups of order n.

5.27. Let G_1, G_2 and G_3 be finite abelian groups, and suppose that $G_1 \times G_2$ is isomorphic to $G_1 \times G_3$. Show that G_2 and G_3 are isomorphic.

5.28. Let a finite abelian group G have invariant factors $n_1, n_2, \ldots, n_k$. What are the invariant factors of $G \times G$?

5.29. Let G be a nontrivial finite abelian 2-group. Show that the number of elements of order 2 in G is $2^k - 1$, for some positive integer k.

5.30. Let G be a finite abelian group. Suppose that, for every $n \in \mathbb{N}$, there are at most n elements $a \in G$ satisfying $a^n = e$. Show that G is cyclic.

5.4 A Word About Infinite Abelian Groups

Unfortunately, that word is "messy". We have seen that finite abelian groups behave very nicely. To be sure, we cannot possibly expect every infinite abelian group to be a direct product of cyclic groups of prime power order. But even if we allow direct products of infinite cyclic groups such as $\mathbb{Z} \times \mathbb{Z}$, that does not come close to covering all of the possibilities. While a deep discussion of infinite abelian groups is beyond the scope of an introductory abstract algebra course, we can make a few remarks.

Definition 5.7. Let G be a nontrivial group. We say that G is **decomposable** if it is the direct product of two proper subgroups. If not, then it is **indecomposable**.

We can easily classify the indecomposable finite abelian groups.

Theorem 5.9. *Let G be a finite abelian group. Then G is indecomposable if and only if G is a cyclic group of order p^n, for some prime p and positive integer n.*

Proof. In view of Theorem 5.3, an indecomposable finite abelian group must indeed be cyclic of prime power order. If G is cyclic of order p^n, then suppose that $G = H \times K$, for some subgroups H and K. Then by Lagrange's theorem, H and K are both p-groups. Furthermore, by Theorem 3.16, they are both cyclic. But since $G = H \times K$ is cyclic, it follows from Theorem 5.4 that $(|H|, |K|) = 1$. As the orders are both powers of p, this means that either H or K is trivial, so either K or H is all of G. Thus, H and K are not both proper and G is indecomposable. $\square$

What about infinite abelian groups?

Example 5.15. The additive group $\mathbb{Q}$ is indecomposable. Indeed, suppose that $\mathbb{Q} = H \times K$, where H and K are proper subgroups. Then neither H nor K is $\{0\}$, so take $a/b \in H$, $c/d \in K$, where a, b, c and d are nonzero integers. Note that $bc(a/b) = ac \in H$ and $ad(c/d) = ac \in K$. Then $H \cap K$ is not trivial, so we do not have a direct product. Also, $\mathbb{Q}$ is not cyclic. Indeed, if $a, b \in \mathbb{Z}$ and $b > 0$, it is clear that $1/(b+1) \notin \langle a/b \rangle$. Thus, $\mathbb{Q} \neq \langle a/b \rangle$.

Now, every element of $\mathbb{Q}$ other than the identity has infinite order. What about infinite abelian groups where every element has finite order?

Example 5.16. Consider the group $\mathbb{Q}/\mathbb{Z}$. Exercise 5.31 asks us to examine some properties of this group. In particular, the distinct elements of the group are precisely of the form $q + \mathbb{Z}$, where $q \in \mathbb{Q}$ and $0 \leq q < 1$. Also, every element has finite order. But this group is decomposable. Indeed, fix any prime p. Then let $H = \{a/b + \mathbb{Z} : a, b \in \mathbb{Z}, b = p^n, n \geq 0\}$ and $K = \{c/d + \mathbb{Z} : c, d \in \mathbb{Z}, (d, p) = 1\}$. In Exercise 5.32, we also demonstrate that $\mathbb{Q}/\mathbb{Z} = H \times K$.

The group H from the preceding example is named for E.P. Heinz Prüfer.

Definition 5.8. Let p be a prime. Then the **Prüfer p-group** is the subgroup $\{a/p^n + \mathbb{Z} : a, n \in \mathbb{Z}, n \geq 0\}$ of the additive group $\mathbb{Q}/\mathbb{Z}$.

Example 5.17. Let H be the Prüfer p-group. We note that H is an abelian p-group; indeed, $p^n(a/p^n + \mathbb{Z}) = a + \mathbb{Z} = 0 + \mathbb{Z}$; thus, the order of $a/p^n + \mathbb{Z}$ divides p^n. But H is not cyclic; indeed, $1/p^n + \mathbb{Z}$ has order p^n, so H has elements of arbitrarily large order. So if it were cyclic, what order could its generator possibly have? However, Exercise 5.36 asks us to show that every nontrivial subgroup of H contains $1/p + \mathbb{Z}$. Thus, H is surely indecomposable.

In fact, $\mathbb{Q}$ and the Prüfer p-group share another interesting property.

Definition 5.9. Let G be an abelian group written additively. We say that G is **divisible** if, for every element a of G and every positive integer n, there exists a $b \in G$ such that $nb = a$.

Note that if G is a nontrivial finite abelian group, then it cannot be divisible. Indeed, if G has order n, then $nb = 0$ for every $b \in G$. Thus, if $0 \neq a \in G$, then $nb = a$ has no solution. So, we must look to infinite abelian groups.

Example 5.18. The group $\mathbb{Q}$ is divisible. Indeed, if $a \in \mathbb{Q}$ and n is a positive integer, then $n(a/n) = a$.

Example 5.19. For any prime p, the Prüfer p-group is divisible. Indeed, to see this, we note that if G is divisible, so is any factor group of G. (See Exercise 5.35.) Thus, $\mathbb{Q}/\mathbb{Z}$ is divisible. As in Example 5.16, write $\mathbb{Q}/\mathbb{Z} = H \times K$, where H is the Prüfer p-group. If $a \in H$, then by the divisibility of $\mathbb{Q}/\mathbb{Z}$, for any positive integer n, there exist $h \in H, k \in K$ such that $n(h, k) = (a, 0)$. But then $nh = a$.

Exercises

5.31. Let $G = \mathbb{Q}/\mathbb{Z}$.

1. Show that the elements of G can be uniquely written in the form $q + \mathbb{Z}$, where $q \in \mathbb{Q}$ and $0 \leq q < 1$.
2. If $a, b \in \mathbb{Z}, b > 0$ and $(a, b) = 1$, what is the order of $a/b + \mathbb{Z}$ in G?

5.32. Show that for any prime p, $\mathbb{Q}/\mathbb{Z} = H \times K$, where H is the Prüfer p-group and $K = \{c/d + \mathbb{Z} : c, d \in \mathbb{Z}, (d, p) = 1\}$.

5.33. Let G be a divisible group, written additively. Show that for every positive integer n, the function $\alpha : G \to G$ given by $\alpha(a) = na$ is an onto homomorphism. Is it necessarily an automorphism?

5.34. Let G and H be abelian groups, written additively. Show that $G \times H$ is divisible if and only if G and H are both divisible.

5.35. Show that if G is a divisible group, then every factor group of G is divisible, but subgroups need not be.

5.36. Let G be the Prüfer p-group, for some prime p. Show that every nontrivial subgroup of G contains $1/p + \mathbb{Z}$.

5.37. Let G be an abelian group having a subgroup N such that G/N is infinite cyclic. Show that G has a subgroup H such that H is infinite cyclic and $G = H \times N$.

5.38. For any prime p, show that every proper subgroup of the Prüfer p-group is finite.

Chapter 6
Symmetric and Alternating Groups

We have seen the definition of the symmetric group S_n, but so far, we do not have too much experience with it. In this chapter, we will introduce the notions of cycles and, in particular, transpositions, which are important elements of the symmetric group. These will help us to understand the group.

We will also construct a subgroup of the symmetric group called the alternating group. If $n \geq 5$, then the alternating group is very special in that it has no nontrivial proper normal subgroups.

6.1 The Symmetric Group and Cycle Notation

Let n be a positive integer. Then we recall that the set of permutations of the set $\{1, 2, \ldots, n\}$ is a group of order $n!$ under composition of functions. It is called the symmetric group and denoted S_n. Why is this group of sufficient interest to merit a chapter on its own? In the earliest years of group theory, the abstract definition of a group had not been written down. Instead, mathematicians worked with groups of permutations. As it turns out, they were not losing much by doing so! If A is any nonempty set, write $P(A)$ for the set of all permutations of A. Then just as we saw that S_n is a group under composition of functions, so is $P(A)$. The following famous result is due to Arthur Cayley.

Theorem 6.1 (Cayley's Theorem). *Let G be any group. Then G is isomorphic to a subgroup of $P(G)$.*

Proof. For each $a \in G$, define $\rho_a : G \to G$ via $\rho_a(g) = ag$, for all $g \in G$. We claim that $\rho_a \in P(G)$. Certainly $\rho_a(g) \in G$. If $\rho_a(g_1) = \rho_a(g_2)$, for $g_1, g_2 \in G$, then $ag_1 = ag_2$, so $g_1 = g_2$. Thus, ρ_a is one-to-one. If $g \in G$, then $\rho_a(a^{-1}g) = g$, so ρ_a is also onto. The claim is proved.

© Springer International Publishing AG, part of Springer Nature 2018
G. T. Lee, *Abstract Algebra*, Springer Undergraduate Mathematics Series,
https://doi.org/10.1007/978-3-319-77649-1_6

Now, define $\rho : G \to P(G)$ via $\rho(a) = \rho_a$. We claim that ρ is a homomorphism. If $a, b \in G$, then $\rho(ab)(g) = \rho_{ab}(g) = abg$ and $(\rho(a) \circ \rho(b))(g) = \rho_a(\rho_b(g)) = \rho_a(bg) = abg$, for all $g \in G$. Thus, $\rho(ab) = \rho(a) \circ \rho(b)$, proving the claim. Also, if $a \in \ker(\rho)$, then ρ_a is the identity permutation. In particular, $\rho_a(e) = e$, and therefore $ae = e$. Thus, $a = e$, and ρ is one-to-one. It now follows that G is isomorphic to $\rho(G)$, which is a subgroup of $P(G)$.   □

Corollary 6.1. *Let G be a group of order $n < \infty$. Then G is isomorphic to a subgroup of S_n.*

Proof. We know that G is isomorphic to a subgroup of $P(G)$, but replacing G with $\{1, 2, \ldots, n\}$ is just a relabelling. Thus, G is isomorphic to a subgroup of S_n.   □

The notation we have been using for elements of S_n is rather cumbersome and tends to hide the properties of the permutations. It is time to introduce something better.

Definition 6.1. Let k be a positive integer. A permutation $\sigma \in S_n$ is called a k-**cycle** if there exist distinct elements $a_1, a_2, \ldots, a_k \in \{1, 2, \ldots, n\}$ such that $\sigma(a_i) = a_{i+1}$, for $1 \leq i < k$, $\sigma(a_k) = a_1$ and if $a \notin \{a_1, \ldots, a_k\}$, then $\sigma(a) = a$. We use the **cycle notation** $\sigma = (a_1 \ a_2 \ \cdots \ a_k)$. A **cycle** means a k-cycle for some k.

Example 6.1. Let us work in S_5. Then $\sigma = \begin{pmatrix} 1 & 2 & 3 & 4 & 5 \\ 1 & 5 & 3 & 2 & 4 \end{pmatrix}$ is a 3-cycle; as $\sigma(2) = 5$, $\sigma(5) = 4$, $\sigma(4) = 2$ and everything else is fixed, we have $\sigma = (2 \ 5 \ 4)$. Note that it would be just as correct to write $\sigma = (5 \ 4 \ 2)$ or $(4 \ 2 \ 5)$ (but not $(2 \ 4 \ 5)$). Similarly, $\tau = \begin{pmatrix} 1 & 2 & 3 & 4 & 5 \\ 3 & 5 & 2 & 1 & 4 \end{pmatrix}$ satisfies $\tau(1) = 3$, $\tau(3) = 2$, $\tau(2) = 5$, $\tau(5) = 4$, $\tau(4) = 1$, and there are no other values to consider, so τ is the 5-cycle $(1 \ 3 \ 2 \ 5 \ 4)$ (or $(3 \ 2 \ 5 \ 4 \ 1)$, and so on).

Note that the only 1-cycle in S_n is the identity permutation, denoted (1).

Theorem 6.2. *Any k-cycle in S_n has order k.*

Proof. Simply note that if $\sigma = (a_1 \ \cdots \ a_k)$, then $\sigma(a_1) = a_2$, $\sigma^2(a_1) = \sigma(a_2) = a_3$, and so on. It takes k steps to reach a_1 again. Similarly for all other a_i.   □

Definition 6.2. We say that cycles $\sigma_1, \ldots, \sigma_r$ are **disjoint** if, whenever $\sigma_i(a) \neq a$, we have $\sigma_j(a) = a$ for all $j \neq i$. If $\sigma \in S_n$ and we write $\sigma = \sigma_1 \sigma_2 \cdots \sigma_r$, where the σ_i are disjoint cycles, then we have a **disjoint cycle decomposition** for σ.

Example 6.2. Let $\sigma = \begin{pmatrix} 1 & 2 & 3 & 4 & 5 & 6 \\ 5 & 6 & 3 & 1 & 4 & 2 \end{pmatrix}$ in S_6. Then noting that $\sigma(1) = 5$, $\sigma(5) = 4$ and $\sigma(4) = 1$, we have a cycle $(1 \ 5 \ 4)$. Also, $\sigma(2) = 6$ and $\sigma(6) = 2$, so we have another cycle $(2 \ 6)$. The remaining number, 3, is fixed by σ, so a disjoint cycle decomposition for σ is $\sigma = (1 \ 5 \ 4)(2 \ 6)$.

Example 6.3. Similarly, consider $\sigma = \begin{pmatrix} 1 & 2 & 3 & 4 & 5 & 6 & 7 & 8 \\ 2 & 1 & 5 & 6 & 8 & 4 & 3 & 7 \end{pmatrix}$. Using the same procedure as above, we find that $\sigma = (1\ 2)(3\ 5\ 8\ 7)(4\ 6)$ is a disjoint cycle decomposition.

In fact, we can always apply the procedure from the last two examples.

Theorem 6.3. *Every element of S_n is a product of disjoint cycles.*

Proof. Take any $\sigma \in S_n$. If σ is the identity, then $\sigma = (1)$ and there is nothing
to do. Assume otherwise, and take $a_1 \in \{1, \ldots, n\}$ such that $\sigma(a_1) = a_2 \neq a_1$. If
$\sigma(a_2) = a_1$, then we have a 2-cycle, $(a_1\ a_2)$. Otherwise, let $\sigma(a_2) = a_3$. Continue
until we find a_k such that $\sigma(a_k) \in \{a_1, \ldots, a_k\}$. Now, if $\sigma(a_k) = a_i$, with $1 < i \leq k$,
then $\sigma^k(a_1) = \sigma^{i-1}(a_1)$. Thus, $\sigma^{k-i+1}(a_1) = a_1$. In other words, $a_{k-i+2} = a_1$. But
this is a contradiction. Therefore, $\sigma(a_k) = a_1$, and we have a k-cycle, $(a_1\ a_2\ \cdots\ a_k)$.
    If $\sigma = (a_1\ a_2\ \cdots\ a_k)$, then we are done. Otherwise, take b_1 which is not in
$\{a_1, \ldots, a_k\}$ such that $\sigma(b_1) = b_2 \neq b_1$. Now repeat the same procedure, obtaining
an l-cycle, $(b_1\ b_2\ \cdots\ b_l)$. We must make sure that these cycles are disjoint; that
is, we cannot have $b_m \in \{a_1, \ldots, a_k\}$, for any m. By choice, $b_1 \notin \{a_1, \ldots, a_n\}$. If
$b_2 = \sigma(b_1) = a_t$, then since $a_t = \sigma(a_s)$, for some s, we have $\sigma(b_1) = \sigma(a_s)$, and as
σ is one-to-one, $b_1 = a_s$, which is impossible. Proceeding in this way, we see that
the cycles are disjoint.
    If $\sigma = (a_1\ \cdots\ a_k)(b_1\ \cdots\ b_l)$, then we are done. Otherwise, take any c_1 that
does not lie in $\{a_1, \ldots, a_k, b_1, \ldots, b_l\}$ such that $\sigma(c_1) \neq c_1$ and repeat. As there are
only n entries in $\{1, \ldots, n\}$, this procedure must stop eventually.   □

We were not too concerned about the order in which we wrote the cycles in the
last proof. But this is ok.

Theorem 6.4. *In S_n, disjoint cycles commute.*

Proof. Let $\sigma = (a_1\ \cdots\ a_k)$ and $\tau = (b_1\ \cdots\ b_m)$ be disjoint cycles. We will
show that $\sigma\tau = \tau\sigma$. Take $c \in \{1, \ldots, n\}$. If $c \in \{a_1, \ldots, a_k\}$, then as σ and τ are
disjoint, τ fixes c. Thus, $\sigma\tau(c) = \sigma(c)$. But $\sigma(c) \in \{a_1, \ldots, a_k\}$ as well. Thus, τ
fixes $\sigma(c)$ too, so $\tau\sigma(c) = \sigma(c)$. By a similar argument, if $c \in \{b_1, \ldots, b_m\}$, then
$\sigma\tau(c) = \tau\sigma(c) = \tau(c)$. If c is not among the a_i or b_i, then both σ and τ fix c, so
$\sigma\tau(c) = \tau\sigma(c) = c$. We are done.   □

Example 6.4. It makes no difference if we write $(1\ 5)(2\ 6\ 4)$ or $(2\ 6\ 4)(1\ 5)$. Both
are the same permutation.

However, it would be wrong to try to extend this to cycles that are not disjoint!

Example 6.5. In S_3, let $\sigma = (1\ 2)$ and $\tau = (1\ 3)$. Let us compute $\sigma\tau$. Now, $\tau(1) = 3$
and $\sigma(3) = 3$, so $\sigma\tau(1) = 3$. Also, $\tau(3) = 1$ and $\sigma(1) = 2$, so $\sigma\tau(3) = 2$. Finally,
$\tau(2) = 2$ and $\sigma(2) = 1$, so $\sigma\tau(2) = 1$. There are no other values to consider, so $\sigma\tau$
is the 3-cycle $(1\ 3\ 2)$. But proceeding in the same way, we find that $\tau\sigma$ is a different
3-cycle, $(1\ 2\ 3)$.

Example 6.6. Let us find a disjoint cycle decomposition for $(2\ 4)(2\ 5\ 3\ 4)(1\ 3)(1\ 5)$. We see (working from right to left) that 1 is mapped by $(1\ 5)$ to 5, which is fixed by $(1\ 3)$, which then goes to 3, which is fixed by $(2\ 4)$. So, 1 goes to 3. Next, 3 is fixed by $(1\ 5)$, then goes to 1, which is fixed by the other cycles, so we have a 2-cycle $(1\ 3)$. Next, 2 is fixed by $(1\ 3)(1\ 5)$, it then goes to 5, which is fixed by $(2\ 4)$, so 2 goes to 5. Now, 5 goes to 1 which goes to 3 which goes to 4 and then back to 2. Thus, we have another 2-cycle, $(2\ 5)$. Finally, 4 goes to 2 then back to 4, so 4 is fixed. Therefore, we have $(2\ 4)(2\ 5\ 3\ 4)(1\ 3)(1\ 5) = (1\ 3)(2\ 5)$.

We can use the disjoint cycle decomposition to find the order of a permutation. Recall that the **least common multiple** of positive integers $a_1, a_2, \ldots, a_r$ is the smallest positive integer m such that $a_i | m$ for all i.

Theorem 6.5. *If $\sigma_1, \ldots, \sigma_r$ are disjoint cycles in S_n, then the order of $\sigma_1 \cdots \sigma_r$ is the least common multiple of the lengths of the σ_i.*

Proof. Let k be a positive integer. Then since the σ_i commute, by Theorem 6.4, we have $(\sigma_1 \cdots \sigma_r)^k = \sigma_1^k \cdots \sigma_r^k$. As the σ_i move disjoint subsets of $\{1, \ldots, n\}$, we have $\sigma_1^k \cdots \sigma_r^k = (1)$ if and only if each $\sigma_i^k = (1)$. In view of Theorem 6.2, this occurs if and only if the length of each σ_i divides k. □

Exercises

6.1. Write each of the following permutations as a product of disjoint cycles.

1. $\begin{pmatrix} 1\ 2\ 3\ 4\ 5\ 6\ 7 \\ 1\ 4\ 5\ 7\ 3\ 2\ 6 \end{pmatrix}$

2. $\begin{pmatrix} 1\ 2\ 3\ 4\ 5\ 6\ 7\ 8 \\ 2\ 5\ 6\ 3\ 1\ 4\ 8\ 7 \end{pmatrix}$

6.2. Write each of the following permutations as a product of disjoint cycles.

1. $(1\ 3\ 2)(1\ 4)(2\ 5\ 3)$
2. $(2\ 5\ 3\ 4)(1\ 2\ 6)(3\ 5\ 4)(1\ 2\ 7)$

6.3. Find the inverse of each of the following permutations. Write the answer as a product of disjoint cycles.

1. $(1\ 2\ 4)(3\ 5\ 7\ 6)$
2. $(1\ 2)(2\ 4\ 3)(2\ 3\ 5)$

6.4. Find all possible orders of elements of S_7.

6.5. How many elements of order 3 are there in S_9?

6.6. Let σ be a k-cycle. If m is a positive integer, show that σ^m is a k-cycle if and only if $(k, m) = 1$.

6.7. Let $\sigma \in S_n$ be a k-cycle. Show that there exists a k-cycle $\tau \in S_n$ such that $\tau^2 = \sigma$ if and only if k is odd.

6.8. If $n \neq 2$, show that $Z(S_n) = \{(1)\}$.

6.9. Find the smallest positive integers m and n such that S_m has an element of order 105 and S_n has an element of order 125.

6.10. Find a subgroup of order 120 in S_8.

6.2   Transpositions and the Alternating Group

While a disjoint cycle decomposition gives us the clearest picture of the action of a permutation, it is often useful to write the permutation as a different sort of product.

Definition 6.3. A **transposition** is a 2-cycle.

Theorem 6.6. *If $n \geq 2$, then every permutation in S_n is a product of transpositions.*

Proof. In view of Theorem 6.3, it is sufficient to show that every cycle is a product of transpositions. The identity is $(1) = (1\ 2)(1\ 2)$. Let us take a k-cycle σ; without loss of generality, say $\sigma = (1\ 2\ 3\ \cdots\ k)$. We claim that $\sigma = (1\ k)(1\ (k-1))\cdots(1\ 2)$. Our proof is by induction on k. If $k = 2$, there is nothing to do. Otherwise, assume that $(1\ 2\ \cdots\ k) = (1\ k)(1\ (k-1))\cdots(1\ 2)$. Then

$$(1\ (k+1))(1\ k)\cdots(1\ 2) = (1\ (k+1))(1\ 2\ \cdots\ k),$$

and performing the calculation, we see that this is $(1\ 2\ \cdots\ (k+1))$, as required. □

Example 6.7. Let us write $(1\ 4\ 5)(1\ 3\ 6\ 4\ 5)$ as a product of transpositions. Using the method described in the above proof,

$$(1\ 4\ 5) = (1\ 5)(1\ 4)$$

and

$$(1\ 3\ 6\ 4\ 5) = (1\ 5)(1\ 4)(1\ 6)(1\ 3),$$

so

$$(1\ 4\ 5)(1\ 3\ 6\ 4\ 5) = (1\ 5)(1\ 4)(1\ 5)(1\ 4)(1\ 6)(1\ 3).$$

It is worth noting that the expression of a permutation as a product of transpositions is by no means unique. For instance, we have seen that $(1\ 2\ 3\ 4) = (1\ 4)(1\ 3)(1\ 2)$. But also, $(1\ 2\ 3\ 4) = (1\ 2)(2\ 3)(3\ 4)$. In fact, the number of transpositions involved does not have to be the same, as both of these are equal to $(5\ 6)(1\ 2)(2\ 3)(3\ 4)(5\ 6)$.

Nevertheless, we note that all of the products we have just calculated involve an odd number of transpositions. It is a very useful fact that this parity is always preserved; that is, a permutation will be a product of either an even or an odd number of transpositions, not both. The following lemma does most of the work in proving this fact.

Lemma 6.1. *In S_n, the identity permutation cannot be written as a product of an odd number of transpositions.*

Proof. Suppose that the lemma is false, and let k be the smallest odd number such that $(1) = \sigma_1\sigma_2\cdots\sigma_k$, where each σ_i is a transposition. Now, choose an element of $\{1,\ldots,n\}$ that is not fixed by all of the σ_i. Without loss of generality, let us say that some $\sigma_i(1) \neq 1$. Let j be such that $\sigma_j(1) \neq 1$ but $\sigma_r(1) = 1$ for all $r > j$. Among all expressions of (1) as a product of k transpositions such that at least one does not fix 1, we proceed by induction on j. If $j = 1$, then we note that $\sigma_2\cdots\sigma_k$ fixes 1, but σ_1 does not, so $\sigma_1\cdots\sigma_k$ does not fix 1, which is a contradiction.

Therefore, assume that $j > 1$ and that our result holds for expressions with a smaller j value. Without loss of generality, say that $\sigma_j = (1\ 2)$. We have four cases to consider for σ_{j-1}. If $\sigma_{j-1} = (1\ 2)$, then since $(1\ 2)(1\ 2)$ is the identity, we can cancel it from our expression. But this contradicts the minimality of k.

Suppose that σ_{j-1} fixes 1 but not 2. Without loss of generality, say $\sigma_{j-1} = (2\ 3)$. Then notice that $(2\ 3)(1\ 2) = (1\ 3\ 2) = (1\ 3)(2\ 3)$. Thus, replacing $\sigma_{j-1}\sigma_j$ with $(1\ 3)(2\ 3)$, we see that the j value has now decreased to $j - 1$. By our inductive hypothesis, it is impossible to write the identity as a product in this way.

Suppose that σ_{j-1} fixes 2 but not 1. Without loss of generality, say $\sigma_{j-1} = (1\ 3)$. Then we see that $(1\ 3)(1\ 2) = (1\ 2\ 3) = (1\ 2)(2\ 3)$. Again, replacing $\sigma_{j-1}\sigma_j$ with $(1\ 2)(2\ 3)$, the j value decreases, and we have a contradiction.

Finally, suppose that σ_{j-1} fixes both 1 and 2. Without loss of generality, say $\sigma_{j-1} = (3\ 4)$. Then by Theorem 6.4, $(3\ 4)(1\ 2) = (1\ 2)(3\ 4)$, so we can once again decrease the j value. Our proof is complete. □

Theorem 6.7. *No permutation in S_n can be written as a product of both an even and an odd number of transpositions.*

Proof. Suppose that

$$\sigma_1\sigma_2\cdots\sigma_k = \tau_1\tau_2\cdots\tau_m,$$

where each σ_i and τ_i is a transposition, k is even and m is odd. Then

$$(1) = \sigma_k^{-1}\cdots\sigma_1^{-1}\tau_1\cdots\tau_m = \sigma_k\cdots\sigma_1\tau_1\cdots\tau_m,$$

since each σ_i has order 2 (by Theorem 6.2) and is therefore its own inverse. Thus, we have written the identity as a product of $k + m$ transpositions. But $k + m$ is odd, contradicting the preceding lemma. □

Definition 6.4. We say that a permutation in S_n is **even** (respectively, **odd**) if it is the product of an even (respectively, odd) number of transpositions.

Example 6.8. In S_5, we note that $(1\ 2\ 3)(4\ 5)$ is odd, as $(1\ 2\ 3)(4\ 5) = (1\ 3)(1\ 2)(4\ 5)$.

Theorem 6.8. *A k-cycle is even if and only if k is odd.*

Proof. If $k = 1$, then we know that the identity is even. If $k > 1$, then refer to the proof of Theorem 6.6, where we wrote a k-cycle as a product of $k - 1$ transpositions.

□

Thus, to determine if a particular permutation is even or odd, we can look at its disjoint cycle decomposition. The preceding theorem tells us whether each cycle is a product of an even or odd number of transpositions, so we can easily determine the answer for the entire permutation.

Definition 6.5. The **alternating group** A_n is the set of all even permutations in S_n.

Example 6.9. We note that S_3 consists of the identity (which is even), three transpositions (which are odd) and two 3-cycles (which are even). Thus,

$$A_3 = \{(1), (1\ 2\ 3), (1\ 3\ 2)\}.$$

Similarly S_4 consists of the identity (even), six transpositions (odd), eight 3-cycles (even), six 4-cycles (odd) and three elements that are products of two disjoint transpositions (even). Thus,

$$A_4 = \{(1), (1\ 2\ 3), (1\ 2\ 4), (1\ 3\ 2), (1\ 3\ 4), (1\ 4\ 2), (1\ 4\ 3), (2\ 3\ 4), (2\ 4\ 3),$$
$$(1\ 2)(3\ 4), (1\ 3)(2\ 4), (1\ 4)(2\ 3)\}.$$

Theorem 6.9. *Let $n \geq 2$. Then A_n is a normal subgroup of S_n, and $[S_n : A_n] = 2$.*

Proof. Define $\alpha : S_n \to \mathbb{Z}_2$ as follows. Let $\alpha(\sigma) = 0$ if σ is even and 1 if σ is odd. We claim that α is a homomorphism. Indeed, as the product of two even or two odd permutations is even, and the product of an even and an odd is odd, this follows immediately. By definition, the kernel is A_n, so A_n is a normal subgroup. Furthermore, $\alpha((1)) = 0$ and $\alpha((1\ 2)) = 1$, so α is onto. Thus, by the First Isomorphism Theorem, S_n/A_n is isomorphic to $\mathbb{Z}_2$. That is, $|S_n/A_n| = 2$, so A_n has index 2.   □

Exercises

6.11. Decide if each of the following permutations is even or odd.

1. $(2\ 3)(1\ 3\ 4)(1\ 4\ 2\ 3)$
2. $(1\ 4\ 3\ 5)(1\ 2)(1\ 3\ 2\ 4)$

6.12. Write each of the following permutations as a product of transpositions.

1. $(1\ 3\ 2)(1\ 4)(2\ 5\ 3)$
2. $(2\ 5\ 3\ 4)(1\ 2\ 6)(3\ 5\ 4)$

6.13. Find every possible order of the product of two transpositions.

6.14. Let $n \geq 2$ and $H \leq S_n$. Show that either every element of H is even, or exactly half of the elements of H are even.

6.15. For which $n \geq 2$ does A_n have a subgroup of order 4? What if we insist that the subgroup be cyclic?

6.16. Find the orders of all the elements of A_8.

6.17. If $n \geq 2$, show that every element of odd order in S_n lies in A_n.

6.18. Show that every permutation other than the identity in S_n is the product of at most $n - 1$ transpositions.

6.19. For which positive integers n does S_n have

1. more elements of even order than odd order;
2. more elements of odd order than even order;
3. the same number of elements of odd order as even order?

6.20. For which integers $n \geq 2$ does there exist a $\sigma \in A_n$ such that $|\sigma| > n$?

6.3   The Simplicity of the Alternating Group

Why are we so interested in the group A_n? In order to explain this, we must start with a definition.

Definition 6.6. A group is **simple** if it is nontrivial and has no nontrivial proper normal subgroups.

If G is abelian, then every subgroup is normal, so we are looking for groups whose only subgroups are G and $\{e\}$. But these were determined in Exercise 3.52. Indeed, we saw that these were precisely the cyclic groups of prime order. By Theorem 4.14, we have the following result.

Theorem 6.10. *Let G be an abelian group. Then G is simple if and only if G is isomorphic to $\mathbb{Z}_p$, for some prime p.*

That was pretty painless! However, the nonabelian case is much much more difficult. Much! The classification of all of the finite simple groups was one of the biggest mathematical projects of the twentieth century. Over one hundred mathematicians contributed to the solution, and the proof consists of many thousands of pages of journal articles. For obvious reasons, we will not be discussing this classification here.

We will content ourselves with proving one of the earliest results on the subject; namely, if $n \geq 5$ then A_n is a nonabelian simple group. (Actually, A_5 is the smallest nonabelian simple group.) The $n = 5$ case was established by Évariste Galois in the early nineteenth century. Decades later, M.E. Camille Jordan provided a proof for all $n \geq 5$.

Why are finite simple groups so interesting? Let us look at it this way. Suppose that G is a nontrivial finite group. Let N_1 be a proper normal subgroup of largest order in G. (If G is simple, this will be $\{e\}$. Otherwise, it will be something larger.) Now, we claim that G/N_1 is simple. Indeed, by Theorem 4.8, the normal subgroups of G/N_1 are precisely of the form H/N_1, where H is a normal subgroup of G containing N_1. But by definition of N_1, $H = N_1$ or G. Thus, G/N_1 has no nontrivial proper normal subgroups, so it is simple.

Now, suppose that $N_1 \neq \{e\}$. Then in the same way, take a proper normal subgroup N_2 of N_1 of largest possible order. Then N_1/N_2 is simple. We can repeat this procedure and obtain

$$G = N_0 \geq N_1 \geq N_2 \geq N_3 \geq \cdots \geq N_{k-1} \geq N_k = \{e\},$$

where each N_{i+1} is normal in N_i and N_i/N_{i+1} is simple. We know the process must end, as each N_{i+1} is properly contained in N_i, and the original group is finite. In a way, then, finite groups can be built up using simple groups.

Let us begin the process of proving that A_n is simple, for $n \geq 5$. We start with a general fact about the conjugation of cycles.

Lemma 6.2. Let $\sigma = (a_1\ a_2\ \cdots\ a_k)$ be a k-cycle in S_n. If $\tau \in S_n$, then $\tau\sigma\tau^{-1} = (\tau(a_1)\ \tau(a_2)\ \cdots\ \tau(a_k))$.

Proof. Suppose that $b = \tau(a_i)$. Then $\tau^{-1}(b) = a_i$; hence, $\sigma(\tau^{-1}(b)) = \sigma(a_i) = a_{i+1}$ (or a_1, if $i = k$). Therefore, $\tau\sigma\tau^{-1}(b) = \tau(a_{i+1})$ (or $\tau(a_1)$, if $i = k$). That is, $\tau\sigma\tau^{-1}$ permutes the $\tau(a_i)$ as described. If b is not among the $\tau(a_i)$, then $\tau^{-1}(b)$ is not equal to any a_i, which means that it is fixed by σ. Thus, $\tau\sigma\tau^{-1}(b) = \tau\tau^{-1}(b) = b$. Therefore, $\tau\sigma\tau^{-1}$ is the k-cycle described in the statement of the lemma. $\qquad\square$

Corollary 6.2. Let n and k be positive integers with $n \geq k$. Then

1. any two k-cycles are conjugate in S_n; and
2. if k is odd and $n \geq k + 2$, then any two k-cycles are conjugate in A_n.

Proof. (1) Let $\sigma = (a_1\ \cdots\ a_k)$ and $\delta = (b_1\ \cdots\ b_k)$ be any two k-cycles. The preceding lemma tells us that in order to show that σ and δ are conjugate, we need only find $\tau \in S_n$ such that $\tau(a_i) = b_i$ for all i; in this case, $\tau\sigma\tau^{-1} = \delta$. But S_n contains every possible permutation of $\{1, \ldots, n\}$. Thus, we can certainly assign $\tau(a_i) = b_i$, and for the $j \notin \{a_1, \ldots, a_k\}$, let the $\tau(j)$ be any distinct values not in $\{b_1, \ldots, b_k\}$.

(2) As k is odd, the k-cycles are even, and therefore lie in A_n. Let σ and δ be any k-cycles. Without loss of generality, let us say that $\delta = (1\ 2\ \cdots\ k)$. Then just as in (1), we can find $\tau \in S_n$ such that $\tau\sigma\tau^{-1} = \delta$. If $\tau \in A_n$, then we are done. Otherwise, τ is odd, so $((k+1)\ (k+2))\tau$ is even. (Note that this is valid, as $n \geq k + 2$.) Thus, letting $\eta = ((k+1)\ (k+2))\tau \in A_n$, we have

$$\eta\sigma\eta^{-1} = ((k+1)\ (k+2))\tau\sigma\tau^{-1}((k+1)\ (k+2))$$
$$= ((k+1)\ (k+2))(1\ 2\ \cdots\ k)((k+1)\ (k+2)).$$

But disjoint cycles commute, so this is

$$((k + 1) \ (k + 2))((k + 1) \ (k + 2))\delta = \delta.$$

We are done.   □

Example 6.10. The preceding lemma tells us that (1 2 3) and (1 3 2) are conjugate in S_3, and the proof suggests how we might demonstrate it. We need to find τ such that $\tau(1) = 1, \tau(2) = 3$ and $\tau(3) = 2$; that is, $\tau = (2 \ 3)$. Then $(1 \ 3 \ 2) = \tau(1 \ 2 \ 3)\tau^{-1}$. However, (1 2 3) and (1 3 2) are not conjugate in A_3; this is obvious, as A_3 is abelian, having order 3, so different elements are not conjugate. It is less obvious that they are not conjugate in A_4 either; however, it is possible to try conjugating (1 2 3) by all of the elements of A_4. None of these conjugates will equal (1 3 2). However, the preceding lemma tells us that (1 2 3) and (1 3 2) are indeed conjugate in A_5, and the proof tells us that if we take $\eta = (4 \ 5)\tau$, then $\eta \in A_5$, and we find that $\eta(1 \ 2 \ 3)\eta^{-1} = (1 \ 3 \ 2)$.

We can now simplify our task by showing that if we have a 3-cycle in a normal subgroup of A_n, then we have all of A_n.

Corollary 6.3. *Let $n \geq 3$. Then*

1. *every element of A_n is a product of 3-cycles; and*
2. *if a normal subgroup N of A_n contains any 3-cycle, then $N = A_n$.*

Proof. (1) We know that an element of A_n is a product of an even number of transpositions. Thus, it is sufficient to show that every product of two transpositions is a product of 3-cycles. (As (1) = (1 2 3)(1 3 2), we need not worry about the identity.) If the two transpositions are equal, then their product is the identity, with which we have just dealt. Suppose they have one number in common. Without loss of generality, say (1 2)(1 3). Then note that (1 2)(1 3) = (1 3 2), which is a 3-cycle. Finally, suppose they have no numbers in common. Without loss of generality, say (1 2)(3 4). Then we observe that (1 2)(3 4) = (1 4 3)(1 2 3), which is a product of 3-cycles.

(2) In view of (1), it is sufficient to show that N contains all of the 3-cycles. But it contains one 3-cycle, so as N is normal, it contains all of its conjugates. If $n \geq 5$, then Corollary 6.2 tells us that these conjugates are all of the 3-cycles, and we are done. If $n = 3$, there is little to do, as the only 3-cycles are (1 2 3) and (1 3 2), and they are squares of each other; thus, if N contains one, it contains the other. The $n = 4$ case requires a little more work, and we leave it as Exercise 6.24.   □

And now, our main result for this section.

Theorem 6.11. *If $n \geq 5$, then A_n is a nonabelian simple group.*

Proof. The fact that (1 2 3)(1 2 4) $\neq$ (1 2 4)(1 2 3) shows that A_n is nonabelian, so we can focus on the simplicity. Let N be a nontrivial normal subgroup of A_n. We must prove that $N = A_n$. In view of Corollary 6.3, it is sufficient to show that N contains a 3-cycle.

Take any $(1) \neq \sigma \in N$, and consider the disjoint cycle decomposition of σ. Suppose, first of all, that there are two or more transpositions in this decomposition. Without loss of generality, say $\sigma = (1\ 2)(3\ 4)\delta$, where δ is a product of disjoint cycles which also fix everything in $\{1, 2, 3, 4\}$ (and $\delta = (1)$ is possible). Let $\tau = (1\ 2\ 4) \in A_n$. Then as N is normal in A_n, we have $\tau\sigma\tau^{-1} \in N$. That is,

$$(1\ 2\ 4)(1\ 2)(3\ 4)\delta(1\ 4\ 2) \in N.$$

(It is easy to check that $(1\ 2\ 4)^{-1} = (1\ 4\ 2)$.) As the cycles in δ are disjoint from all the other cycles in the product, we see from Theorem 6.4 that N contains

$$(1\ 2\ 4)(1\ 2)(3\ 4)(1\ 4\ 2)\delta = (1\ 3)(2\ 4)\delta.$$

But N also contains σ^{-1}, and therefore

$$\sigma^{-1}(1\ 3)(2\ 4)\delta = \delta^{-1}(3\ 4)(1\ 2)(1\ 3)(2\ 4)\delta \in N.$$

Again, δ commutes with these other cycles, so we have

$$\delta^{-1}\delta(3\ 4)(1\ 2)(1\ 3)(2\ 4) = (1\ 4)(2\ 3) \in N.$$

Let $\eta = (1\ 4\ 5) \in A_n$ (since $n \geq 5$). Then N must contain

$$\eta(1\ 4)(2\ 3)\eta^{-1} = (1\ 4\ 5)(1\ 4)(2\ 3)(1\ 5\ 4) = (2\ 3)(4\ 5).$$

Thus, N also contains
$$(1\ 4)(2\ 3)(2\ 3)(4\ 5) = (1\ 4\ 5).$$

But when N contains a 3-cycle, we know that $N = A_n$. Thus, from this point on, we may assume that the disjoint cycle decomposition of σ contains at most one transposition.

Now, let us consider the length k of the longest cycle appearing in the disjoint cycle decomposition of σ. If $k = 2$, then σ is a product of an even number of disjoint transpositions, and we have already dealt with this case.

Suppose that $k = 3$. Then σ is a product of some 3-cycles and, possibly, some transpositions. But the product of some 3-cycles and a single transposition is odd, and therefore not in A_n. Furthermore, multiple transpositions are not allowed. Therefore, we may assume that σ is a product of one or more 3-cycles. If it is just one 3-cycle, then we are done. So assume that it is a product of two or more disjoint 3-cycles. Without loss of generality, say $\sigma = (1\ 2\ 3)(4\ 5\ 6)\delta$, where either $\delta = (1)$ or δ is a product of disjoint 3-cycles, all of which fix everything in $\{1, 2, 3, 4, 5, 6\}$. Let $\tau = (3\ 4\ 5) \in A_n$. Then as N is normal, it contains

$$\tau\sigma\tau^{-1} = (3\ 4\ 5)(1\ 2\ 3)(4\ 5\ 6)\delta(3\ 5\ 4) = (1\ 2\ 4)(3\ 6\ 5)\delta,$$

since δ commutes with the other cycles. But N also contains σ^{-1}, so we have

$$\sigma^{-1}(1\ 2\ 4)(3\ 6\ 5)\delta = \delta^{-1}(4\ 6\ 5)(1\ 3\ 2)(1\ 2\ 4)(3\ 6\ 5)\delta = (2\ 6\ 4\ 3\ 5) \in N,$$

again, since δ is disjoint from the other cycles. Replacing σ with $(2\ 6\ 4\ 3\ 5)$, we can move to our final case.

Let us suppose that $k \geq 4$. Then without loss of generality, we may write $\sigma = (1\ 2\ 3\ \cdots\ k)\delta$, where $k \geq 4$ and δ is some product of disjoint cycles, all of which fix everything in $\{1, 2, \ldots, k\}$. Let $\tau = (1\ 2\ 3) \in A_n$. Then by normality, N contains

$$\tau\sigma\tau^{-1} = (1\ 2\ 3)(1\ 2\ 3\ \cdots\ k)\delta(1\ 3\ 2) = (1\ 4\ 5\ \cdots\ k\ 2\ 3)\delta.$$

But N also contains σ^{-1}, so noting that $(1\ 2\ 3\ \cdots\ k)^{-1} = (1\ k\ (k-1)\ \cdots\ 2)$, we have

$$\sigma^{-1}(1\ 4\ 5\ \cdots\ k\ 2\ 3)\delta = \delta^{-1}(1\ k\ (k-1)\ \cdots\ 2)(1\ 4\ 5\ \cdots\ k\ 2\ 3)\delta = (1\ 3\ k) \in N,$$

again using the fact that δ commutes with everything else. Thus, N contains a 3-cycle, and the proof is complete.   □

We might well ask about A_n when $n < 5$. For $n = 2$, A_2 is the trivial group; hence, by definition, not simple. When $n = 3$, A_3 has order 3 and by Corollary 4.2, it is isomorphic to $\mathbb{Z}_3$. By Theorem 6.10, it is an abelian simple group. The big exception is the $n = 4$ case, as illustrated in the following example.

Example 6.11. The alternating group A_4 is not simple. To see, this let

$$N = \{(1), (1\ 2)(3\ 4), (1\ 3)(2\ 4), (1\ 4)(2\ 3)\}.$$

It simply requires some computation to see that N is a nontrivial proper normal subgroup of A_4.

With the exception of S_2, which is abelian of order 2, and hence isomorphic to $\mathbb{Z}_2$, the symmetric groups are not simple. Indeed, A_n is a nontrivial proper normal subgroup of S_n, whenever $n \geq 3$. However, we can state the following result.

Corollary 6.4. *If $n \geq 5$, then the only nontrivial proper normal subgroup of S_n is A_n.*

Proof. Let N be a normal subgroup of S_n. Then $N \cap A_n$ is a normal subgroup of A_n. As A_n is simple, this means that $N \cap A_n = A_n$ or $\{(1)\}$. If $N \cap A_n = A_n$, then $A_n \leq N$. But by Lagrange's theorem, this implies that $|A_n|$ divides $|N|$ and $|N|$ divides $|S_n|$. As $|S_n| = 2|A_n|$ (because A_n is of index 2), this can only mean that $|N| = |A_n|$ or $|S_n|$. Thus, $N = A_n$ or S_n, as desired.

On the other hand, suppose that $N \cap A_n = \{(1)\}$. Then by Theorem 4.4, $|NA_n| = |N||A_n|$. As $|A_n| = |S_n|/2$ and $|NA_n| \leq |S_n|$, we see that $|N| = 1$ or 2. If $|N| = 1$, we are done, so suppose that $|N| = 2$. But by Exercise 4.3, a normal subgroup of order 2 in a group is central. However, Exercise 6.8 tells us that the centre of S_n is trivial. Thus, we have a contradiction, and the proof is complete.   □

Exercises

6.21. Show that A_5 has no subgroup of order 30.

6.22. In S_7, describe the conjugates of $(1\ 2)(3\ 4\ 5)$.

6.23. Can a nonabelian simple group have a nonabelian simple proper subgroup? Either prove that it cannot, or construct an explicit example.

6.24. Let N be a normal subgroup of A_4 containing a 3-cycle. Show that $N = A_4$.

6.25. Show that the only nontrivial proper normal subgroup of A_4 is the one exhibited in Example 6.11.

6.26. Let $n \geq 2$. Show that every element of S_n can be written as a product of transpositions of the form $(1\ i)$, for various i.

6.27. If $n \geq 2$, show that every element of S_n can be written as a product of the transpositions $(1\ 2), (2\ 3), \ldots, ((n-1)\ n)$.

6.28. If $n \geq 2$, let $\sigma = (1\ 2)$ and $\tau = (1\ 2\ 3\ \cdots\ n)$. Show that every element of S_n can be written in the form $\sigma^{i_1}\tau^{j_1}\sigma^{i_2}\tau^{j_2}\cdots\sigma^{i_k}\tau^{j_k}$, where the exponents are any integers and $k \in \mathbb{N}$.

Chapter 7
The Sylow Theorems

In this chapter, we will prove the Sylow theorems. These are difficult results, but fundamental to our understanding of the structure of finite groups. In particular, we will show that if p^n is the largest power of a prime p dividing the order of a finite group G, then G has at least one subgroup of order p^n. Furthermore, we will discover that any two such subgroups are conjugate to each other, and determine a restriction upon the number of such subgroups. We will then explore various applications of these theorems, and conclude the chapter by classifying all groups of order smaller than 16.

7.1 Normalizers and Centralizers

We are very familiar with the centre of a group, which consists of all elements that commute with everything. Let us generalize.

Definition 7.1. Let G be a group, $a \in G$ and H a subgroup of G. Then the **centralizer** of a is the set of all elements of G that commute with a. We write $C(a) = \{g \in G : ag = ga\}$. Also, the **centralizer** of H is $C(H) = \{g \in G : gh = hg$ for all $h \in H\}$.

Example 7.1. If $a \in Z(G)$, then $C(a) = G$. If $H \leq Z(G)$, then $C(H) = G$.

In particular, $C(e) = G$, so we cannot assume that centralizers are necessarily abelian.

Example 7.2. Let $G = D_8$. Then we find that $C(R_{270}) = \langle R_{90} \rangle$, $C(R_{180}) = G$ and $C(F_1) = \{R_0, R_{180}, F_1, F_2\}$.

© Springer International Publishing AG, part of Springer Nature 2018
G. T. Lee, *Abstract Algebra*, Springer Undergraduate Mathematics Series,
https://doi.org/10.1007/978-3-319-77649-1_7

Theorem 7.1. *Let G be a group, $a \in G$ and H a subgroup of G. Then*

1. $C(H) = \bigcap_{h \in H} C(h)$;
2. $C(a)$ and $C(H)$ are both subgroups of G;
3. if H is a normal subgroup of G, then so is $C(H)$;
4. $Z(G)$ is a subgroup of both $C(a)$ and $C(H)$; and
5. $C(a) = C(\langle a \rangle)$.

Proof. (1) This follows from the definition.

(2) Clearly $ae = a = ea$, so $e \in C(a)$. Suppose that $b, c \in C(a)$. Then $bca = bac = abc$, so $bc \in C(a)$. Also, $ba = ab$, so $b^{-1}(ba)b^{-1} = b^{-1}(ab)b^{-1}$. Thus, $ab^{-1} = b^{-1}a$, so $b^{-1} \in C(a)$. Hence, $C(a) \leq G$. Furthermore, combining this fact with (1) and Exercise 3.37, we see that $C(H) \leq G$.

(3) See Exercise 4.4.

(4) Central elements commute with everything hence, in particular, they commute with a and elements of H.

(5) If $b \in C(\langle a \rangle)$, then since $a \in \langle a \rangle$, we see that b commutes with a. Thus, $b \in C(a)$. Conversely, if $b \in C(a)$, then $ab = ba$. Therefore, $a \in C(b)$. As $C(b) \leq G$ by (2), we see that $a^i \in C(b)$ for all integers i. That is, $a^i b = ba^i$, for all $i \in \mathbb{Z}$. In other words, $b \in C(\langle a \rangle)$. □

Suppose we have a subgroup H of G that is not normal. Of course, H is a normal subgroup of H. Furthermore, it is easy to see that H is normal in $HZ(G)$. How big a subgroup of G can we find in which H is a normal subgroup? This is where normalizers come in.

Definition 7.2. Let G be a group and H a subgroup. Then the **normalizer** of H is the set $N(H) = \{a \in G : a^{-1}Ha = H\}$. If K is another subgroup of G, then we write $N_K(H) = N(H) \cap K$, and call it the normalizer of H in K.

Remember, if $a \in C(H)$, then $a^{-1}ha = h$, for all $h \in H$. But if $a \in N(H)$, then $a^{-1}Ha = H$. In particular, $a^{-1}ha = h_1$, for some (possibly different) $h_1 \in H$. Thus, the normalizer and centralizer are different concepts.

Example 7.3. If H is a normal subgroup of G, then $N(H) = G$. See Theorem 4.3.

Example 7.4. Let $G = S_4$ and $H = \langle (1\ 2\ 3\ 4) \rangle$. Then notice that $(2\ 4) \notin C(H)$, as $(2\ 4)(1\ 2\ 3\ 4) = (1\ 4)(2\ 3)$, but $(1\ 2\ 3\ 4)(2\ 4) = (1\ 2)(3\ 4)$. However,

$$(2\ 4)^{-1}(1\ 2\ 3\ 4)(2\ 4) = (2\ 4)(1\ 2\ 3\ 4)(2\ 4) = (1\ 4\ 3\ 2) = (1\ 2\ 3\ 4)^3 \in H.$$

Thus,

$$(2\ 4)^{-1}(1\ 2\ 3\ 4)^i(2\ 4) = ((2\ 4)^{-1}(1\ 2\ 3\ 4)(2\ 4))^i = (1\ 4\ 3\ 2)^i \in H,$$

for all $i \in \mathbb{Z}$. Therefore, $(2\ 4)^{-1}H(2\ 4) \leq H$. By Theorem 4.2, $|(2\ 4)^{-1}H(2\ 4)| = |H|$, so we conclude that $(2\ 4)^{-1}H(2\ 4) = H$. Thus, $(2\ 4) \in N(H)$.

Theorem 7.2. *Let H be a subgroup of G. Then $N(H)$ is a subgroup of G containing H. Furthermore, if K is a subgroup of G containing H, then H is normal in K if and only if K is a subgroup of $N(H)$.*

Proof. Take any $h \in H$. Then for any $c \in H$, we have $h^{-1}ch \in H$, so $h^{-1}Hh \leq H$. Also, $h^{-1}(hch^{-1})h = c$, and $hch^{-1} \in H$. Thus, every element of H is in $h^{-1}Hh$, so $H = h^{-1}Hh$, and $H \subseteq N(H)$. In particular, $e \in N(H)$. Now, take any $a, b \in N(H)$. Then $(ab)^{-1}Hab = b^{-1}(a^{-1}Ha)b = b^{-1}Hb = H$; thus, $ab \in N(H)$. Also, as $a^{-1}Ha = H$, we have $aa^{-1}Haa^{-1} = aHa^{-1}$; that is, $H = (a^{-1})^{-1}Ha^{-1}$, hence $a^{-1} \in N(H)$, and $N(H) \leq G$.

Let $H \leq K \leq G$. Then in view of Theorem 4.3, H is a normal subgroup of K if and only if $k^{-1}Hk = H$ for all $k \in K$. By definition of $N(H)$, this occurs if and only if $K \leq N(H)$. $\qquad\square$

It is clear that if H is a subgroup of G, then $C(H) \leq N(H)$. They could, of course be equal; indeed, if $H \leq Z(G)$, then $C(H) = N(H) = G$. But as we saw in Example 7.4, they need not be. An interesting fact about the relationship between these two subgroups is given in the following result.

Theorem 7.3 (*N/C* **Theorem**). *Let G be a group and H a subgroup. Then $C(H)$ is a normal subgroup of $N(H)$, and $N(H)/C(H)$ is isomorphic to a subgroup of* $\mathrm{Aut}(H)$.

Proof. We will show that $C(H)$ is a normal subgroup of $N(H)$ by illustrating that it is the kernel of a homomorphism from $N(H)$ to $\mathrm{Aut}(H)$. Define $\alpha : N(H) \to \mathrm{Aut}(H)$ via $\alpha(a)(h) = aha^{-1}$, for all $a \in N(H)$, $h \in H$. If a is in the normalizer, then so is a^{-1}, and therefore $aHa^{-1} = H$. Thus, we see immediately that $\alpha(a)$ is an onto function from H to H. Also, if $ah_1a^{-1} = ah_2a^{-1}$, then $h_1 = h_2$ by cancellation, so $\alpha(a)$ is one-to-one as well. Furthermore, for any $h_1, h_2 \in H$, we have

$$\alpha(a)(h_1h_2) = ah_1h_2a^{-1} = ah_1a^{-1}ah_2a^{-1} = (\alpha(a)(h_1))(\alpha(a)(h_2)).$$

Therefore, $\alpha(a) \in \mathrm{Aut}(H)$.

We need to show that α is a homomorphism. But if $a, b \in N(H)$, then for any $h \in H$, we have

$$\alpha(ab)(h) = abh(ab)^{-1} = abhb^{-1}a^{-1} = a(\alpha(b)(h))a^{-1} = \alpha(a)(\alpha(b)(h)).$$

Thus, $\alpha(ab) = \alpha(a) \circ \alpha(b)$, as required. Now, the kernel of α is the set of all $c \in N(H)$ such that $\alpha(c)$ acts as the identity on H; specifically, we must have $chc^{-1} = h$, for all $h \in H$. But this is precisely the definition of $C(H)$. The First Isomorphism Theorem now tells us that $N(H)/C(H)$ is isomorphic to $\alpha(N(H))$, which is a subgroup of $\mathrm{Aut}(H)$, as required. $\qquad\square$

The following example illustrates a cute application of the *N/C* Theorem.

Example 7.5. Suppose that G is a nonabelian group of order 39. Let us demonstrate that G cannot possibly have a normal subgroup H of order 3. Suppose such a normal subgroup exists. Then $N(H) = G$. Also, what can the centralizer of H be? As H has prime order, Corollary 3.6 tells us that it is cyclic, hence abelian. In particular, H centralizes itself. Thus, $H \leq C(H) \leq G$. By Lagrange's theorem, we see that 3 divides $|C(H)|$, which in turn divides 39. The only possibilities are $|C(H)| = 3$ or 39. Suppose that $|C(H)| = 39$. Then $C(H) = G$, so H is central. In particular, $H \leq Z(G) \leq G$. But again, looking at the orders, we see that $Z(G) = H$ or G. As G is not abelian, $Z(G) = H$, but that cannot be the case either, as then $Z(G)$ has prime index which, by Corollary 4.1, is impossible.

Therefore, $|C(H)| = 3$, so $C(H) = H$. By the preceding theorem, $N(H)/C(H)$ is isomorphic to a subgroup of $\text{Aut}(H)$. That is, G/H is isomorphic to a subgroup of $\text{Aut}(H)$. But $|G/H| = |G|/|H| = 13$. As H is cyclic, the structure of $\text{Aut}(H)$ is given by Theorem 4.22. But even if we did not have that resource, H is a set of 3 elements, so there are only $3! = 6$ ways to permute them (and not all of those are automorphisms). Thus, we are trying to fit a group of order 13 inside one that is simply too small. Hence, H cannot exist.

Exercises

7.1. Which matrices lie in the centralizer of $\begin{pmatrix} 1 & 1 \\ 0 & 1 \end{pmatrix}$ in $GL_2(\mathbb{R})$?

7.2. Which permutations lie in the centralizer of $(1\ 2\ 3)$ in S_5?

7.3. In $GL_2(\mathbb{R})$, let H be the subgroup generated by $\begin{pmatrix} 2 & 3 \\ 5 & 6 \end{pmatrix}$. Show that $C(H) = N(H)$.

7.4. Let $H_1 \leq G_1$ and $H_2 \leq G_2$. Show that in $G_1 \times G_2$, $C(H_1 \times H_2) = C(H_1) \times C(H_2)$.

7.5. If G is a group having a subgroup H of order 2, show that $C(H) = N(H)$.

7.6. If G is a nonabelian group, show that G has a subgroup H such that $Z(G) \subsetneq H \subsetneq G$. (Yes, this is the same as Exercise 4.20. Solve it using the results in this section.)

7.7. In any group, show that $C(a) = C(a^{-1})$.

7.8. If $a \in G$ and a has odd order, show that $C(a) = C(a^4)$.

7.9. Let G be a group of order 77 having a normal subgroup H of order 11.

1. If G is not abelian, show that $C(H) = H$.
2. Conclude that G must, in fact, be abelian.

7.10. Let G be a group of order 77.

1. Show that G has a subgroup H of order 11.
2. Show that H is unique, and hence normal.
3. Conclude that G is isomorphic to $\mathbb{Z}_{77}$.

7.2 Conjugacy and the Class Equation

We are already familiar with the notion of conjugacy in groups. To reiterate, we say that a and b in G are conjugate if there exists a $g \in G$ such that $g^{-1}ag = b$. Here is a simple fact that we have not mentioned.

Theorem 7.4. *If G is any group, then conjugacy is an equivalence relation on the elements of G.*

Proof. Reflexivity: For any $a \in G$, we have $e^{-1}ae = a$, so a is conjugate to itself. Symmetry: Suppose that a is conjugate to b, say $g^{-1}ag = b$. Then $a = (g^{-1})^{-1}bg^{-1}$, and therefore b is conjugate to a. Transitivity: Suppose that a is conjugate to b, and b to c, say $g^{-1}ag = b$ and $h^{-1}bh = c$. Then $c = h^{-1}(g^{-1}ag)h = (gh)^{-1}a(gh)$. Thus, a is conjugate to c. $\square$

We know that equivalence classes partition a set; thus we can break a group down into disjoint sets of elements, all elements in each set being conjugate to each other.

Definition 7.3. Let G be a group and $a \in G$. Then the **conjugacy class** of a is the set $C_a = \{g^{-1}ag : g \in G\}$.

Conjugacy classes are subsets of G, not subgroups. Indeed, the only one that will contain the identity is C_e.

Example 7.6. Note that C_a contains only a if and only if $a \in Z(G)$. (This happens if and only if $g^{-1}ag = a$ for all $g \in G$.)

Example 7.7. Let k and n be positive integers with $n \geq k$. If $G = S_n$ and $\sigma = (1\ 2\ 3\ \cdots\ k)$, then we see from Lemma 6.2 that C_σ is the set of all k-cycles in G.

It is important to know the size of a conjugacy class.

Lemma 7.1. *Let G be a finite group and $a \in G$. Then the number of elements in C_a is the index of the centralizer, $[G : C(a)]$.*

Proof. Take $g, h \in G$. Then notice that $g^{-1}ag = h^{-1}ah$ if and only if $gh^{-1}a = agh^{-1}$. That is, $g^{-1}ag$ and $h^{-1}ah$ produce the same conjugate of a if and only if $gh^{-1} \in C(a)$ or, equivalently, if and only if the right cosets $C(a)g$ and $C(a)h$ are equal. In other words, we get a distinct conjugate of a for each right coset of $C(a)$, so the number of distinct conjugates is the index, $[G : C(a)]$, as required. $\square$

This allows us to establish an important equation, called the **class equation**.

Theorem 7.5 (Class Equation). *Let G be a finite group, and let $a_1, \ldots, a_k$ be representatives of the conjugacy classes in G with more than one element. Then*

$$|G| = |Z(G)| + [G : C(a_1)] + \cdots + [G : C(a_k)].$$

Proof. As G is partitioned into conjugacy classes, we know that $|G|$ is the sum of the sizes of these classes. We noted in Example 7.6 that the conjugacy classes of size 1 are precisely those of the central elements. Collecting them together, we obtain $|Z(G)|$ elements. For the remaining classes, we now apply the preceding lemma. $\square$

This has powerful consequences!

Corollary 7.1. *Let G be a group of order p^n, for some prime p and positive integer n. Then the centre of G is not trivial.*

Proof. In the class equation, each $[G : C(a_i)]$ is the size of a conjugacy class with more than one element. But it is also $|G|/|C(a_i)|$, and therefore a divisor of p^n. Thus, each $[G : C(a_i)]$ is a multiple of p. But $|G|$ is also a multiple of p, and therefore the one remaining term in the equation, $|Z(G)|$, is a multiple of p. In particular, it is not 1. $\square$

Corollary 7.2. *Let G be a group of order p^2, for some prime p. Then G is isomorphic to either $\mathbb{Z}_{p^2}$ or $\mathbb{Z}_p \times \mathbb{Z}_p$.*

Proof. By Lagrange's theorem, $|Z(G)| \in \{1, p, p^2\}$. But by the preceding corollary, it cannot be 1. Suppose it is p. Then $[G : Z(G)] = p^2/p = p$. By Corollary 4.1, this is impossible. Therefore, $Z(G) = G$, and G is abelian. By Corollary 5.1, we are done. $\square$

Theorem 5.6 tells us that $\mathbb{Z}_{p^2}$ and $\mathbb{Z}_p \times \mathbb{Z}_p$ are not isomorphic, so we now have a complete picture for groups of order p^2.

We also need to know about conjugacy of subgroups.

Definition 7.4. Let G be a group and H a subgroup. We say that subgroups K and L of G are H-**conjugate** if there exists an $h \in H$ such that $h^{-1}Kh = L$. When $H = G$, we simply say that K and L are **conjugate**.

Example 7.8. Let G be S_5 and $H = \langle (1\ 3)(2\ 5\ 4) \rangle$. Take $\sigma = (1\ 3)(2\ 4\ 5) = ((1\ 3)(2\ 5\ 4))^{-1} \in H$. Then we notice that

$$\sigma^{-1}(1\ 2\ 3\ 4)\sigma = (1\ 3)(2\ 5\ 4)(1\ 2\ 3\ 4)(1\ 3)(2\ 4\ 5) = (1\ 2\ 3\ 5).$$

Therefore, for any integer i, $\sigma^{-1}(1\ 2\ 3\ 4)^i\sigma = (\sigma^{-1}(1\ 2\ 3\ 4)\sigma)^i = (1\ 2\ 3\ 5)^i$. This means that $\sigma^{-1}\langle (1\ 2\ 3\ 4) \rangle\sigma = \langle (1\ 2\ 3\ 5) \rangle$. Thus, $\langle (1\ 2\ 3\ 4) \rangle$ and $\langle (1\ 2\ 3\ 5) \rangle$ are H-conjugate (and, therefore, conjugate).

Theorem 7.6. *Let G be a group and H a subgroup of G. Then H-conjugacy is an equivalence relation on the set of all subgroups of G.*

Proof. Reflexivity: Let $K \leq G$. Then $e \in H$ and $e^{-1}Ke = K$. Therefore, K is H-conjugate to itself. Symmetry: Suppose that $h^{-1}Kh = L$, with $h \in H$. Then

$$K = hh^{-1}Khh^{-1} = hLh^{-1} = (h^{-1})^{-1}Lh^{-1}.$$

Since $h^{-1} \in H$, we see that H-conjugacy is symmetric. Transitivity: Suppose that $h_1^{-1} K h_1 = L$ and $h_2^{-1} L h_2 = M$, where $h_1, h_2 \in H$. Then

$$M = h_2^{-1}(h_1^{-1} K h_1)h_2 = (h_1 h_2)^{-1} K (h_1 h_2).$$

Since $h_1 h_2 \in H$, we are done. $\qquad\square$

Thus, the subgroups of G are partitioned into equivalence classes, in which those in each class are H-conjugate to each other. In a similar fashion to Lemma 7.1, we have a formula for the number of H-conjugates of a subgroup.

Theorem 7.7. *Let G be a finite group and H a subgroup. Then for any subgroup K of G, the number of H-conjugates of K in G is $[H : N_H(K)]$.*

Proof. Take $h_1, h_2 \in H$. Then $h_1^{-1} K h_1 = h_2^{-1} K h_2$ if and only if $h_2 h_1^{-1} K h_1 h_2^{-1} = K$; that is, if and only if $(h_1 h_2^{-1})^{-1} K (h_1 h_2^{-1}) = K$. But this means precisely that $h_1 h_2^{-1} \in N_H(K)$ or, in other words, that $N_H(K)h_1 = N_H(K)h_2$. Thus, we get one distinct conjugate for each right coset of $N_H(K)$ in H, so the number of such conjugates is the index, $[H : N_H(K)]$. $\qquad\square$

Exercises

7.11. What are the conjugacy classes of D_8?

7.12. What are the conjugacy classes of S_4?

7.13. Let G be a group having subgroups H and K. Suppose that two subgroups of G are both H-conjugate and K-conjugate. Does it follow that they are $(H \cap K)$-conjugate? Either prove that it does or construct a counterexample.

7.14. Let G be a finite group with normal subgroup N. Show that there are at least as many conjugacy classes in G as in G/N.

7.15. Let G be a group of order p^n, where p is a prime and $n \geq 2$. Suppose that $|Z(G)| = p$. Show that there exists an $a \in G$ such that $|C(a)| = p^{n-1}$.

7.16. If G is a group of order p^n, for some prime p and positive integer n, show that G has a subgroup of order p^m for each positive integer $m \leq n$.

7.17. For each of the following lists, determine if it is the list of sizes of the conjugacy classes of some finite group. If it is, provide such a group. If not, explain why not.

1. $1, 1, 1, 1, 1, 5, 5, 5, 5$
2. $1, 2, 3$
3. $2, 4, 6$

7.18. Let G be a group and H the set of elements of G having only finitely many conjugates. Show that H is a subgroup of G.

7.19. Suppose that G is a finite group and there exists $e \neq a \in G$ such that $a^{-1} \in C_a$. Show that G has even order.

7.20. Let G be a group of order $n > 1$. Show that no conjugacy class can have order greater than $n/2$.

7.3   The Three Sylow Theorems

We can now present the three major theorems due to P. Ludwig Sylow concerning subgroups of prime power order in a finite group. We will give the statements and proofs in this section, and some applications in the following sections.

Definition 7.5. Let G be a finite group, and suppose that $|G| = p^n r$, where p is a prime, $n \geq 0$ and r is a positive integer such that $(p, r) = 1$. Then a **Sylow p-subgroup** of G is any subgroup of order p^n.

By Lagrange's theorem, if H is a subgroup of G of order p^k, for some k, then the order of H cannot possibly be any larger than that of a Sylow p-subgroup.

Example 7.9. If the p-elements of G form a subgroup H, then that is the unique Sylow p-subgroup. By Lemma 5.3, this happens whenever G is a finite abelian group. But it can also occur for certain nonabelian groups. As an obvious example, consider D_8. The entire group is the Sylow 2-subgroup.

Example 7.10. As $|A_4| = 12$, a Sylow 2-subgroup has to have order 4 and a Sylow 3-subgroup has to have order 3. In fact, there is just one Sylow 2-subgroup, namely $\{(1), (1\ 2)(3\ 4), (1\ 3)(2\ 4), (1\ 4)(2\ 3)\}$ (discussed in Example 6.11). However, there are four different Sylow 3-subgroups, namely $\langle(1\ 2\ 3)\rangle$, $\langle(1\ 2\ 4)\rangle$, $\langle(1\ 3\ 4)\rangle$ and $\langle(2\ 3\ 4)\rangle$.

The First Sylow Theorem says that we can always find a Sylow p-subgroup.

Theorem 7.8 (First Sylow Theorem). *Let G be a finite group. Then for every prime p, G has at least one Sylow p-subgroup.*

Proof. We will proceed by strong induction on $|G|$. If $|G| = 1$, then $\{e\}$ is the Sylow p-subgroup for any prime p. Therefore, let $|G| > 1$ and suppose that the theorem holds for smaller groups. Fix a prime p, and let $|G| = p^n r$, where $n \geq 0$ and $(p, r) = 1$. If $n = 0$, then again, the Sylow p-subgroup is $\{e\}$, so assume that $n \geq 1$.

Suppose there exists a noncentral element $a \in G$ such that p does not divide $[G : C(a)]$. Then as $|G| = |C(a)|[G : C(a)]$, we see that p^n divides $|C(a)|$ (and certainly no higher power of p can do so, as $C(a) \leq G$). But a is not central, so $C(a) \neq G$. Therefore, by our inductive hypothesis, $C(a)$ has a subgroup of order p^n. But this is also a subgroup of G, completing this case.

Therefore, assume that for every noncentral $a \in G$, we have $p|[G : C(a)]$. Also, $p||G|$. Therefore, p divides every term in the class equation except for $|Z(G)|$, which means that p must divide $|Z(G)|$ as well. By Cauchy's theorem for abelian groups, $Z(G)$ has an element z of order p. Then $\langle z \rangle$ is a central, hence normal, subgroup of order p in G. Furthermore, $|G/\langle z \rangle| = |G|/|\langle z \rangle| = p^n r/p = p^{n-1} r$. By our inductive hypothesis, $G/\langle z \rangle$ has a subgroup of order p^{n-1}. But Theorem 4.8 tells us that the subgroups of $G/\langle z \rangle$ are of the form $H/\langle z \rangle$, where H is a subgroup of G containing $\langle z \rangle$. However, $|H| = |H/\langle z \rangle||\langle z \rangle| = p^{n-1} p = p^n$. Therefore, H is a Sylow p-subgroup.   □

The Second Sylow Theorem says that Sylow p-subgroups are always conjugate to each other.

Theorem 7.9 (Second Sylow Theorem). *Fix a prime p. Let G be a finite group and P a Sylow p-subgroup of G. If H is a subgroup of G of order p^k, for some $k \geq 0$, then H is conjugate to a subgroup of P. In particular, all Sylow p-subgroups of G are conjugate.*

Proof. By Theorem 7.7, there are $[G : N(P)]$ different conjugates of P in G. Also, by Theorem 7.2, $P \leq N(P)$. Therefore,

$$|G|/|P| = (|G|/|N(P)|)(|N(P)|/|P|) = [G : N(P)][N(P) : P].$$

Hence, $[G : N(P)]$ divides $|G|/|P|$. But by definition of a Sylow p-subgroup, $|G|/|P|$ is relatively prime to p; thus, the number of conjugates of P is relatively prime to p.

Among all of these subgroups conjugate to P (and hence to each other), let us consider those that are H-conjugate. We know that H-conjugacy is an equivalence relation, and the H-conjugacy classes partition the set of all conjugates of P. If all of these H-conjugacy classes contained numbers of elements that are divisible by p, then the total number of conjugates would be divisible by p, which is impossible. Therefore, there is a subgroup K of G, conjugate to P, such that the number of H-conjugates of K is not divisible by p. Now conjugate subgroups have the same order, so K is also a Sylow p-subgroup of G.

By Theorem 7.7, the number of H-conjugates of K is $[H : N_H(K)]$. But $|H| = p^k$, and $[H : N_H(K)] = |H|/|N_H(K)|$ is a divisor of $|H|$. The only way we can avoid having $[H : N_H(K)]$ be a multiple of p is if it is 1. Thus, $H = N_H(K)$. That is, $H = H \cap N(K)$, which means that $H \leq N(K)$.

Theorem 7.2 tells us that $N(K)$ contains K as a normal subgroup. Thus, by Theorem 4.5, HK is also a subgroup of $N(K)$. But by Theorem 4.4, $|HK| = |H||K|/|H \cap K|$. However, $|K|$ is the largest power of p dividing $|G|$, and since $|HK|$ must divide $|G|$, we conclude that p does not divide $|H|/|H \cap K|$. As H has order p^k, this means that $H \cap K = H$; thus, $H \leq K$. But K is a conjugate of P! That is, $H \leq g^{-1}Pg$, for some $g \in G$. Equivalently, $(g^{-1})^{-1}Hg^{-1} \leq P$, as required.

The fact that Sylow subgroups are conjugate now follows immediately from the fact that conjugate subgroups have the same order. $\qquad\square$

The Third Sylow Theorem imposes restrictions upon the possible numbers of Sylow p-subgroups in a group.

Theorem 7.10 (Third Sylow Theorem). *Let p be a prime and G a group of order $p^n r$, where $n \geq 0$ and $(p, r) = 1$. Then the number of Sylow p-subgroups of G is congruent to 1 modulo p and divides r.*

Proof. Fix a Sylow p-subgroup P. By the Second Sylow Theorem, every Sylow p-subgroup of G is conjugate to P. Also, as conjugate subgroups have the same order, only Sylow p-subgroups can be conjugate to P. Therefore, the set of Sylow p-subgroups of G is precisely the set of conjugates of P. By Theorem 7.7, there are $[G : N(P)]$ such conjugates. But $P \leq N(P)$, which means that

$$[G : P] = |G|/|P| = (|G|/|N(P)|)(|N(P)|/|P|).$$

By definition of the Sylow p-subgroup, $[G : P] = r$, which means that $[G : N(P)]$ divides r, giving us the last part of the theorem.

Now, we know that P-conjugacy is an equivalence relation on the set of all Sylow p-subgroups. Thus, the number of Sylow p-subgroups is the sum of the sizes of the P-conjugacy classes. But if H is a Sylow p-subgroup, then by Theorem 7.7, it has precisely $[P : N_P(H)]$ P-conjugates. As P has order p^k, we see that $[P : N_P(H)] = |P|/|N_P(H)|$ is also a power of p. Thus, it is in particular a multiple of p, unless it is 1. So, to determine the number of Sylow p-subgroups modulo p, we have only to consider those H such that $P = N_P(H)$. But this happens if and only if $P = N(H) \cap P$; that is, if and only if $P \leq N(H)$. Now proceed as in the proof of the Second Sylow Theorem; we see that this happens if and only if $P \leq H$. However, P and H have the same order, so this means that $P = H$.

That is, modulo p, the number of Sylow p-subgroups is $[P : N_P(P)]$. But P is normal in itself, so this is $[P : P] = 1$. The proof is complete.   □

Exercises

7.21. Find all Sylow 2-, 5- and 7-subgroups of $\mathbb{Z}_{100} \times \mathbb{Z}_{14}$.

7.22. Find all Sylow 2- and 3-subgroups of A_4.

7.23. Let G be a group of order 294. Show that G has exactly one Sylow 7-subgroup.

7.24. Let G be a finite group. Explain why it is impossible for G to have one Sylow 2-subgroup isomorphic to $\mathbb{Z}_4$ and another Sylow 2-subgroup isomorphic to $\mathbb{Z}_2 \times \mathbb{Z}_2$.

7.25. Suppose that p is a prime and p^n divides $|G|$, for some $n \in \mathbb{N}$. Show that G has a subgroup of order p^n.

7.26. Find a Sylow 2-subgroup of S_4. To what familiar group is it isomorphic?

7.27. Let G be a finite group having a normal subgroup N. If H is a Sylow p-subgroup of G, show that HN/N is a Sylow p-subgroup of G/N.

7.28. If G is a finite group with normal subgroup N, and H is a Sylow p-subgroup of G, show that $H \cap N$ is a Sylow p-subgroup of N.

7.4 Applying the Sylow Theorems

Let us discuss some interesting consequences of the Sylow theorems. For one thing, we can now complete Cauchy's theorem, which we previously discussed for abelian groups.

Theorem 7.11 (Cauchy's Theorem). *Let G be a finite group, and suppose that a prime p divides its order. Then G has an element of order p.*

Proof. By the First Sylow Theorem, G has a Sylow p-subgroup P, the order of which is p^n, for some positive integer n. Take any $e \neq a \in P$. By Lagrange's theorem, $|a| = p^k$, $1 \leq k \leq n$. Then by Corollary 3.2, $|a^{p^{k-1}}| = p$.   □

As a consequence, we can now extend Corollary 5.2 to nonabelian groups.

Corollary 7.3. *Let p be a prime. Then a finite group G is a p-group if and only if $|G| = p^n$, for some $n \geq 0$.*

Proof. If $|G| = p^n$, then by Lagrange's theorem, every element has order dividing p^n, so G is a p-group. On the other hand, if some prime q different from p divides $|G|$, then by Cauchy's theorem, G has an element of order q, so it is not a p-group. □

The Third Sylow Theorem tells us about the possible numbers of Sylow p-subgroups. But from the Second Sylow Theorem, we can deduce when there is just one such subgroup.

Corollary 7.4. *Let p be a prime and G a finite group. Then G has just one Sylow p-subgroup if and only if the Sylow p-subgroup is normal in G.*

Proof. Let P be a Sylow p-subgroup of G. Then P is normal if and only if $a^{-1}Pa = P$ for all $a \in G$; in other words, if and only if P has only itself as a conjugate. But Theorem 7.9 tells us that every Sylow p-subgroup is conjugate to P, and since conjugates have the same order, this means that nothing that is not a Sylow p-subgroup can be conjugate to P. That is, P is normal if and only if it is conjugate only to itself, if and only if there is only one Sylow p-subgroup.   □

This corollary is highly useful in finding normal subgroups of groups of a particular order. In particular, if we are asked to show that groups of some particular order cannot be simple, then our first step is often to see if some Sylow p-subgroup must be normal. For instance:

Theorem 7.12. *Let G be a group of order pq, where p and q are primes with $p < q$. Then the Sylow q-subgroup of G is normal. In particular, G is not simple.*

Proof. By the Third Sylow Theorem, the number of Sylow q-subgroups is of the form $1 + kq$, with $k \in \mathbb{Z}$, and divides p. As $q > p$, the only possibility is $k = 0$ and $1 + kq = 1$. Now apply the preceding corollary.   □

Let us try something slightly more complicated.

Example 7.11. Let us show that there are no simple groups of order 351. As $351 = 3^3 \cdot 13$, we note that the number of Sylow 3-subgroups is $1 + 3k$ and divides 13 and the number of Sylow 13-subgroups is $1 + 13l$ and divides 27, with $k, l \in \mathbb{Z}$. The only solutions for l are 0 and 2; that is, the number of Sylow 13-subgroups is either 1 or 27. If it is 1, then we know that the Sylow 13-subgroup is normal, and we are done. So let us assume that it is 27. Now, each Sylow 13-subgroup is of order 13. In a group of prime order, everything but the identity has order equal to that of the group; thus, each Sylow 13-subgroup has 12 elements of order 13. Furthermore, if P and Q are different Sylow 13-subgroups, then since $|P \cap Q|$ must divide $|P| = 13$, either $P = Q$ (which is impossible) or $P \cap Q = \{e\}$. Thus, each of the 27 Sylow 13-subgroups contributes 12 elements of order 13, and there is no overlap. We have now used up $12 \cdot 27 = 324$ elements of the group. This leaves only $351 - 324 = 27$ elements. But that is the size of one Sylow 3-subgroup! Thus, there is only room for one such subgroup. In order words, either the Sylow 13-subgroup or the Sylow 3-subgroup must be normal.

We do have to be a bit careful in solving problems like the one in the preceding example. It would not have worked well if we had considered the Sylow 3-subgroups first. To be sure, we would have found that the number of such subgroups is 1 or 13, and if it is 1, we are done. But if it is 13, we would have a problem counting the 3-elements we have used, because the Sylow 3-subgroups do not have prime order and, therefore, do not necessarily intersect trivially.

Let us consider groups with order the product of three primes.

Theorem 7.13. *Let G be a group of order pqr, where p, q and r are distinct primes. Then G is not simple.*

Proof. Without loss of generality, let us say that $p < q < r$. Then the number of Sylow r-subgroups is of the form $1 + kr$, with $k \in \mathbb{Z}$, and divides pq. Now, the only positive divisors of pq are 1, p, q and pq. Since $r > q > p$, we cannot have $1 + kr = p$ or q. If there is only one Sylow r-subgroup, then by Corollary 7.4, we are done. Thus, let us assume that there are pq of them. Now, these Sylow r-subgroups have prime order, so just as in Example 7.11, they intersect trivially, and provide us with $pq(r - 1)$ elements of order r.

Similarly, the number of Sylow q-subgroups is $1 + lq$, with $l \in \mathbb{Z}$, and divides pr. As $q > p$, it cannot be p. If it is 1, then once again, we are done. So it is either r or pr. In any case, it is at least r. Therefore, by the same argument, we obtain at least $r(q - 1)$ elements of order q.

Finally, the number of Sylow p-subgroups is $1 + mp$, with $m \in \mathbb{Z}$, and divides qr. If it is 1, then we are done, so we may assume that it is at least q. Thus, we obtain at least $q(p - 1)$ elements of order p.

Adding in the identity, we now have at least

$$pq(r-1)+r(q-1)+q(p-1)+1 = pqr+qr-q-r+1 > pqr+qr-2r+1$$

elements (since $q < r$). But as $q > p$, and p is a prime, we have $q \geq 3$, so $qr > 2r$, and we have accounted for more than pqr group elements, which is impossible. $\square$

In the special case where all of the Sylow p-subgroups are normal, we are in an even better position.

Theorem 7.14. *Let G be a group of order $p_1^{n_1} \cdots p_k^{n_k}$, where the p_i are distinct primes and the n_i are positive integers. If, for each i, G has a unique Sylow p_i-subgroup P_i, then $G = P_1 \times \cdots \times P_k$.*

Proof. Let a_i be a p_i-element of G. Then by the Second Sylow Theorem, there exists a $g_i \in G$ such that $g_i^{-1} \langle a_i \rangle g_i \leq P_i$; say $g_i^{-1} a_i g_i = h_i \in P_i$. Then $a_i = g_i h_i g_i^{-1} \in P_i$ since, by Corollary 7.4, P_i is normal. In particular, each P_i is the set of all p_i-elements of G.

By Lemma 5.4, every element of G can be written as a product of p_i-elements, $1 \leq i \leq k$. Thus, $G = P_1 P_2 \cdots P_k$. By Exercise 5.10, G is the internal direct product of the P_i. $\square$

Example 7.12. Suppose we wish to classify the groups of order 45. The number of Sylow 3-subgroups is $1 + 3k$ and divides 5, for some $k \in \mathbb{Z}$. Thus, it can only be 1. The number of Sylow 5-subgroups is $1 + 5l$ and divides 9, for some $l \in \mathbb{Z}$. Therefore, the Sylow 5-subgroup is unique as well. According to the preceding theorem, a group of order 45 must be the direct product of its Sylow subgroups. By Corollary 7.2, a group of order 9 is isomorphic to either $\mathbb{Z}_3 \times \mathbb{Z}_3$ or $\mathbb{Z}_9$, and Corollary 4.2 tells us that a group of order 5 is isomorphic to $\mathbb{Z}_5$. Hence, every group of order 45 is isomorphic to either $\mathbb{Z}_3 \times \mathbb{Z}_3 \times \mathbb{Z}_5$ or $\mathbb{Z}_9 \times \mathbb{Z}_5$ (and these are not isomorphic to each other, by Theorem 5.6).

Exercises

7.29. Show that there are no simple groups of order 84.

7.30. Show that there are no simple groups of order 56.

7.31. Let G be a group of order $4352 = 2^8 \cdot 17$. Show that either a Sylow 2-subgroup or a Sylow 17-subgroup of G must be normal.

7.32. Let G be a group of order $870 = 2 \cdot 3 \cdot 5 \cdot 29$. Show that at least one of the Sylow p-subgroups of G must be normal, for some prime p dividing $|G|$.

7.33. Let G be a group of order $p^2 q$, for some distinct primes p and q. If $q \nmid (p^2 - 1)$ and $p \nmid (q - 1)$, show that G is abelian.

7.34. Show that Theorem 7.13 is still true even if the primes p, q and r are not assumed to be distinct.

7.35. Let G be a group of order 57. There are only two possible numbers of elements of order 3 in G. What are they?

7.36. Let G be a group of order $935 = 5 \cdot 11 \cdot 17$. Show that the Sylow 17-subgroup of G is central.

7.37. Let G be a group of order $595 = 5 \cdot 7 \cdot 17$. Show that G has a subgroup of order 119.

7.38. Let G be a nontrivial finite p-group. If H is a proper subgroup of G, show that H is a proper subgroup of $N(H)$.

7.5 Classification of the Groups of Small Order

We conclude our discussion of groups by classifying the groups of order up to 15. Why 15 in particular? Because the classification of the groups of order 16 is a confounded nuisance! There are, in fact, 14 different nonisomorphic groups of that order. We are aware of the five abelian groups (see Example 5.7), the dihedral group D_{16}, and $D_8 \times \mathbb{Z}_2$, but constructing the other seven would be a lot of work.

Let G be a group of order n. We already know all of the possibilities for most of the values $n < 16$. If $n = 1$, there is only the trivial group, $\{e\}$. When $n \in \{2, 3, 5, 7, 11, 13\}$, Corollary 4.2 tells us that G is isomorphic to $\mathbb{Z}_n$. If $n \in \{4, 9\}$, we rely upon Corollary 7.2, which says that if p is a prime and $n = p^2$, then G is isomorphic to $\mathbb{Z}_{p^2}$ or $\mathbb{Z}_p \times \mathbb{Z}_p$. Also, if $n \in \{6, 10, 14\}$, then we use Theorem 4.15; when $n = 2p$, for some odd prime p, we find that G is isomorphic to $\mathbb{Z}_{2p}$ or D_{2p}.

We are left with groups of order 8, 12 and 15. With the aid of the Sylow theorems, the $n = 15$ case is a piece of cake.

Theorem 7.15. *Every group of order 15 is isomorphic to $\mathbb{Z}_{15}$.*

Proof. By the Third Sylow Theorem, the number of Sylow 3-subgroups is $1 + 3k$, for some $k \in \mathbb{Z}$, and divides 5. Thus, there is only one Sylow 3-subgroup. By Theorem 7.12, the same is true for the Sylow 5-subgroup. Therefore, by Theorem 7.14, our group is the direct product of these Sylow subgroups. But the Sylow subgroups have prime order and, therefore, are cyclic. By Theorem 5.4, the direct product of cyclic groups of relatively prime order is also cyclic. Thus, by Theorem 4.14, our group is isomorphic to $\mathbb{Z}_{15}$. □

Unfortunately, when it comes to groups of order 8, the Sylow theorems cannot help us. Indeed, for any finite p-group, the unique Sylow p-subgroup is the whole group. We can, nevertheless, classify the groups of order 8 up to isomorphism. In view of Corollary 5.1, we know that the abelian groups of order 8 are all isomorphic to one of $\{\mathbb{Z}_8, \mathbb{Z}_4 \times \mathbb{Z}_2, \mathbb{Z}_2 \times \mathbb{Z}_2 \times \mathbb{Z}_2\}$ (and, by Theorem 5.6, these groups are not isomorphic to each other).

Let G be a nonabelian group of order 8. By Lagrange's theorem, every nonidentity element of G has order 2, 4 or 8. If there is an element of order 8 then G is cyclic, hence abelian, which is not the case. Also, if every nonidentity element has order 2, then G is abelian, by Exercise 3.32. Therefore, we may assume the existence of an element a of order 4. Then $\langle a \rangle$ has index 2 and, by Theorem 4.1, is normal in G.

Take any element b in G that is not in $\langle a \rangle$. Then we observe that the elements of G are precisely a^i and ba^i, $0 \le i \le 3$. Also, by the normality of $\langle a \rangle$, we have $b^{-1}ab = a^j$, for some j. Now, conjugate elements have the same order, so $j = 1$ or 3. If $j = 1$, then a and b commute. But this means that all elements of the form a^i and ba^i commute as well, so G is abelian, which is not the case. Thus, $b^{-1}ab = a^3$.

What is the order of b? We know it is 2 or 4. Suppose that it is 2. We can now follow the final part of the proof of Theorem 4.15, and we see that G is isomorphic to D_8.

Therefore, let $|b| = 4$. Now, $G/\langle a \rangle$ has order 2. Thus, $(b\langle a \rangle)^2 = e\langle a \rangle$, so $b^2 \in \langle a \rangle$. Furthermore, $|b^2| = 2$ (by Corollary 3.2), so $b^2 = a^2$. But now we know everything about the group. We know what the elements are. Furthermore, $a^k a^l = a^{k+l}, ba^k a^l = ba^{k+l}$,

$$a^k ba^l = b(b^{-1}a^k b)a^l = b(b^{-1}ab)^k a^l = b(a^3)^k a^l = ba^{3k+l}$$

and

$$(ba^k)(ba^l) = b(ba^{3k+l}) = a^2 a^{3k+l} = a^{3k+l+2},$$

for any $k, l \in \mathbb{Z}$.

What this means is, we can completely fill in the group table so, up to isomorphism, there can be at most one group meeting this description. This does not, however, mean that such a group necessarily exists. As it happens, it does!

Example 7.13. The **quaternion group** is the group $Q_8 = \{\pm 1, \pm i, \pm j \pm k\}$, where $i^2 = j^2 = k^2 = -1, ij = k = -ji, jk = i = -kj$ and $ki = j = -ik$. The element 1 is the identity, and it is easy to see that the group is closed and every element has an inverse (for instance, $i^{-1} = -i$). Checking associativity involves verifying a lot of cases, but it does work. Furthermore, $ij \ne ji$, so Q_8 is not abelian. Also, we note that the only element of order 2 is -1, whereas D_8 has many elements of order 2. Thus, we have a new group, and it must be the one we described above. (In the notation we used, let $a = i$ and $b = j$.)

We now record our classification of the groups of order 8.

Theorem 7.16. *Every group of order* 8 *is isomorphic to one of the following, namely* $\mathbb{Z}_8, \mathbb{Z}_4 \times \mathbb{Z}_2, \mathbb{Z}_2 \times \mathbb{Z}_2 \times \mathbb{Z}_2, D_8$ *or* Q_8.

Finally, suppose that G has order 12. The number of Sylow 3-subgroups is $1+3m$, $m \in \mathbb{Z}$, and divides 4, so it is 1 or 4. As a group of order 3 is cyclic, let $H = \langle d \rangle$ be a Sylow 3-subgroup. The number of Sylow 2-subgroups is $1 + 2l, l \in \mathbb{Z}$, and divides 3, so it is 1 or 3. Let K be a Sylow 2-subgroup, which we know is isomorphic either to $\mathbb{Z}_4$ or $\mathbb{Z}_2 \times \mathbb{Z}_2$. Let us break our discussion down into cases.

CASE I: The Sylow 2- and 3-subgroups are both unique. (Note that this must be the case if G is abelian, as every subgroup is normal.) By Theorem 7.14, G is the direct product of its Sylow subgroups. Thus, G is isomorphic to $\mathbb{Z}_4 \times \mathbb{Z}_3$ or $\mathbb{Z}_2 \times \mathbb{Z}_2 \times \mathbb{Z}_3$. By Theorem 5.6, these groups are not isomorphic.

CASE II: There are three Sylow 2-subgroups and four Sylow 3-subgroups. Now, we proceed as in Example 7.11. As the Sylow 3-subgroups have prime order, they intersect trivially, and every element other than the identity has order 3. Thus, we have $4 \cdot 2 = 8$ elements of order 3. But this leaves only four elements unaccounted for. Thus, there is only room for one Sylow 2-subgroup. This case cannot occur.

CASE III: The Sylow 2-subgroup is unique, but there are four Sylow 3-subgroups. Let us further break this case down. CASE IIIa: K is isomorphic to $\mathbb{Z}_2 \times \mathbb{Z}_2$. Notice that $|HK| = |H||K|/|H \cap K| = 4 \cdot 3/1 = 12$, using Theorem 4.4 and the fact that $H \cap K$ must have order dividing both $|H|$ and $|K|$. Thus, $HK = G$. Also, H and K are abelian, and if $d \in C(K)$, then $H \le C(K)$, and we see that G is abelian which, as we noted above, means we must be in Case I. Thus, there exists $e \ne a \in K$ such that $d^{-1}ad = b \ne a$. As K is normal, $b \in K$. Now, if $d^{-1}ad = d^{-1}bd$, then $a = b$, which is impossible, so $d^{-1}bd \ne b$. Also, if $d^{-1}bd = a$, then $d^{-2}ad^2 = d^{-1}(d^{-1}ad)d = d^{-1}bd = a$, so $d^2 \in C(a)$, and therefore $d = (d^2)^2 \in C(a)$, which is not the case. As $K = \{e, a, b, ab\}$, we must have $d^{-1}bd = ab$. This means that $d^{-1}abd = d^{-1}add^{-1}bd = bab = a$, as the Sylow 2-subgroup is isomorphic to $\mathbb{Z}_2 \times \mathbb{Z}_2$. Now, we know that the elements of G are precisely hk, with $h \in H$ and $k \in K$. We also know how products work and can construct a group table. For example,

$$(d^2 ab)(d^2 b) = d(d^{-2}abd^2)b = d(d^{-1}(d^{-1}abd)d)b = d(d^{-1}ad)b = db^2 = d.$$

Thus, there is at most one group in this case, up to isomorphism.

The question that remains is, can such a group be constructed? In fact it can, and we have already seen it. If we let $G = A_4$, $a = (1\ 2)(3\ 4)$, $b = (1\ 3)(2\ 4)$ and $d = (1\ 2\ 3)$, we find that all of our conditions are met.

CASE IIIb: $K = \langle c \rangle$ is cyclic of order 4. As K is normal, we have $d^{-1}cd \in \langle c \rangle$, say $d^{-1}cd = c^i$. As conjugates have the same order, $i = 1$ or 3. If $i = 1$, then $d \in C(c)$, so $K \le C(c)$. As in CASE IIIa, $G = HK$, and we see that G is abelian, which is not permitted. Therefore, $d^{-1}cd = c^3$. But then

$$d^{-2}cd^2 = d^{-1}(d^{-1}cd)d = d^{-1}c^3 d = (d^{-1}cd)^3 = (c^3)^3 = c.$$

Thus, $d^{-3}cd^3 = d^{-1}cd = c^3$. But $d^3 = e$, so we have a contradiction. Therefore, this case cannot occur.

CASE IV: The Sylow 3-subgroup is unique, but G has three Sylow 2-subgroups. Again, let us break this down further. CASE IVa: G has a Sylow 2-subgroup isomorphic to $\mathbb{Z}_2 \times \mathbb{Z}_2$. Now, if $K \le C(d)$, then we see that elements of K commute with elements of H and once again, G is abelian, which is not permitted. Therefore, take $a \in K$ such that $a^{-1}da \ne d$. Now, $\langle d \rangle$ is normal, and given that the only nonidentity elements are d and d^2, we have $a^{-1}da = d^2$. If b is another nonidentity element of K, we must also have $b^{-1}db = d$ or d^2. In the latter case,

$$(ba)^{-1}d(ba) = a^{-1}(b^{-1}db)a = a^{-1}d^2a = (a^{-1}da)^2 = (d^2)^2 = d.$$

Thus, one of the nonidentity elements of K centralizes d. Without loss of generality, say $bd = db$.

What is the order of bd? If $(bd)^j = e$, then as b and d commute, we have $b^j d^j = e$, so $b^j = d^{-j} \in H \cap K = \{e\}$, since H and K have relatively prime orders. Thus, j must be divisible by 2 and 3, and hence 6. On the other hand, $(bd)^6 = b^6 d^6 = e$, so $|bd| = 6$. Also, a has order 2 and

$$a^{-1}bda = (a^{-1}ba)(a^{-1}da) = bd^2 = (bd)^{-1},$$

since $(bd^2)(bd) = b^2 d^3 = e$. Since a does not commute with bd, $a \notin \langle bd \rangle$. Hopefully this situation rings a bell! Refer to the proof of Theorem 4.15. It is at this point that we can conclude that we have constructed D_{12}. Note that D_{12} and A_4 are not isomorphic, since D_{12} has an element of order 6 and A_4 does not.

Finally, we have CASE IVb: $K = \langle c \rangle$ is cyclic of order 4. As H is normal, we have $c^{-1}dc = d^j$, for some j. Since the identity is only conjugate to itself, this means that $c^{-1}dc = d$ or d^2. If $c^{-1}dc = d$, then we see immediately that all powers of c and d commute. But once again, $G = KH$, so G is abelian, which is not the case. Thus, $c^{-1}dc = d^2$. But since $G = KH$, the elements of G are precisely $c^r d^s$, $0 \le i \le 3$ and $0 \le s \le 2$. And we now know how to take a product of any two group elements. For instance,

$$(cd)(cd^2) = c^2(c^{-1}dc)d^2 = c^2 d^2 d^2 = c^2 d.$$

In particular, we can fill in the group table. This means that there is at most one more group of order 12 that is not isomorphic to any of the ones we have constructed so far. In fact, such a group exists.

Example 7.14. Let $G = S_3 \times \mathbb{Z}_4$. This is a group of order 24. Let H be the set of all elements $(\sigma, t) \in G$ such that the permutation σ and the number t are either both even or both odd. In Exercise 7.40, we are asked to show that H is a subgroup of G of order 12 and that it is not isomorphic to any of the other groups of order 12 that we have found. Thus, it must be the group from CASE IVb. In fact, using $c = ((1\ 2), 1)$ and $d = ((1\ 2\ 3), 0)$, we find that it has the desired properties.

We have now completed the classification of groups of order 12.

Theorem 7.17. *Let G be a group of order 12. Then G is isomorphic either to $\mathbb{Z}_4 \times \mathbb{Z}_3$, $\mathbb{Z}_2 \times \mathbb{Z}_2 \times \mathbb{Z}_3$, A_4, D_{12}, or the group H from Example 7.14.*

Exercises

7.39. To which of the groups listed in Theorem 7.17 is $D_6 \times \mathbb{Z}_2$ isomorphic?

7.40. Let H be the subset of $S_3 \times \mathbb{Z}_4$ described in Example 7.14. Show that H is a subgroup of order 12 in $S_3 \times \mathbb{Z}_4$, and that H is not isomorphic to $\mathbb{Z}_4 \times \mathbb{Z}_3$, $\mathbb{Z}_2 \times \mathbb{Z}_2 \times \mathbb{Z}_3$, A_4 or D_{12}.

7.41. Show that every subgroup of Q_8 is normal.

7.42. Let H be a finite abelian group. Show that every subgroup of $Q_8 \times H$ is normal if and only if H has no elements of order 4.

7.43. Let p be a prime. If $a, b \in \mathbb{Z}_p$ and $a \neq 0$, define $\alpha_{a,b} : \mathbb{Z}_p \to \mathbb{Z}_p$ via $\alpha_{a,b}(x) = ax + b$.

1. Show that these $\alpha_{a,b}$ form a group G under composition and, if $p > 2$, that this group is not abelian.
2. If $p = 7$, find a nonabelian subgroup H of G such that $|H| = 21$.

7.44. Show that every group of order 21 is isomorphic either to $\mathbb{Z}_{21}$ or to the group H from the second part of the preceding exercise.

7.45. Generalize Theorem 7.15 as follows. If p and q are primes, with $p > q$ and $q \nmid (p - 1)$, show that every group of order pq is isomorphic to $\mathbb{Z}_{pq}$.

7.46. We know from Theorem 6.11 that A_5 is a nonabelian simple group of order 60. Show that there are no nonabelian simple groups with order smaller than 60. (The methods we have discussed up to this point are sufficient to deal with every order except for 24, 36 and 48. Here is a hint if $|G| = 36$: Suppose that G has distinct Sylow 3-subgroups H and K. What is $|H \cap K|$? What can you say about $|N(H \cap K)|$? Find a nontrivial proper normal subgroup of G.)

Part III
Rings

Chapter 8
Introduction to Rings

We now move on to the second major type of algebraic object that we are considering: the ring. At first blush, rings look a bit more complicated than groups. Indeed, a ring is an abelian group written additively, and we must still impose a multiplication operation along with several new rules. But in another sense, rings are easier to deal with, because they are more familiar. Indeed, when we think of a ring, we tend to think of the integers (although, as we shall see, the integers are actually a special sort of ring).

In this chapter, we will define a ring and prove some properties of rings and subrings. We shall also discuss two well-behaved types of rings; namely, integral domains and fields.

8.1 Rings

Let us now define a ring.

Definition 8.1. A **ring** is a set R together with two binary operations, written as addition and multiplication, such that

1. R is an abelian group under addition;
2. if $a, b \in R$, then $ab \in R$ (closure under multiplication);
3. if $a, b, c \in R$, then $(ab)c = a(bc)$ (associativity of multiplication);
4. if $a, b, c \in R$, then $a(b + c) = ab + ac$ (distributive law); and
5. if $a, b, c \in R$, then $(a + b)c = ac + bc$ (distributive law).

As usual when we have an additive group, we will use additive notation. In particular, we write 0 for the additive identity of a ring, and $-a$ for the additive inverse of a. Notice that we do not insist that the multiplication operation be commutative.

© Springer International Publishing AG, part of Springer Nature 2018

G. T. Lee, *Abstract Algebra*, Springer Undergraduate Mathematics Series,

https://doi.org/10.1007/978-3-319-77649-1_8

135

Definition 8.2. A ring R is said to be a **commutative ring** if $ab = ba$ for all $a, b \in R$.

Also, while there is an identity for the addition operation, there does not have to be one for the multiplication operation.

Definition 8.3. A ring R is said to be a **ring with identity** if R has an element, denoted 1, such that $1a = a1 = a$ for all $a \in R$. In this case, we call 1 the **identity** of R.

Note that if we refer to the identity in a ring, we mean the multiplicative identity 1 (if it exists), not the additive identity 0.

Example 8.1. As we observed in Section 2.4, the sets $\mathbb{Z}$, $\mathbb{Q}$, $\mathbb{R}$ and $\mathbb{C}$ are all commutative rings with identity, under the usual addition and multiplication operations. Also, we saw in Section 2.5 that the same can be said for $\mathbb{Z}_n$, for any positive integer $n \geq 2$.

Example 8.2. The set of even integers, $2\mathbb{Z}$, can easily be seen to be a commutative ring without an identity. There is no even integer that can be multiplied by 2 to get 2.

Example 8.3. The set of all polynomials with real coefficients is a commutative ring with identity, using the usual polynomial addition and multiplication operations. We denote it by $\mathbb{R}[x]$. The same can be said for the polynomials with integer coefficients, $\mathbb{Z}[x]$. In each case, the identity is the constant polynomial, 1.

How about an example of a noncommutative ring?

Example 8.4. Let n be a positive integer. Then the $n \times n$ matrices with real entries form a ring under matrix addition and multiplication. The identity matrix is the identity of the ring. However, if $n > 1$, then it is not a commutative ring as, for instance, $\begin{pmatrix} 1 & 1 \\ 0 & 1 \end{pmatrix} \begin{pmatrix} 1 & 0 \\ 1 & 1 \end{pmatrix} \neq \begin{pmatrix} 1 & 0 \\ 1 & 1 \end{pmatrix} \begin{pmatrix} 1 & 1 \\ 0 & 1 \end{pmatrix}$. We denote this ring by $M_n(\mathbb{R})$. In fact, as we observe in Appendix B, we can substitute entries from any ring R in place of the real numbers, and we obtain a new ring, $M_n(R)$. If R is a ring with identity, then we can form the identity matrix, so $M_n(R)$ is also a ring with identity. The conditions under which it is a commutative ring are discussed in Exercise 8.10.

We also have a way of constructing new rings from old, simply extending the idea of the direct product of groups.

Definition 8.4. Let R and S be rings. Then the **direct sum** of R and S, denoted $R \oplus S$, is the Cartesian product $R \times S$ under the operations

$$(r_1, s_1) + (r_2, s_2) = (r_1 + r_2, s_1 + s_2) \text{ and } (r_1, s_1)(r_2, s_2) = (r_1 r_2, s_1 s_2),$$

for all $r_i \in R$, $s_i \in S$.

Theorem 8.1. *Let R and S be rings. Then $R \oplus S$ is a ring. Furthermore, if R and S are commutative rings, then so is $R \oplus S$. Also, if R and S are rings with identity, then so is $R \oplus S$.*

Proof. The proof is very similar to that of Theorem 3.1. The ring properties all hold in the direct sum because they hold in R and S. We will prove one of the distributive laws, and leave the rest as Exercise 8.6.

Take $r_i \in R$, $s_i \in S$. Then

$$
\begin{aligned}
(r_1, s_1)((r_2, s_2) + (r_3, s_3)) &= (r_1, s_1)(r_2 + r_3, s_2 + s_3) \\
&= (r_1(r_2 + r_3), s_1(s_2 + s_3)) \\
&= (r_1 r_2 + r_1 r_3, s_1 s_2 + s_1 s_3) \\
&= (r_1 r_2, s_1 s_2) + (r_1 r_3, s_1 s_3) \\
&= (r_1, s_1)(r_2, s_2) + (r_1, s_1)(r_3, s_3).
\end{aligned}
$$

$\square$

Example 8.5. In $\mathbb{Z}_5 \oplus \mathbb{Z}_6$, we have $(3, 5) + (4, 2) = (7, 7) = (2, 1)$ and $(3, 5)(4, 2) = (12, 10) = (2, 4)$.

One additional point is important to keep in mind. A ring is a group under addition, not under multiplication! While the multiplication operation satisfies the closure and associativity properties, a ring does not have to have an identity. And even if it does, elements do not have to have inverses. For instance, $\mathbb{Z}$ has an identity, but there is nothing we can multiply by 2 to obtain 1.

Exercises

8.1. Write the addition and multiplication tables for the ring $\mathbb{Z}_5$.

8.2. Write the addition and multiplication tables for the ring $\mathbb{Z}_3 \oplus \mathbb{Z}_2$.

8.3. Let $R = \{0, 3, 6, 9, 12\}$ with addition and multiplication in $\mathbb{Z}_{15}$. Is R a ring? If so, is it commutative, and does it have an identity?

8.4. Let R be the set of all functions from $\mathbb{R}$ to $\mathbb{R}$, under addition and multiplication of functions. Is R a ring? If so, is it commutative, and does it have an identity?

8.5. Let R be the set of all functions from $\mathbb{R}$ to $\mathbb{R}$. Let the addition operation be the usual addition of functions, but let the multiplication operation be composition. That is, the product of α and β is $\alpha \circ \beta$. Is R a ring? If so, is it commutative, and does it have an identity?

8.6. Complete the proof of Theorem 8.1.

8.7. Let R be the set of matrices of the form $\begin{pmatrix} a & b \\ 0 & c \end{pmatrix}$, for all $a, b, c \in \mathbb{Z}$. Is R a ring under matrix addition and multiplication? If so, is it commutative, and does it have an identity?

8.8. Show that every ring with a prime number of elements is commutative.

8.9. Must a ring with a prime number of elements be a ring with identity?

8.10. Let R be a ring and n a positive integer. Under what conditions is $M_n(R)$ commutative?

8.2 Basic Properties of Rings

Let us mention a few straightforward properties of rings.

Theorem 8.2. *Let R be a ring. Then the additive identity, 0, is unique. If R has a multiplicative identity 1, then it too is unique.*

Proof. As R is a group under addition, we see from Theorem 3.2 that 0 is unique. Suppose that a and b are both multiplicative identities for R. As a is an identity, $ab = b$. But as b is an identity, $ab = a$. Thus, $a = b$. $\qquad\square$

Theorem 8.3. *Let R be a ring. If $a, b \in R$, then*

1. $0a = a0 = 0$;
2. $(-a)b = a(-b) = -(ab)$; *and*
3. $(-a)(-b) = ab$.

Proof. (1) As $0 = 0 + 0$, we have $0a = (0+0)a = 0a + 0a$. Adding $-0a$ to both sides, we get $0 = 0a$. The proof that $a0 = 0$ is similar.

(2) Notice that $ab + (-a)b = (a + (-a))b = 0b = 0$, by (1). As adding $(-a)b$ to ab gives 0, we have $(-a)b = -(ab)$. The proof that $a(-b) = -(ab)$ is similar.

(3) By (2), we have $(-a)(-b) = -(a(-b)) = -(-(ab))$. But remember that R is a group under addition, and hence $-(-(ab)) = ab$, as required. $\qquad\square$

Corollary 8.1. *If R is a ring with identity, then $(-1)a = -a$, for any $a \in R$.*

Proof. By the preceding theorem, $(-1)a = -(1a) = -a$. $\qquad\square$

As a ring is a group under addition, we know from Theorem 3.3 that an expression such as $a_1 + a_2 + \cdots + a_n$ is unambiguous, without the need for brackets. Even though the ring is not a group under multiplication, we can apply precisely the same proof as that of Theorem 3.3 to show that the expression $a_1 a_2 \cdots a_n$ also does not require brackets.

Theorem 8.4. *Let R be any ring, and $a_1, a_2, \ldots, a_n \in R$. Then regardless of how the product $a_1 a_2 \cdots a_n$ is bracketed, the result equals $(\cdots (((a_1 a_2) a_3) a_4) \cdots a_{n-1}) a_n$.*

In order to avoid mistakes, it is also important to recognize which rules cannot be applied in general. For instance, in ordinary arithmetic using the real numbers, we take for granted that if $ab = 0$, then $a = 0$ or $b = 0$. This is simply not the case in an arbitrary ring.

Example 8.6. In $\mathbb{Z}_6$, we have $2 \cdot 3 = 0$, but $2 \neq 0$ and $3 \neq 0$.

Example 8.7. In $M_2(\mathbb{R})$, we have

$$\begin{pmatrix} 1 & 2 \\ 3 & 6 \end{pmatrix} \begin{pmatrix} -2 & 4 \\ 1 & -2 \end{pmatrix} = \begin{pmatrix} 0 & 0 \\ 0 & 0 \end{pmatrix},$$

but

$$\begin{pmatrix} 1 & 2 \\ 3 & 6 \end{pmatrix} \neq \begin{pmatrix} 0 & 0 \\ 0 & 0 \end{pmatrix} \neq \begin{pmatrix} -2 & 4 \\ 1 & -2 \end{pmatrix}.$$

In dealing with groups, we have the cancellation law. We are used to something similar happening in ordinary arithmetic; that is, if $ab = ac$ and $a \neq 0$, then $b = c$. Again, this does not have to hold in rings.

Example 8.8. In $\mathbb{Z}_{12}$, we have $3 \cdot 1 = 3 \cdot 5$, but $3 \neq 0$ and $1 \neq 5$.

Finally, in a group G, we note that if there exists a $b \in G$ such that $ab = b$, then a is the identity. (Just multiply on the right by b^{-1}.) But even if a ring has an identity, the fact that $ab = b$ does not mean that $a = 1$. Indeed, the previous example points us in the right direction.

Example 8.9. In $\mathbb{Z}_{12}$, we have $5 \cdot 3 = 3$, but $5 \neq 1$.

Thus, to check that a ring element a is the identity, we must make sure that $ab = b = ba$ for every $b \in R$, not just for one such b.

Exercises

8.11. Let a and b be elements of a ring R. Simplify the following expressions as far as possible.

1. $(a + b)(a - b)$
2. $(a - b)^3$

8.12. Let R be a ring with identity. Suppose that there exist $a, b, c \in R$ such that $ab = ba = 1$ and $ac = 0$. Show that $c = 0$.

8.13. Let R be a ring with identity. Suppose there exist $a, b, c \in R$ such that $ba = ac = 1$. Does it follow that $b = c$? Show that it does, or find an explicit counterexample.

8.14. Let R be a ring and $n > 2$ a positive integer. Show that if there exists $0 \neq a \in R$ such that $a^n = 0$, then there exists $0 \neq b \in R$ such that $b^2 = 0$.

8.15. Let R be a ring with identity. Suppose that $a(a - 1) = 0$ for every $a \in R$. Does it follow that $a \in \{0, 1\}$ for every $a \in R$? Either prove that it does, or construct an explicit counterexample.

8.16. Let R be a ring in which $a^2 = a$ for every $a \in R$.

1. Show that $a + a = 0$ for every $a \in R$.
2. Show that R is commutative.

8.3   Subrings

Just as we have the notion of a subgroup, we can discuss subrings.

Definition 8.5. Let R be a ring. Then a subset S of R is said to be a **subring** if S is a ring under the same addition and multiplication operations as in R.

Example 8.10. We see that $\mathbb{Z}$ is a subring of $\mathbb{Q}$, and both are subrings of $\mathbb{R}$.

Example 8.11. The matrix ring $M_2(\mathbb{Q})$ is a subring of $M_2(\mathbb{R})$.

Example 8.12. For any ring R, $\{0\}$ and R are subrings of R.

How can we test if a subset is a subring?

Theorem 8.5. *Let R be a ring and S a subset of R. Then S is a subring of R if and only if*

1. *$0 \in S$;*
2. *if $a, b \in S$, then $a - b \in S$; and*
3. *if $a, b \in S$, then $ab \in S$.*

Proof. Suppose that S is a subring of R. Then it is an additive subgroup. By Theorem 3.13, (1) and (2) hold. As a ring is closed under multiplication, (3) holds as well. Conversely, suppose that (1)–(3) hold. Then by Theorem 3.13, S is an additive subgroup of R. By (3), S is closed under multiplication. The remaining ring properties (associativity and the distributive laws) hold in R, hence in any subset of R. Thus, S is indeed a subring. $\square$

Note that for condition (1), it is actually sufficient to check that S is not the empty set.

Example 8.13. Let us show that $2\mathbb{Z}$ is a subring of $\mathbb{Z}$. Certainly $0 \in 2\mathbb{Z}$. If $2a, 2b \in 2\mathbb{Z}$, for some $a, b \in \mathbb{Z}$, then $2a - 2b = 2(a - b) \in 2\mathbb{Z}$. Also, $(2a)(2b) = 2(2ab) \in 2\mathbb{Z}$.

Example 8.14. Let $S = \left\{ \begin{pmatrix} a & 0 \\ 0 & 0 \end{pmatrix} : a \in \mathbb{R} \right\}$. Then letting $a = 0$, we see that S contains the zero matrix. Also, if $a, b \in \mathbb{R}$, then

$$\begin{pmatrix} a & 0 \\ 0 & 0 \end{pmatrix} - \begin{pmatrix} b & 0 \\ 0 & 0 \end{pmatrix} = \begin{pmatrix} a-b & 0 \\ 0 & 0 \end{pmatrix} \in S$$

and

$$\begin{pmatrix} a & 0 \\ 0 & 0 \end{pmatrix} \begin{pmatrix} b & 0 \\ 0 & 0 \end{pmatrix} = \begin{pmatrix} ab & 0 \\ 0 & 0 \end{pmatrix} \in S.$$

Thus, S is a subring of $M_2(\mathbb{R})$.

We recall that the centre of every group is a subgroup. A similar thing happens for rings.

Definition 8.6. Let R be a ring. Then the **centre** of R is the set $\{z \in R : az = za \text{ for all } a \in R\}$; that is, it is the set of elements of R that commute with everything in R.

Theorem 8.6. *The centre of any ring is a subring.*

Proof. Let R be a ring and Z its centre. If $a \in R$, then $0a = 0 = a0$, so $0 \in Z$. Take any $y, z \in Z$. Then for any $a \in R$, we have $a(y-z) = ay-az = ya-za = (y-z)a$, since y and z are central. Thus, $y - z \in Z$. Also, $ayz = yaz = yza$, and hence $yz \in Z$. By Theorem 8.5, we are done. $\square$

Example 8.15. If R is a commutative ring, then its centre is all of R.

Example 8.16. The centre of $M_2(\mathbb{R})$ is the set of all matrices of the form $\begin{pmatrix} r & 0 \\ 0 & r \end{pmatrix}$, for all real numbers r. See Exercise 8.26.

One particular type of subring deserves special mention.

Definition 8.7. Let R be a ring with identity 1. Then a subring S of R is said to be a **unital** subring if $1 \in S$.

Example 8.17. We observe that $\mathbb{Z}$ is a unital subring of $\mathbb{Q}$, but $2\mathbb{Z}$ is not a unital subring.

Note that a subring can fail to be a unital subring because it does not have an identity (as is the case with $2\mathbb{Z}$ above), but it can also have an identity which is not the same as that for R.

Example 8.18. Let $R = \mathbb{Z}_6$ and $S = \{0, 3\}$. Theorem 8.5 shows us that S is a subring of R. It does not contain 1, so it is not a unital subring. However, S is still a ring with identity, as $3 \cdot 0 = 0$ and $3 \cdot 3 = 3$. That is, 3 is the identity of S.

Exercises

8.17. Let $R = \{a + bi : a, b \in \mathbb{Z}\}$. Show that R is a subring of $\mathbb{C}$. Is it a ring with identity? If so, is it unital?

8.18. Let $R = \left\{ \begin{pmatrix} 0 & a & b \\ 0 & 0 & c \\ 0 & 0 & 0 \end{pmatrix} : a, b, c \in \mathbb{R} \right\}$. Show that R a subring of $M_3(\mathbb{R})$. Is it a ring with identity? If so, is it a unital subring?

8.19. Let R be the set of matrices of the form $\begin{pmatrix} 0 & 0 \\ 0 & a \end{pmatrix}$ for all real numbers a. Show that R is a subring of $M_2(\mathbb{R})$. Is it a ring with identity? If so, is it a unital subring?

8.20. Let R be a ring with subrings S and T. Show that $S \cap T$ is a subring. Extend this to show the intersection of any collection of subrings of R is also a subring.

8.21. Let R and S be rings. Show that $T = \{(r, 0) : r \in R\}$ is a subring of $R \oplus S$.

8.22. Find a ring R and an additive subgroup S of R such that S is not a subring of R.

8.23. Let R be a ring and $a \in R$. Show that $S = \{ra : r \in R\}$ is a subring of R.

8.24. Let R be a ring and $a \in R$. Let $S = \{r \in R : ra = 0\}$. Is S necessarily a subring of R? Prove that it is, or find an explicit counterexample.

8.25. Let R be a ring and $a \in R$. Fix a subring S of R, and let $T = \{r \in R : ra \in S\}$. Is T necessarily a subring of R? Prove that it is, or find an explicit counterexample.

8.26. Show that the centre of $M_2(\mathbb{R})$ is the set of matrices of the form $\begin{pmatrix} r & 0 \\ 0 & r \end{pmatrix}$, for all $r \in \mathbb{R}$.

8.4 Integral Domains and Fields

Let us discuss a couple of special sorts of rings.

Definition 8.8. Let R be a commutative ring. Then a nonzero element $a \in R$ is said to be a **zero divisor** if there exists a nonzero $b \in R$ such that $ab = 0$.

Example 8.19. In $\mathbb{Z}_6$, we note that 4 is a zero divisor, as $4 \cdot 3 = 0$. On the other hand, 5 is not a zero divisor.

Example 8.20. The ring of integers has no zero divisors.

As we mentioned at the beginning of the chapter, while we tend to think of the integers when we work with rings, they are actually rather special, and this is the reason why.

Definition 8.9. An **integral domain** is a commutative ring R with identity $1 \neq 0$ having no zero divisors.

The condition that $1 \neq 0$ may seem a bit curious. In fact, if $1 = 0$, then for any $a \in R$, we have $a = 1a = 0a = 0$. Thus, $R = \{0\}$. So we are only ruling out one ring with that restriction.

Example 8.21. The rings $\mathbb{Z}, \mathbb{Q}, \mathbb{R}$ and $\mathbb{C}$ are all integral domains.

Example 8.22. The polynomial ring $\mathbb{R}[x]$ is an integral domain. Indeed, we know that it is a commutative ring with identity. Also, if $f(x) = a_0 + a_1x + \cdots + a_nx^n$ and $g(x) = b_0 + b_1x + \cdots + b_mx^m$, with $a_i, b_i \in \mathbb{R}$ and $a_n \neq 0 \neq b_m$, then the unique term of highest degree in $f(x)g(x)$ is $a_nb_mx^{m+n}$. As $\mathbb{R}$ is an integral domain, $a_nb_m \neq 0$. Thus, $f(x)g(x)$ is not the zero polynomial.

Example 8.23. The rings $2\mathbb{Z}, \mathbb{Z}_6$ and $M_2(\mathbb{R})$ all fail to be integral domains. The first lacks an identity, the second has zero divisors and the third is not commutative.

As we discussed in Section 8.2, rings in general do not enjoy a cancellation law. However, integral domains do.

Theorem 8.7 (Cancellation Law). *Let R be an integral domain. Suppose that $a, b, c \in R$ and $ab = ac$. If $a \neq 0$, then $b = c$.*

Proof. If $ab = ac$, then $ab - ac = 0$, and hence $a(b-c) = 0$. Since R is an integral domain, either $a = 0$ (which is not true), or $b - c = 0$, as required. $\square$

We also wish to discuss a stronger restriction on the ring. We need a definition first.

Definition 8.10. Let R be a ring with identity. Then we say that an element $a \in R$ is a **unit** if there exists an element $b \in R$ such that $ab = ba = 1$. In this case, we call b the **inverse** of a and write $b = a^{-1}$. We write $U(R)$ for the set of all units of R, and call it the **unit group** of R.

Theorem 8.8. *Let R be a ring with identity. Then $U(R)$ is a group under multiplication.*

Proof. Let $a, b \in U(R)$. Then $abb^{-1}a^{-1} = a1a^{-1} = aa^{-1} = 1$, and $b^{-1}a^{-1}ab = b^{-1}1b = b^{-1}b = 1$. Thus, $b^{-1}a^{-1} = (ab)^{-1}$, and $ab \in U(R)$. Multiplication in a ring is associative. Plainly, $1 \in U(R)$, as $1 \cdot 1 = 1$. Also, if $a \in U(R)$, then $aa^{-1} = a^{-1}a = 1$. That is, a is the inverse of a^{-1}, hence $a^{-1} \in U(R)$. We are done. $\square$

Example 8.24. By definition, $U(M_n(\mathbb{R})) = GL_n(\mathbb{R})$.

Example 8.25. The unit group of $\mathbb{Z}$ is $\{\pm 1\}$.

Example 8.26. The unit group of $\mathbb{Z}_n$ is $U(n)$. See Exercise 8.30.

Example 8.27. Every element other than 0 in $\mathbb{R}$ is a unit. The same can be said for $\mathbb{Q}$ and $\mathbb{C}$.

This last example leads us to our next definition.

Definition 8.11. Let F be a commutative ring with identity $1 \neq 0$. Then F is said to be a **field** if $U(F)$ consists of every element of F other than 0.

Example 8.28. As we noted above, $\mathbb{Q}$, $\mathbb{R}$ and $\mathbb{C}$ are fields.

Lemma 8.1. *Let R be a commutative ring with identity. Then a unit in R cannot be a zero divisor.*

Proof. See Exercise 8.12. □

This immediately yields the following result.

Theorem 8.9. *Every field is an integral domain.*

Of course, the integers are an integral domain, but not a field. However, we can say something for finite integral domains. As we might expect, if $a \in R$, and n is a positive integer, we write

$$a^n = \underbrace{aa \cdots a}_{n \text{ times}}.$$

Theorem 8.10. *Let R be a finite integral domain. Then R is a field.*

Proof. By definition, R is a commutative ring with identity $1 \neq 0$. It remains only to check that each nonzero element is a unit. Take $0 \neq a \in R$. Consider the set $\{a^i : i > 0\}$. It consists of infinitely many powers of a. But R is finite. Thus, there cannot be infinitely many distinct powers. Let us say that $a^i = a^j$ with $i > j > 0$. Then $a^j a^{i-j} = a^i = a^j$. More importantly, $a^j a^{i-j} = a^j \cdot 1$. Now, a is a nonzero element of an integral domain, and products of nonzero elements in such a domain do not become zero. Thus, $a^j \neq 0$. By the cancellation law, $a^{i-j} = 1$. If $i - j = 1$, then $a = 1$, which is surely a unit. Otherwise, $a a^{i-j-1} = 1$. Since $i - j - 1$ is a positive integer, $a^{i-j-1} \in R$, and we have an inverse for a. □

We can now handle a particular collection of finite rings of interest.

Theorem 8.11. *Let $n \geq 2$ be a positive integer. Then the following are equivalent:*

1. $\mathbb{Z}_n$ *is an integral domain;*
2. $\mathbb{Z}_n$ *is a field; and*
3. n *is prime.*

Proof. In view of Theorems 8.9 and 8.10, we know that (1) and (2) are equivalent. We need only show that they are equivalent to (3). If n is composite, then write $n = kl$, where k and l are positive integers smaller than n. Then k and l are not 0 in $\mathbb{Z}_n$, and yet $kl = 0$ in $\mathbb{Z}_n$. Thus, $\mathbb{Z}_n$ is not an integral domain. On the other hand, suppose that n is prime. Surely $\mathbb{Z}_n$ is a commutative ring with identity $1 \neq 0$. Suppose we have integers i and j such that $ij = 0$ in $\mathbb{Z}_n$. Then $n|ij$. By Theorem 2.7, $n|i$ or $n|j$. That is, $i = 0$ or $j = 0$ in $\mathbb{Z}_n$. Thus, $\mathbb{Z}_n$ is an integral domain. $\qquad \square$

Just as we have subrings, it will also be necessary to know about subfields.

Definition 8.12. Let F be a field. Then a subring K of F is said to be a **subfield** if it is a field using the same addition and multiplication operations.

Example 8.29. $\mathbb{Q}$ is a subfield of $\mathbb{R}$, which in turn is a subfield of $\mathbb{C}$.

But how do we test if a subset is a subfield?

Theorem 8.12. *Let F be a field. Then a subset S of F is a subfield of F if and only if*

1. $1 \in S$;
2. *if $a, b \in S$, then $a - b \in S$; and*
3. *if $a, b \in S$, and $b \neq 0$, then $ab^{-1} \in S$.*

Proof. Suppose that S is a subfield of F. Then S contains an identity $f \neq 0$. We must check that f is 1, the identity of F. But as f is the identity for S, we have $ff = f$. Now, f is a unit in F, so multiplying by f^{-1}, we get $f = 1$. Thus, (1) is proved. Since S is a subring of F, (2) follows from Theorem 8.5. As S is a field, every element except 0 has an inverse. This inverse is unique, as $U(F)$ is a group. Therefore, if $0 \neq b \in S$, then $b^{-1} \in S$. Since S is a subring, we get (3) as well.

Conversely, suppose that (1)–(3) hold. In view of (1) and (2), we see that $0 = 1 - 1 \in S$. Take any $a, b \in S$. By (2), $a - b \in S$. If $b = 0$, then $ab = 0 \in S$. Otherwise, we have $b^{-1} = 1b^{-1} \in S$, and therefore $ab = a(b^{-1})^{-1} \in S$. By Theorem 8.5, S is a subring of F. It certainly has an identity $1 \neq 0$, and it is commutative, since F is. Thus, it remains only to check that every nonzero element has an inverse in S. But we just did that! If $0 \neq b \in S$, then $b^{-1} = 1b^{-1} \in S$. Therefore, S is indeed a subfield of F. $\qquad \square$

A small word of caution. It is not sufficient to replace (1) with the condition that S is not empty; indeed, if we did so, then we would accept $\{0\}$ as a field, which is wrong. It would be sufficient to assume that S contains a nonzero element b, for then (3) would give $1 = bb^{-1} \in S$.

Example 8.30. Let $F = \{a + b\sqrt{2} : a, b \in \mathbb{Q}\}$. We claim that F is a subfield of $\mathbb{R}$. Let us check the conditions. Certainly $1 = 1 + 0\sqrt{2} \in F$, so (1) holds. If $a_i, b_i \in \mathbb{Q}$, then $(a_1 + b_1\sqrt{2}) - (a_2 + b_2\sqrt{2}) = (a_1 - a_2) + (b_1 - b_2)\sqrt{2} \in F$, and we have (2). Let us check the final condition. To begin with, we shall show that F is closed under multiplication. But $(a_1 + b_1\sqrt{2})(a_2 + b_2\sqrt{2}) = (a_1a_2 + 2b_1b_2) + (a_1b_2 + a_2b_1)\sqrt{2} \in$

F. Thus, if we can show that every nonzero element of F has an inverse in F, then we will be done, as we can obtain (3). Take $0 \neq a + b\sqrt{2} \in F$. If $b = 0$, then $0 \neq a \in \mathbb{Q}$, and certainly $a^{-1} \in \mathbb{Q} \subseteq F$. Assume that $b \neq 0$. Notice that $(a + b\sqrt{2})(a - b\sqrt{2}) = a^2 - 2b^2 \in \mathbb{Q}$. Also, $a^2 - 2b^2 \neq 0$. Otherwise, we would have $(ab^{-1})^2 = 2$, meaning that $\sqrt{2}$ is rational, which is not the case. Thus, $a^2 - 2b^2$ has an inverse $c \in \mathbb{Q}$. But then $(a + b\sqrt{2})(ac - bc\sqrt{2}) = (a^2 - 2b^2)c = 1$. Hence, $a + b\sqrt{2}$ has an inverse in F, and F is a subfield of $\mathbb{R}$.

Exercises

8.27. Let $R = \{a + bi : a, b \in \mathbb{Q}\}$. Show that R is a subfield of $\mathbb{C}$.

8.28. For each of the following rings, which elements are units? Which are zero divisors?

1. $\mathbb{Z}_{18}$
2. $\mathbb{Z}_3 \oplus \mathbb{Z}_9$

8.29. Let R and S be rings with identity. Show that $U(R \oplus S) = U(R) \times U(S)$.

8.30. Let $n \geq 2$ be a positive integer. Show that $U(\mathbb{Z}_n) = U(n)$.

8.31. Show that every integral domain contains exactly two elements a satisfying $a^2 = a$.

8.32. Let R and S be rings. Under precisely what circumstances is $R \oplus S$ an integral domain?

8.33. Let F be a field with subfields K and L. Show that $K \cap L$ is a subfield of F. Extend this to show that the intersection of any collection of subfields is a subfield.

8.34. Let p be a prime and F a field with p^2 elements. Show that F cannot have more than one proper subfield.

8.35. Let R be an integral domain. Suppose that we have $a, b \in R$ such that $a^{13} = b^{13}$ and $a^{10} = b^{10}$. Show that $a = b$.

8.36. Let R be a finite commutative ring having no zero divisors. Show that R is $\{0\}$ or an integral domain.

8.5   The Characteristic of a Ring

One rather important property of a ring is its characteristic. Letting R be a ring, recall that using additive notation, if we have $a \in R$ and some positive integer n, then

$$na = \underbrace{a + a + \cdots + a}_{n \text{ times}} .$$

Definition 8.13. Let R be a ring. Then the **characteristic** of R, denoted char R, is the smallest positive integer n such that $na = 0$ for all $a \in R$. If no such n exists, then char $R = 0$.

Example 8.31. The characteristic of $\mathbb{Z}_n$ is n, as clearly $na = 0$ for any $a \in \mathbb{Z}_n$, whereas no smaller value than n will work if we take $a = 1$.

Example 8.32. The ring of integers has characteristic zero.

In fact, for rings with identity, we only need to look at the identity.

Theorem 8.13. *Let R be a ring with identity. Regarding R as an additive group, if the order of 1 is $n < \infty$, then R has characteristic n. If 1 has infinite order, then R has characteristic zero.*

Proof. If 1 has infinite order, then there is no positive integer n such that $n1 = 0$, and therefore char $R = 0$. Suppose 1 has order $n < \infty$. Then no number $1 \le m < n$ can be the characteristic, as $m1 \ne 0$. But on the other hand, if $a \in R$, then

$$na = \underbrace{a + a + \cdots + a}_{n \text{ times}} = \underbrace{1a + 1a + \cdots + 1a}_{n \text{ times}} = \underbrace{(1 + 1 + \cdots + 1)}_{n \text{ times}}a = 0a = 0.$$

Thus, n is the characteristic.   □

Corollary 8.2. *Let R be a ring with identity. Then every unital subring of R has the same characteristic as R.*

Proof. The same identity has the same order.   □

The corollary does not apply to subrings that are not unital. For instance, if $R = \mathbb{Z}_6$, then char $R = 6$, but taking the subring $S = \{0, 2, 4\}$, we see that char $S = 3$.

In a commutative ring of prime characteristic, we have the following interesting fact.

Theorem 8.14 (Freshman's Dream). *Let R be a commutative ring of prime characteristic p. Then for any $a, b \in R$, we have*

$$(a + b)^p = a^p + b^p.$$

Proof. Let us apply the Binomial Theorem. (We are really only familiar with it for real numbers, but the proof in any commutative ring is the same.) We have

$$(a + b)^p = a^p + \binom{p}{1}a^{p-1}b + \binom{p}{2}a^{p-2}b^2 + \cdots + \binom{p}{p-1}ab^{p-1} + b^p.$$

Now, if $1 < k < p$, then

$$\binom{p}{k} = \frac{p!}{(p-k)!k!}.$$

148   8 Introduction to Rings

Notice that the numerator is divisible by p. However, p does not divide any of the terms in the denominator and, therefore, it does not divide the denominator. Thus, p divides each $\binom{p}{k}$, with $1 < k < p$. As our ring has characteristic p, multiplying any element by p, and hence by any multiple of p, gives 0. We have our result. □

We tend to encounter commutative rings with prime characteristic a lot in the context of integral domains.

Theorem 8.15. *The characteristic of an integral domain is either zero or a prime.*

Proof. Let R be an integral domain. There is nothing to do if char $R = 0$, so let char $R = n > 0$. We cannot have $n = 1$, for then 1 has additive order 1, but only 0 has that order. The only remaining problem is if n is composite. Suppose that $n = kl$, with $1 < k, l < n$. Then we have

$$0 = n1 = (kl)1 = \underbrace{1 + \cdots + 1}_{kl \text{ times}} = \underbrace{(1 + \cdots + 1)}_{k \text{ times}}\underbrace{(1 + \cdots + 1)}_{l \text{ times}} = (k1)(l1).$$

Since R is an integral domain, $k1 = 0$ or $l1 = 0$. But the additive order of 1 is n, and we have a contradiction. □

Exercises

8.37. Find the characteristic of each of the following rings.

1. $3\mathbb{Z}_{21} = \{0, 3, \ldots, 18\}$
2. $\mathbb{R}[x]$

8.38. Find the characteristic of each of the following rings.

1. $\mathbb{Z}_4 \oplus \mathbb{Z}_{10}$
2. $M_2(\mathbb{Z}_3)$

8.39. Show that a finite integral domain R must have order p^n for some prime p and positive integer n.

8.40. Let F be a field of prime characteristic p. Show that for every positive integer n, $\{a \in F : a^{p^n} = a\}$ is a subfield of F.

8.41. Let R be a commutative ring with identity, and suppose that $a \in R$ satisfies $a^n = 0$ for some positive integer n.

1. Show that $1 + a \in U(R)$.
2. If char R is prime, show that $1 + a$ has finite order in $U(R)$.

8.42. Let $F = \{0, 1, a, b\}$ be a field with four elements. Write the addition and multiplication tables for F.

Chapter 9
Ideals, Factor Rings and Homomorphisms

We saw in Chapter 4 that for some purposes, subgroups are not quite good enough; we needed to consider normal subgroups. There is a similar concept in ring theory. In this chapter, we introduce the notion of an ideal, which is a subring with one additional condition imposed. We can then define factor rings and discuss ring homomorphisms and isomorphisms. Along the way, we will mention several important sorts of ideals, including principal, maximal and prime ideals.

9.1 Ideals

When we discussed normal subgroups of a group, our main concern was to find a condition that we could impose in order to make the group operation on a factor group well-defined. Our motivation is the same here. Of course, a subring is necessarily an additive subgroup of a ring, and as the additive group is abelian, we do not have to worry about normality. However, we need an additional condition to make the multiplication operation work properly.

Definition 9.1. Let R be a ring. Then a subring I of R is said to be an **ideal** if $ir, ri \in I$ for all $i \in I$ and $r \in R$. We call this the **absorption property**.

Note that closure under multiplication is not enough. We need to be able to multiply an element of the ideal by any element of the ring and stay within the ideal. Combining the definition with Theorem 8.5, we immediately obtain the following.

Theorem 9.1. *Let R be a ring and I a subset of R. Then I is an ideal if and only if*

1. *$0 \in I$;*
2. *$i - j \in I$ for all $i, j \in I$; and*
3. *$ir, ri \in I$ for all $i \in I, r \in R$.*

© Springer International Publishing AG, part of Springer Nature 2018
G. T. Lee, *Abstract Algebra*, Springer Undergraduate Mathematics Series,
https://doi.org/10.1007/978-3-319-77649-1_9

Example 9.1. Let n be any integer. Then $n\mathbb{Z}$ is an ideal of $\mathbb{Z}$. Indeed, we already know that it is a subring. But also, if $nk \in n\mathbb{Z}$, then for any integer r, $r(nk) = n(rk) \in n\mathbb{Z}$.

Example 9.2. Let I be the set of all polynomials $f(x) \in \mathbb{R}[x]$ such that $f(0) = 0$. We claim that I is an ideal in $\mathbb{R}[x]$. Certainly I contains the zero polynomial. Also, if $f(0) = g(0) = 0$, then $(f - g)(0) = f(0) - g(0) = 0$, hence $f(x) - g(x) \in I$. Also, if $f(0) = 0$ and $h(x) \in \mathbb{R}[x]$, then $h(0)f(0) = h(0)0 = 0$. Hence, $h(x)f(x) \in I$.

Example 9.3. Let I be the set of all polynomials in $\mathbb{Z}[x]$ whose constant term is a multiple of 5. Then I is an ideal. See Exercise 9.2.

Example 9.4. For any ring R, $\{0\}$ and R are ideals of R.

Example 9.5. In $\mathbb{Q}$, the ring of integers is a subring but not an ideal. Indeed, $3 \in \mathbb{Z}$, but $3(1/5) \notin \mathbb{Z}$. Thus, $\mathbb{Z}$ does not have the absorption property.

Actually, this last example is not particularly surprising. Fields do not have interesting ideals, as the following results illustrate.

Theorem 9.2. *Let R be a ring with identity. If an ideal I of R contains a unit, then $I = R$.*

Proof. Let $u \in I$ be a unit. Then by the absorption property, $1 = uu^{-1} \in I$. But then for any $a \in R$, we also have $a = 1a \in I$. Thus, $I = R$. $\qquad\qquad\square$

Corollary 9.1. *Let F be a field. Then the only ideals of F are $\{0\}$ and F.*

Proof. Let I be an ideal of F. If $I = \{0\}$, there is nothing to do, so assume that $0 \neq a \in F$. Then a is a unit, and by the preceding theorem, $I = F$. $\qquad\qquad\square$

There are a number of ways in which we can obtain new ideals from old ones. For instance, let I and J be ideals of a ring R. Then we write $I + J = \{i + j : i \in I, j \in J\}$.

Example 9.6. In $\mathbb{Z}$, if we let $I = 4\mathbb{Z}$ and $J = 6\mathbb{Z}$, then $I + J = 2\mathbb{Z}$. Indeed, if $m \in I + J$, then $m = 4a + 6b = 2(2a + 3b)$, for some integers a and b. In particular, $I + J \subseteq 2\mathbb{Z}$. On the other hand, for any $c \in \mathbb{Z}$, $2c = 4(-c) + 6c \in I + J$, and hence $2\mathbb{Z} \subseteq I + J$. We can say something more general here. See Exercise 9.4.

Theorem 9.3. *If I and J are ideals of a ring R, then so is $I + J$.*

Proof. As I and J are subgroups of the abelian additive group R, and hence normal, we know from Theorem 4.5 that $I + J$ is an additive subgroup of R. It remains only to check the absorption property. Take $i \in I$, $j \in J$ and $r \in R$. Then $r(i + j) = ri + rj$. Now, $ri \in I$ and $rj \in J$. Thus, $r(i + j) \in I + J$. Similarly, $(i + j)r = ir + jr \in I + J$. $\qquad\qquad\square$

Also, we can define $IJ = \{i_1 j_1 + i_2 j_2 + \cdots + i_n j_n : i_k \in I, j_k \in J, n \in \mathbb{N}\}$. (We cannot simply take terms of the form ij with $i \in I$ and $j \in J$, as sums of terms of that form cannot necessarily be written in the same form.)

Theorem 9.4. *Let I and J be ideals in a ring R. Then IJ is also an ideal.*

Proof. Clearly $0 = 0 \cdot 0 \in IJ$. If we have $i_k \in I$, $j_k \in J$, then

$$(i_1 j_1 + \cdots + i_m j_m) - (i_{m+1} j_{m+1} + \cdots + i_n j_n)$$
$$= i_1 j_1 + \cdots + i_m j_m + (-i_{m+1}) j_{m+1} + \cdots + (-i_n) j_n \in IJ.$$

Also, for any $r \in R$,

$$r(i_1 j_1 + \cdots + i_m j_m) = (r i_1) j_1 + \cdots + (r i_m) j_m \in IJ$$

since $r i_1, \ldots, r i_m \in I$ and, similarly,

$$(i_1 j_1 + \cdots + i_m j_m) r = i_1 (j_1 r) + \cdots + i_m (j_m r) \in IJ.$$

$\square$

Notice that I and J are both subsets of $I + J$, as we can take $i + 0$ and $0 + j$ for any $i \in I$, $j \in J$. But by the absorption property, $IJ \subseteq I \cap J$.

One type of ideal is particularly important.

Definition 9.2. Let R be a commutative ring with identity and $a \in R$. Then the **principal ideal** generated by a, denoted (a), is the set $\{ra : r \in R\}$.

Example 9.7. In $\mathbb{Z}$, we have $(n) = n\mathbb{Z}$ for any $n \in \mathbb{Z}$.

Example 9.8. The ideal from Example 9.2 is (x). Indeed, if $f(x) \in \mathbb{R}[x]$, then $f(0) = 0$ if and only if the constant term is 0; that is, if and only if the polynomial is a multiple of x.

Theorem 9.5. *If R is a commutative ring with identity, and $a \in R$, then (a) is an ideal of R; indeed, it is the intersection of all ideals of R containing a.*

Proof. We have $0 = 0a \in (a)$. If $r, s \in R$, then $ra - sa = (r - s)a \in (a)$. Also, if $ra \in (a)$ and $s \in R$, then $s(ra) = (ra)s = (sr)a \in (a)$. Furthermore, if I is an ideal of R containing a, then by the absorption property, $ra \in I$ for all $r \in R$. Thus, (a) is a subset of every ideal containing a. As (a) is an ideal containing a, our result is proved. $\square$

Notice that the preceding proof does not work if R is not a commutative ring with identity. Indeed, if R is not commutative, then $\{ra : r \in R\}$ need not be an ideal; if R does not have an identity, then it may not contain a. See Exercise 9.5.

In a similar fashion, if R is a commutative ring with identity, and $a_1, \ldots, a_n \in R$, we can construct the ideal generated by $a_1, \ldots, a_n$, namely, the set of all elements $r_1 a_1 + \cdots + r_n a_n$, with $r_i \in R$.

Exercises

9.1. List all of the elements in each of the following principal ideals of $\mathbb{Z}_4 \oplus \mathbb{Z}_6$.

1. $((0, 5))$
2. $((2, 3))$

9.2. Let I be the set of all polynomials in $\mathbb{Z}[x]$ whose constant term is a multiple of 5. Show that I is an ideal of $\mathbb{Z}[x]$.

9.3. Show that the intersection of ideals I and J in a ring R is also an ideal. Extend this to the intersection of an arbitrary collection of ideals.

9.4. Let m and n be positive integers. Show that $m\mathbb{Z} + n\mathbb{Z} = (m, n)\mathbb{Z}$.

9.5. Given a ring R and an element a, let $S = \{ra : r \in R\}$. Show by example that

1. S need not contain a, even if R is commutative; and
2. S need not be an ideal, even if R has an identity.

9.6. Let R be a commutative ring. If $I = \{a \in R : a^n = 0 \text{ for some } n \in \mathbb{N}\}$, show that I is an ideal of R.

9.7. Find ideals I and J in a ring R such that $IJ \neq I \cap J$.

9.8. Let R be a commutative ring with identity having exactly two ideals. Show that R is a field.

9.9. Consider the additive group G and subgroup H from Exercise 3.42. Define a multiplication operation on G via $(a_1, a_2, \ldots)(b_1, b_2, \ldots) = (a_1 b_1, a_2 b_2, \ldots)$. Show that G is a ring and H is an ideal.

9.10. In the preceding exercise, show that H is not a principal ideal.

9.2   Factor Rings

Let R be a ring and I an ideal. Then R is an abelian group under addition, and I is a subgroup. Thus, we can consider the left cosets $a + I$, for all $a \in R$. We use these to form a factor ring. Remember that $a + I = b + I$ if and only if $a - b \in I$.

Definition 9.3. Let R be a ring and I an ideal of R. Then the **factor ring** (or **quotient ring**), R/I, is the set of all left cosets $\{a + I : a \in R\}$ together with the operations $(a + I) + (b + I) = a + b + I$ and $(a + I)(b + I) = ab + I$, for all $a, b \in R$.

Theorem 9.6. *For any ring R and ideal I, the factor ring R/I is a ring.*

Proof. Since R is an abelian group under addition, I is necessarily a normal subgroup. Thus, we know from Theorem 4.6 that R/I is a group under addition. By Theorem 4.7, it is abelian.

Let us show that the multiplication operation is well-defined. Suppose that $a_1 + I = a_2 + I$ and $b_1 + I = b_2 + I$. Then notice that

$$a_1 b_1 - a_2 b_2 = a_1 b_1 - a_1 b_2 + a_1 b_2 - a_2 b_2 = a_1 (b_1 - b_2) + (a_1 - a_2) b_2.$$

Now, $b_1 - b_2 \in I$. Thus, by absorption, $a_1 (b_1 - b_2) \in I$. Similarly, $a_1 - a_2 \in I$, and hence $(a_1 - a_2) b_2 \in I$. Thus, $a_1 b_1 - a_2 b_2 \in I$, and therefore $a_1 b_1 + I = a_2 b_2 + I$. That is, the multiplication operation is well-defined.

We must check the remaining properties of a ring. Take any $a, b, c \in R$. Then since $ab + I \in R/I$, we have closure under multiplication. Also,

$$
\begin{aligned}
(a + I)((b + I)(c + I)) &= (a + I)(bc + I) \\
&= a(bc) + I \\
&= (ab)c + I \\
&= (ab + I)(c + I) \\
&= ((a + I)(b + I))(c + I),
\end{aligned}
$$

and associativity holds. Similarly,

$$
\begin{aligned}
(a + I)((b + I) + (c + I)) &= (a + I)(b + c + I) \\
&= a(b + c) + I \\
&= ab + ac + I \\
&= (ab + I) + (ac + I) \\
&= (a + I)(b + I) + (a + I)(c + I).
\end{aligned}
$$

The other distributive law is proved in the same fashion. □

Let us discuss a few examples of factor rings.

Example 9.9. Let $R = \mathbb{Z}$ and $I = (5) = 5\mathbb{Z}$. Then $R/I = \{0 + I, 1 + I, 2 + I, 3 + I, 4 + I\}$ and, for instance, $(2 + I) + (4 + I) = 6 + I = 1 + I$ and $(3 + I)(4 + I) = 12 + I = 2 + I$.

Example 9.10. Let $R = M_2(\mathbb{Z})$ and let I be the ideal consisting of all matrices whose entries are even. Then notice that for any $a_{ij} \in \mathbb{Z}$, we have

$$\begin{pmatrix} a_{11} & a_{12} \\ a_{21} & a_{22} \end{pmatrix} + I = \begin{pmatrix} b_{11} & b_{12} \\ b_{21} & b_{22} \end{pmatrix} + I,$$

where b_{ij} is 0 if a_{ij} is even and 1 if a_{ij} is odd. Thus, R/I consists of the sixteen different elements $\begin{pmatrix} b_{11} & b_{12} \\ b_{21} & b_{22} \end{pmatrix} + I, b_{ij} \in \{0, 1\}$. We perform arithmetic in the following fashion:

$$\left(\begin{pmatrix} 1 & 0 \\ 1 & 1 \end{pmatrix} + I \right) + \left(\begin{pmatrix} 1 & 1 \\ 0 & 1 \end{pmatrix} + I \right) = \begin{pmatrix} 2 & 1 \\ 1 & 2 \end{pmatrix} + I = \begin{pmatrix} 0 & 1 \\ 1 & 0 \end{pmatrix} + I$$

and

$$\left(\begin{pmatrix} 1 & 0 \\ 1 & 1 \end{pmatrix} + I\right)\left(\begin{pmatrix} 1 & 1 \\ 0 & 1 \end{pmatrix} + I\right) = \begin{pmatrix} 1 & 1 \\ 1 & 2 \end{pmatrix} + I = \begin{pmatrix} 1 & 1 \\ 1 & 0 \end{pmatrix} + I.$$

Example 9.11. Let $R = \mathbb{R}[x]$ and $I = (x^2 + 3)$. Readers familiar with polynomial long division will know that if $f(x) \in R$, then $f(x) = (x^2 + 3)q(x) + r(x)$, where q and r are polynomials, with $r(x) = a + bx$, for some $a, b \in \mathbb{R}$. (Those who are unfamiliar with polynomial long division can peek ahead to Section 10.1, where it will be discussed in more generality.) Since $(x^2 + 3)q(x) \in I$ by absorption, we know that elements of R/I are of the form $a + bx + I$, with $a, b \in \mathbb{R}$. Addition behaves as expected; for instance, $(2 + 3x + I) + (7 - 4x + I) = 9 - x + I$. To deal with multiplication, observe that $x^2 - (-3) \in I$; thus, $x^2 + I = -3 + I$. Therefore, we have calculations such as

$$(5 + 4x + I)(-7 + 2x + I) = -35 - 18x + 8x^2 + I$$
$$= -35 - 18x + 8(-3) + I$$
$$= -59 - 18x + I.$$

Let us also record a few basic facts about factor rings.

Theorem 9.7. *Let R be a ring and I an ideal. Then*

1. *if R is commutative, then so is R/I;*
2. *if R has an identity, then so does R/I; and*
3. *if u is a unit of R, then $u + I$ is a unit of R/I.*

Proof. (1) If $a, b \in R$, then $(a + I)(b + I) = ab + I = ba + I = (b + I)(a + I)$.
   (2) If $a \in R$, then $(1 + I)(a + I) = a + I = (a + I)(1 + I)$. Hence, $1 + I$ is the identity of R/I.
   (3) Observe that $(u + I)(u^{-1} + I) = 1 + I = (u^{-1} + I)(u + I)$; thus, $(u + I)^{-1} = u^{-1} + I$. $\square$

Theorem 9.8. *Let R be a ring with ideals I and J, such that $I \subseteq J$. Then J/I is an ideal of R/I.*

Proof. We see that I is a subring of J, and since I enjoys the absorption property in R, it enjoys it in J as well. Thus, I is an ideal of J, and so J/I makes sense. Now, $0 \in J$, and therefore $0 + I \in J/I$. If $j_1, j_2 \in J$, then $(j_1 + I) - (j_2 + I) = (j_1 - j_2) + I \in J/I$. Also, if $j \in J$ and $r \in R$, then $(r + I)(j + I) = rj + I \in J/I$, since $rj \in J$. Similarly, $(j + I)(r + I) = jr + I \in J/I$. The proof is complete. $\square$

Exercises

9.11. Let $\mathbb{R} = \mathbb{Z}$ and $I = (5)$. Write the addition and multiplication tables for R/I.

9.12. Let $\mathbb{R} = \mathbb{Z}_6 \oplus \mathbb{Z}_4$ and $I = ((4, 2))$. Write the addition and multiplication tables for R/I.

9.13. Let $R = \mathbb{R}[x]$ and $f(x) = x^3 + 6x^2 + 2$. If $I = (f(x))$, calculate the product $(4x^2 + 3x + 1 + I)(2x^2 - x + 2 + I)$ in R/I. Reduce the answer to the form $ax^2 + bx + c + I$, for some $a, b, c \in \mathbb{R}$.

9.14. Let R be a ring and I an ideal. Show that R/I is commutative if and only if $ab - ba \in I$ for all $a, b \in R$.

9.15. Let I and J be ideals in a ring R. Show that R/I and R/J are both commutative rings if and only if $R/(I \cap J)$ is commutative.

9.16. Let R be a ring and I a proper ideal.

1. If R is an integral domain, does it follow that R/I is an integral domain? Prove that it does, or find a counterexample.
2. If R/I is an integral domain, does it follow that R is an integral domain? Prove that it does, or find a counterexample.

9.17. If F is a field of order 81, what are the possible orders of F/I, where I is an ideal of F?

9.18. Let $I_1 \subseteq I_2 \subseteq I_3 \subseteq \cdots$ be ideals of R. Let $I = \bigcup_{n=1}^{\infty} I_n$.

1. Show that I is an ideal of R.
2. Suppose that R/I is commutative. Show that for every $a, b \in R$, there exists an $n \in \mathbb{N}$ such that $ab - ba \in I_n$.

9.19. Let R and I be as in Exercise 9.6. Define the analogous ideal for R/I, namely $\{a + I \in R/I : (a + I)^n = 0 + I \text{ for some } n \in \mathbb{N}\}$. Show that this ideal is $\{0 + I\}$.

9.20. If I is an ideal of a ring R, show that the subrings of R/I are precisely of the form S/I, where S is a subring of R containing I. Further show that S/I is an ideal of R/I if and only if S is an ideal of R.

9.3 Ring Homomorphisms

We recall that a group homomorphism is a function from one group to another that respects the group operation. There is a similar concept for rings, but both of the ring operations must be respected.

Definition 9.4. Let R and S be rings. Then a **ring homomorphism** (or, simply, **homomorphism**) from R to S is a function $\alpha : R \rightarrow S$ satisfying

$$\alpha(r_1 + r_2) = \alpha(r_1) + \alpha(r_2)$$

and

$$\alpha(r_1 r_2) = \alpha(r_1)\alpha(r_2)$$

for all $r_1, r_2 \in R$.

Thus, a ring homomorphism is a homomorphism of additive groups, with the additional property that it respects the multiplication operation. The kernel is the same as the kernel of the additive group homomorphism.

Definition 9.5. Let $\alpha : R \to S$ be a ring homomorphism. Then the **kernel** of α is

$$\ker(\alpha) = \{r \in R : \alpha(r) = 0\}.$$

Example 9.12. Let $n \geq 2$ be a positive integer. Then $\alpha : \mathbb{Z} \to \mathbb{Z}_n$ given by $\alpha(a) = [a]$ (where we insert the square brackets for clarity) is a homomorphism. Indeed, by Example 4.10, it respects the addition operation. Also, for any $a, b \in \mathbb{Z}$, $\alpha(ab) = [ab] = [a][b] = \alpha(a)\alpha(b)$. Furthermore, by Example 4.10, $\ker(\alpha) = n\mathbb{Z}$.

Example 9.13. Define $\alpha : \mathbb{C} \to \mathbb{C}$ via $\alpha(a + bi) = a - bi$, for all $a, b \in \mathbb{R}$. Then notice that
$$\alpha((a + bi) + (c + di)) = \alpha((a + c) + (b + d)i)$$
$$= a + c - (b + d)i$$
$$= (a - bi) + (c - di)$$
$$= \alpha(a + bi) + \alpha(c + di),$$

for all $a, b, c, d \in \mathbb{R}$. Similarly,

$$\alpha((a + bi)(c + di)) = \alpha((ac - bd) + (ad + bc)i)$$
$$= ac - bd - (ad + bc)i$$
$$= (a - bi)(c - di)$$
$$= \alpha(a + bi)\alpha(c + di).$$

Thus, α is a homomorphism. Also, if $\alpha(a + bi) = 0$, then $a - bi = 0$, and hence $a = b = 0$. Thus, $\ker(\alpha) = \{0\}$.

Example 9.14. If R and S are any rings, then $\alpha : R \to S$ given by $\alpha(r) = 0$ for all $r \in R$ is a homomorphism. Indeed, by Example 4.13, it is an additive group homomorphism, and $\alpha(r_1 r_2) = 0 = 0 \cdot 0 = \alpha(r_1)\alpha(r_2)$, for all $r_1, r_2 \in R$. The kernel of α is R.

Let us record a few basic properties of ring homomorphisms. If $\alpha : R \to S$ is a ring homomorphism, then as with group homomorphisms, we write $\alpha(M) = \{\alpha(m) : m \in M\}$ and $\alpha^{-1}(N) = \{r \in R : \alpha(r) \in N\}$, for any subring M of R, and any subring N of S.

Theorem 9.9. *Let $\alpha : R \to S$ be a ring homomorphism. Then*

1. $\ker(\alpha)$ *is an ideal of R;*
2. α *is one-to-one if and only if $\ker(\alpha) = \{0\}$;*
3. $\alpha(0) = 0$; *and*
4. *if R is a ring with identity, then so is $\alpha(R)$, and $\alpha(1)$ is the identity of $\alpha(R)$.*

Proof. (1) By Theorem 4.11, ker(α) is an additive subgroup of R. It remains to check the absorption property. But if $k \in$ ker(α) and $r \in R$, then $\alpha(rk) = \alpha(r)\alpha(k) = \alpha(r)0 = 0$, and hence $rk \in$ ker(α). Similarly, $kr \in$ ker(α).

(2) See Theorem 4.11.

(3) This follows immediately from Theorem 4.10.

(4) If $r \in R$, then $\alpha(1)\alpha(r) = \alpha(1r) = \alpha(r) = \alpha(r1) = \alpha(r)\alpha(1)$. $\qquad\square$

We must be a bit careful with the final part of the last theorem. It certainly does not follow that $\alpha(1)$ is the identity of S, as the following example illustrates.

Example 9.15. Define $\alpha : \mathbb{R} \to M_2(\mathbb{R})$ via

$$\alpha(r) = \begin{pmatrix} r & 0 \\ 0 & 0 \end{pmatrix}$$

for all $r \in \mathbb{R}$. It is easy to verify that α is a homomorphism. Now, $\alpha(1)$ is the identity of $\alpha(\mathbb{R})$, but not of $M_2(\mathbb{R})$.

Theorem 9.10. *Let $\alpha : R \to S$ be a ring homomorphism. Let M be a subring of R and N a subring of S. Then*

1. *$\alpha(M)$ is a subring of S;*
2. *if M is an ideal of R, then $\alpha(M)$ is an ideal of $\alpha(R)$;*
3. *$\alpha^{-1}(N)$ is a subring of R; and*
4. *if N is an ideal of S, then $\alpha^{-1}(N)$ is an ideal of R.*

Proof. (1) By Theorem 4.12, $\alpha(M)$ is an additive subgroup of S. If $m_1, m_2 \in M$, then $\alpha(m_1)\alpha(m_2) = \alpha(m_1m_2) \in \alpha(M)$, since $m_1m_2 \in M$.

(2) By (1), it remains only to check the absorption property. If $m \in M, r \in R$, then $\alpha(r)\alpha(m) = \alpha(rm) \in \alpha(M)$, since $rm \in M$. Similarly, $\alpha(m)\alpha(r) \in \alpha(M)$.

(3) By Theorem 4.12, $\alpha^{-1}(N)$ is an additive subgroup of R. If $r_1, r_2 \in \alpha^{-1}(N)$, then $\alpha(r_1r_2) = \alpha(r_1)\alpha(r_2) \in N$, since $\alpha(r_1), \alpha(r_2) \in N$. Thus, $r_1r_2 \in \alpha^{-1}(N)$.

(4) In view of (3), we only need to check the absorption property. Take $a \in \alpha^{-1}(N)$ and $r \in R$. Then $\alpha(ra) = \alpha(r)\alpha(a) \in N$, since $\alpha(a) \in N$, and N is an ideal of S. Therefore, $ra \in \alpha^{-1}(N)$. Similarly, $ar \in \alpha^{-1}(N)$. $\qquad\square$

Once again, we note that the second part of the preceding theorem does not say that $\alpha(M)$ is necessarily an ideal of S.

One more homomorphism will prove useful later.

Theorem 9.11. *Let R be a ring with identity of characteristic n. Then there is a homomorphism $\alpha : \mathbb{Z} \to R$ with kernel (n).*

Proof. Define $\alpha : \mathbb{Z} \to R$ via $\alpha(k) = k1$, for all $k \in \mathbb{Z}$. Let us check that α is a homomorphism. If $j, k \in \mathbb{Z}$, then $\alpha(j+k) = (j+k)1 = j1+k1 = \alpha(j)+\alpha(k)$. (This is Theorem 3.6, using additive notation.) Also, $\alpha(jk) = (jk)1$, whereas $\alpha(j)\alpha(k) = (j1)(k1)$. Again by Theorem 3.6, $(jk)1 = j(k1)$. If $j > 0$, then

$$(j1)(k1) = \underbrace{(1 + \cdots + 1)}_{j \text{ times}}(k1) = \underbrace{k1 + \cdots + k1}_{j \text{ times}} = j(k1).$$

If $j < 0$, then $(jk)1 = ((-j)(-k))1$ and $(j1)(k1) = (-j1)(-k1)$, and we can use the $j > 0$ argument. If $j = 0$, then $(jk)1 = (j1)(k1) = 0$. Thus, in any case, $\alpha(jk) = \alpha(j)\alpha(k)$, and α is a homomorphism.

We note that $k \in \ker(\alpha)$ if and only if $k1 = 0$. If $n = 0$, then by Theorem 8.13, 1 has infinite additive order. Therefore, the only solution is $k = 0$; thus, $\ker(\alpha) = (0)$. If $n > 0$, then by Theorem 8.13, n is the additive order of 1. Furthermore, by Corollary 3.2, $k1 = 0$ if and only if the additive order of 1 divides k; that is, if and only if n divides k. In other words, the kernel is the set of multiples of n. The proof is complete.   □

Exercises

9.21. Decide if each of the following functions is a ring homomorphism.

1. $\alpha : \mathbb{Z} \to \mathbb{R}, \alpha(a) = 2a$.
2. $\alpha : \mathbb{R}[x] \to \mathbb{R}, \alpha(f(x)) = f(2)$.

9.22. Decide if each of the following functions is a ring homomorphism.

1. $\alpha : M_2(\mathbb{R}) \to M_2(\mathbb{R}), \alpha(A) = \begin{pmatrix} 1 & 1 \\ 0 & 1 \end{pmatrix} A \begin{pmatrix} 1 & -1 \\ 0 & 1 \end{pmatrix}$.
2. $\alpha : M_2(\mathbb{R}) \to \mathbb{R}, \alpha(A) = \det(A)$.

9.23. Let $\alpha : R \to S$ and $\beta : S \to T$ be ring homomorphisms. Show that $\beta\alpha : R \to T$ is also a ring homomorphism.

9.24. Let $\alpha : R \to S$ and $\beta : S \to T$ be ring homomorphisms. Show that $\ker(\alpha) \subseteq \ker(\beta\alpha)$. If β is one-to-one, show that $\ker(\beta\alpha) = \ker(\alpha)$.

9.25. Define $\alpha : \mathbb{Z} \oplus \mathbb{Z} \to \mathbb{Z} \oplus \mathbb{Z}$ via $\alpha((a, b)) = (a, 0)$. Is this a ring homomorphism? If so, find $\ker(\alpha)$ and $\alpha^{-1}(2\mathbb{Z} \oplus 3\mathbb{Z})$.

9.26. Define $\alpha : \mathbb{Z}_8 \to \mathbb{Z}_{16}$ via $\alpha([a]) = [a]$, for all $a \in \{0, 1, \ldots, 7\}$, where the square brackets represent the congruence classes. Is this a ring homomorphism? If so, find $\ker(\alpha)$ and $\alpha^{-1}([3])$.

9.27. Let R be a ring and I an ideal. Show that there exist a ring S and a homomorphism $\alpha : R \to S$ such that $\ker(\alpha) = I$.

9.28. Let R be a commutative ring with prime characteristic p. Show that $\alpha : R \to R$ given by $\alpha(a) = a^p$ is a ring homomorphism.

9.29. Let F be a field of order 16 and K a field of order 4. Find all homomorphisms from F to K.

9.30. Find all homomorphisms from $\mathbb{Z}$ to $\mathbb{Q}$.

9.4 Isomorphisms and Automorphisms

As with groups, we use isomorphisms to establish if two rings have the same structure.

Definition 9.6. Let R and S be rings. Then a **ring isomorphism** (or, simply, **isomorphism**) is a bijective homomorphism from R to S. When such an isomorphism exists, we say that R and S are **isomorphic** rings.

Example 9.16. Consider the function $\alpha : \mathbb{Z}_{15} \to \mathbb{Z}_3 \oplus \mathbb{Z}_5$ given by $\alpha(a) = (a, a)$ for all a. By Example 4.16, this is an isomorphism of additive groups. We claim that it is, in fact, a ring isomorphism. All that remains is to show that α respects multiplication. But if $a, b \in \mathbb{Z}_{15}$, then $\alpha(ab) = (ab, ab) = (a, a)(b, b) = \alpha(a)\alpha(b)$. Thus, $\mathbb{Z}_{15}$ and $\mathbb{Z}_3 \oplus \mathbb{Z}_5$ are isomorphic rings.

We must not, however, make the mistake of thinking that rings that are isomorphic as additive groups are necessarily isomorphic as rings!

Example 9.17. Let $R = \mathbb{Z}$ and $S = 5\mathbb{Z}$. As additive groups, these are isomorphic, since $5\mathbb{Z}$ is infinite cyclic, being generated by 5, and we can apply Theorem 4.14. However, as rings, there cannot even be an onto homomorphism from R to S. Why not? If there were, by Theorem 9.9, 1 would have to map to the identity of S, which is sadly lacking. Thus, R and S are not isomorphic rings.

Example 9.18. Define $\alpha : \mathbb{C} \to M_2(\mathbb{R})$ via

$$\alpha(a + bi) = \begin{pmatrix} a & -b \\ b & a \end{pmatrix},$$

for all $a, b \in \mathbb{R}$. Let us check that α is a homomorphism. If $a, b, c, d \in \mathbb{R}$, then

$$\alpha((a + bi) + (c + di)) = \alpha((a + c) + (b + d)i)$$
$$= \begin{pmatrix} a + c & -(b + d) \\ b + d & a + c \end{pmatrix}$$
$$= \alpha(a + bi) + \alpha(c + di).$$

Also,

$$\alpha((a + bi)(c + di)) = \alpha((ac - bd) + (ad + bc)i) = \begin{pmatrix} ac - bd & -(ad + bc) \\ ad + bc & ac - bd \end{pmatrix}$$

whereas

$$\alpha(a + bi)\alpha(c + di) = \begin{pmatrix} a & -b \\ b & a \end{pmatrix} \begin{pmatrix} c & -d \\ d & c \end{pmatrix},$$

and these are the same. Clearly, $\ker(\alpha) = \{0\}$, so α is one-to-one. Now, α is not onto, but we see that $\mathbb{C}$ is isomorphic to the image $\alpha(\mathbb{C})$, namely

$$\left\{ \begin{pmatrix} a & -b \\ b & a \end{pmatrix} : a, b \in \mathbb{R} \right\}.$$

Let us discuss a few properties of isomorphisms.

Theorem 9.12. *On any collection of rings, isomorphism is an equivalence relation.*

Proof. Reflexivity: The function $\alpha : R \to R$ given by $\alpha(a) = a$ for all a is an isomorphism. Symmetry: Let $\alpha : R \to S$ be an isomorphism. By Theorem 4.13, the inverse $\alpha^{-1} : S \to R$ is an isomorphism of additive groups. We only need to check that it respects multiplication. Take any $s_1, s_2 \in S$, and suppose that $\alpha^{-1}(s_i) = r_i$. Then $\alpha(r_1 r_2) = \alpha(r_1)\alpha(r_2) = s_1 s_2$; that is, $\alpha^{-1}(s_1 s_2) = r_1 r_2 = \alpha^{-1}(s_1)\alpha^{-1}(s_2)$. Transitivity: Let $\alpha : R \to S$ and $\beta : S \to T$ be isomorphisms. By Theorem 4.13, $\beta \circ \alpha : R \to T$ is an isomorphism of additive groups. Again, we must check that it respects multiplication. Take $r_1, r_2 \in R$. Then

$$(\beta \circ \alpha)(r_1 r_2) = \beta(\alpha(r_1 r_2)) = \beta(\alpha(r_1)\alpha(r_2)) = \beta(\alpha(r_1))\beta(\alpha(r_2)).$$

The proof is complete.   □

Theorem 9.13. *Let $\alpha : R \to S$ be a ring isomorphism. Then*

1. *if R is commutative, then so is S;*
2. *if R has an identity, then so does S;*
3. *if R is an integral domain, then so is S; and*
4. *if R is a field, then so is S.*

Proof. (1) Take $s_1, s_2 \in S$. Then $s_i = \alpha(r_i)$, for some $r_i \in R$. Thus,

$$s_1 s_2 = \alpha(r_1)\alpha(r_2) = \alpha(r_1 r_2) = \alpha(r_2 r_1) = \alpha(r_2)\alpha(r_1) = s_2 s_1.$$

(2) Use Theorem 9.9 and the fact that α is onto.

(3) By (1) and (2), S is a commutative ring with identity. If $1 = 0$ in S, then $S = \{0\}$. As α is bijective, $R = \{0\}$, which is impossible. It remains only to check for zero divisors. Suppose that s_1 and s_2 are nonzero elements of S with $s_1 s_2 = 0$. Let us say $\alpha(r_i) = s_i$. Then $0 = s_1 s_2 = \alpha(r_1)\alpha(r_2) = \alpha(r_1 r_2)$. As α is one-to-one, $r_1 r_2 = 0$. But R is an integral domain, so either $r_1 = 0$ or $r_2 = 0$, which means that $s_1 = 0$ or $s_2 = 0$, giving us a contradiction.

(4) Once again, S is a commutative ring with identity and $1 \neq 0$. Suppose that $0 \neq s \in S$. Then $s = \alpha(r)$, for some $r \in R$. Now, $r \neq 0$, so r has an inverse in R. Since α is onto, we know that $\alpha(1) = 1$. Thus,

$$1 = \alpha(1) = \alpha(rr^{-1}) = \alpha(r)\alpha(r^{-1}) = s\alpha(r^{-1}).$$

That is, $\alpha(r^{-1}) = s^{-1}$, and every nonzero element of S has an inverse.   □

Example 9.19. The rings $2\mathbb{Z}$, $M_2(\mathbb{R})$, $\mathbb{Z}_6$, $\mathbb{Z}$ and $\mathbb{Q}$ are all nonisomorphic. Indeed, $2\mathbb{Z}$ does not have an identity, so it cannot be isomorphic to any of the others, which do have an identity. Also, all of the rings are commutative except for $M_2(\mathbb{R})$, so it is ruled out. Next, $\mathbb{Z}_6$ is not an integral domain, but the remaining two are. (Also, it is finite and the others are infinite.) Finally, $\mathbb{Q}$ is a field but $\mathbb{Z}$ is not.

Given a ring R, we might well ask if it is a subring of a field. For example, $\mathbb{Z}$ is a subring of $\mathbb{Q}$. But not every ring can be a subring of a field. Indeed, a noncommutative ring or a ring containing zero divisors cannot exist inside a field. So, it seems that integral domains are a good place to start. In fact, if R is an integral domain, then we can construct a field F containing an isomorphic copy of R. The method we will use may seem somewhat familiar; actually, it is exactly the way in which $\mathbb{Q}$ is constructed from $\mathbb{Z}$. We will need something comparable to numerators and denominators. Also, the denominator must not be zero. Furthermore, we need some way of recognizing that $10/25 = 8/20$, for instance. This gives us an idea of how to proceed.

Let R be an integral domain. Then let S be the Cartesian product $R \times (R\backslash\{0\})$. Note that S is only a set, not a ring. (If $R = \mathbb{Z}$, for instance, when we look at $(10, 25) \in S$, we are thinking of the fraction $10/25$.) Let us define a relation $\sim$ on S via $(a, b) \sim (c, d)$ if and only if $ad = bc$. (Continuing our parenthetical thought, $(10, 25) \sim (8, 20)$.)

We claim that $\sim$ is an equivalence relation. Reflexivity: As R is commutative, we see that $(a, b) \sim (a, b)$. Symmetry: Suppose that $(a, b) \sim (c, d)$. Then $ad = bc$ and again, by commutativity, $(c, d) \sim (a, b)$. Transitivity: Suppose that $(a, b) \sim (c, d)$ and $(c, d) \sim (e, f)$. Then $ad = bc$, and hence $adf = bcf$. Also, $cf = de$, and hence $bcf = bde$. Thus, $adf = bde$. Now, $d \neq 0$ and R is an integral domain. Thus, we have $af = be$, and hence $(a, b) \sim (e, f)$. For the sake of simplicity, write $[a, b]$ for the equivalence class of (a, b). (In our example in $\mathbb{Z}$, we have $[10, 25] = [8, 20]$, and this is the set of all pairs (a, b), with $a, b \in \mathbb{Z}$, $b \neq 0$ and $10b = 25a$.) Let us write F for the set of all equivalence classes of S.

The addition and multiplication operations on F work precisely as we would expect with fractions. Specifically,

$$[a, b] + [c, d] = [ad + bc, bd]$$

and

$$[a, b][c, d] = [ac, bd].$$

We must verify that these operations are well-defined. Suppose that $[a_1, b_1] = [a, b]$ and $[c_1, d_1] = [c, d]$. Then $[a_1, b_1] + [c_1, d_1] = [a_1d_1 + b_1c_1, b_1d_1]$. But

$$(ad + bc)(b_1d_1) - (a_1d_1 + b_1c_1)bd = dd_1(ab_1 - a_1b) + bb_1(cd_1 - c_1d)$$
$$= dd_1(0) + bb_1(0)$$
$$= 0;$$

thus, $[a, b] + [c, d] = [a_1, b_1] + [c_1, d_1]$. Similarly, $[a_1, b_1][c_1, d_1] = [a_1c_1, b_1d_1]$, and

$$acb_1d_1 - a_1c_1bd = (acb_1d_1 - a_1cbd_1) + (a_1cbd_1 - a_1c_1bd)$$
$$= cd_1(ab_1 - a_1b) + a_1b(cd_1 - c_1d)$$
$$= cd_1(0) + a_1b(0)$$
$$= 0;$$

thus, $[a, b][c, d] = [a_1, b_1][c_1, d_1]$. Also, we must note that if b and d are nonzero, then so is bd, since R is an integral domain.

Definition 9.7. Let R be an integral domain. Then the **field of fractions** (or **field of quotients**) of R is the field F constructed above.

Of course, the fact that F is indeed a field needs proving!

Theorem 9.14. *Let R be an integral domain. Then the field of fractions, F, is a field.*

Proof. The proof of this theorem is not difficult. However, there are many steps to complete, as we must verify that F is an abelian group under addition, then that it has all of the remaining properties of a ring, and finally that it is a field. We will prove a few selected properties, and leave the rest to the reader[1] as Exercise 9.38.

Take any $a, b, c, d, e, f \in R$ with b, d and f all nonzero. Let us show that the addition operation on F is associative. We have

$$([a, b] + [c, d]) + [e, f] = [ad + bc, bd] + [e, f]$$
$$= [adf + bcf + bde, bdf]$$
$$= [a, b] + [cf + de, df]$$
$$= [a, b] + ([c, d] + [e, f]),$$

as required.

Next, let us prove a distributive law. Observe that

$$[a, b]([c, d] + [e, f]) = [a, b][cf + de, df] = [acf + ade, bdf]$$

whereas

$$[a, b][c, d] + [a, b][e, f] = [ac, bd] + [ae, bf] = [abcf + abde, b^2df].$$

But as $(acf + ade)b^2df = (abcf + abde)bdf$, these are equal.

Notice that $[0, 1]$ is the additive identity of F and $[1, 1]$ is the multiplicative identity. Let us show that every nonzero element of F has an inverse. Take a nonzero $[a, b] \in F$. Note that $[0, b] = [0, 1]$, so $a \neq 0$. But then $[b, a] \in F$ as well, and $[a, b][b, a] = [ab, ab] = [1, 1]$. Thus, $[b, a] = [a, b]^{-1}$. $\square$

---

[1] Aren't you lucky!

What is the connection between R and F? The idea is that F contains a copy of R and it is, in a sense, the smallest field that does.

Theorem 9.15. *Let R be an integral domain and F its field of fractions. Then F has a subring isomorphic to R. Furthermore, if K is any field having R as a subring, then K has a subfield isomorphic to F, and this subfield contains R.*

Proof. Define $\alpha : R \to F$ via $\alpha(r) = [r, 1]$, for all $r \in R$. We claim that α is a homomorphism. If $r_1, r_2 \in R$, then

$$\alpha(r_1 + r_2) = [r_1 + r_2, 1] = [r_1, 1] + [r_2, 1] = \alpha(r_1) + \alpha(r_2)$$

and

$$\alpha(r_1 r_2) = [r_1 r_2, 1] = [r_1, 1][r_2, 1] = \alpha(r_1)\alpha(r_2),$$

proving the claim. Furthermore, if $r \in \ker(\alpha)$, then $[r, 1] = [0, 1]$, and therefore $r = 0$. Thus, α is one-to-one, and hence R is isomorphic to $\alpha(R)$ which, by Theorem 9.10, is a subring of F.

Also, define $\beta : F \to K$ via $\beta([a, b]) = ab^{-1}$, for all $a, b \in R$ with $b \neq 0$. (Since K is a field, b has an inverse in K. But we still need to check that β is well-defined. Suppose that $[a, b] = [a_1, b_1]$. Then $ab_1 = a_1 b$, and hence $ab^{-1} = a_1 b_1^{-1}$.) Now, let us show that β is a homomorphism. But

$$\begin{aligned}
\beta([a, b] + [c, d]) &= \beta([ad + bc, bd]) \\
&= (ad + bc)(bd)^{-1} \\
&= ab^{-1} + cd^{-1} \\
&= \beta([a, b]) + \beta([c, d]).
\end{aligned}$$

Similarly,

$$\beta([a, b][c, d]) = \beta([ac, bd]) = ac(bd)^{-1} = ab^{-1}cd^{-1} = \beta([a, b])\beta([c, d]).$$

Thus, β is a homomorphism. Now, $\ker(\beta)$ is an ideal of F. By Corollary 9.1, $\ker(\beta) = \{0\}$ or F. But $\beta([1, 1]) = 1 \neq 0$, and therefore β is one-to-one. Thus, K has a subfield $\beta(F)$, which is isomorphic to F. Also, for any $r \in R$, we have $r = \beta([r, 1]) \in \beta(F)$. Thus, R is a subring of the isomorphic copy of the field of fractions of R contained in K. $\square$

Example 9.20. As we mentioned above, the field of fractions of $\mathbb{Z}$ is $\mathbb{Q}$.

Example 9.21. Let $R = \{a + b\sqrt{2} : a, b \in \mathbb{Z}\}$. Then R is an integral domain, and its field of fractions is isomorphic to $\{a + b\sqrt{2} : a, b \in \mathbb{Q}\}$. See Exercise 9.36.

One particular type of isomorphism deserves special mention.

Definition 9.8. Let R be a ring. Then an **automorphism** of R is an isomorphism from R to R.

Example 9.22. For any ring R, the function $\alpha : R \to R$ given by $\alpha(a) = a$ for all a is an automorphism.

Example 9.23. The function $\alpha : \mathbb{C} \to \mathbb{C}$ given by $\alpha(a+bi) = a-bi$ for all $a, b \in \mathbb{R}$ is an automorphism. Indeed, we saw in Example 9.13 that it is a homomorphism. It is immediately obvious that α is one-to-one, and if $a+bi \in \mathbb{C}$, then $\alpha(a-bi) = a+bi$, and therefore α is onto.

We have something similar to an inner automorphism of a group as well.

Theorem 9.16. *Let R be a ring with identity and u a unit of R. Then $\alpha : R \to R$, given by $\alpha(a) = u^{-1}au$ for all $a \in R$, is an automorphism of R.*

Proof. Take $a, b \in R$. Then

$$\alpha(a + b) = u^{-1}(a + b)u = u^{-1}au + u^{-1}bu = \alpha(a) + \alpha(b).$$

Also,

$$\alpha(ab) = u^{-1}abu = u^{-1}auu^{-1}bu = \alpha(a)\alpha(b).$$

Thus, α is a homomorphism. If $\alpha(a) = 0$, then $u^{-1}au = 0$, and therefore $a = uu^{-1}auu^{-1} = u0u^{-1} = 0$. Therefore, α is one-to-one. Finally, take any $a \in R$. Then $\alpha(uau^{-1}) = u^{-1}uau^{-1}u = a$. Thus, α is onto, and the proof is complete.   □

Exercises

9.31. Explain why the following pairs of rings are not isomorphic.

1. $\mathbb{Z}_4 \oplus \mathbb{Z}_4$ and $\mathbb{Z}_4 \oplus \mathbb{Z}_2 \oplus \mathbb{Z}_2$
2. $\mathbb{Z}[x]$ and $2\mathbb{Z}[x]$

9.32. Explain why the following pairs of rings are not isomorphic.

1. $\mathbb{R}$ and $M_2(\mathbb{R})$
2. $\mathbb{R}$ and $\mathbb{R} \oplus \mathbb{R}$

9.33. Show that if two rings are isomorphic, then their centres are isomorphic.

9.34. Let R and S be any rings. Show that $R \oplus S$ is isomorphic to $S \oplus R$.

9.35. Let F be a field. Show that F is isomorphic to its field of fractions by constructing an explicit isomorphism.

9.36. Let $R = \{a + b\sqrt{2} : a, b \in \mathbb{Z}\}$. Show that R is an integral domain, and that its field of fractions is isomorphic to $\{a + b\sqrt{2} : a, b \in \mathbb{Q}\}$.

9.37. Let R and S be integral domains. If the fields of fractions of R and S are isomorphic, does it follow that R and S are isomorphic? Prove that it does, or give an explicit counterexample.

9.38. Complete the proof of Theorem 9.14.

9.39. Let R be a ring. An involution on R is a function $\alpha : R \to R$ such that, for all $r_i \in R$, we have $\alpha(r_1 + r_2) = \alpha(r_1) + \alpha(r_2)$, $\alpha(r_1 r_2) = \alpha(r_2)\alpha(r_1)$ and $\alpha(\alpha(r_1)) = r_1$. Show that the following functions α are involutions on $M_2(\mathbb{R})$.

1. $\alpha\left(\begin{pmatrix} a & b \\ c & d \end{pmatrix}\right) = \begin{pmatrix} a & c \\ b & d \end{pmatrix}$ (called the transpose involution)

2. $\alpha\left(\begin{pmatrix} a & b \\ c & d \end{pmatrix}\right) = \begin{pmatrix} d & -b \\ -c & a \end{pmatrix}$ (called the symplectic involution)

9.40. Let R be a ring. Using the definition of an involution from the preceding question,

1. determine under what circumstances an involution on R is an automorphism; and
2. show that the composition of two involutions on R is an automorphism.

9.5 Isomorphism Theorems for Rings

We recall that the three isomorphism theorems for groups were presented in Section 4.5. Let us now state the analogues for rings. The first is certainly the most important.

Theorem 9.17 (First Isomorphism Theorem for Rings). *Let $\alpha : R \to S$ be a ring homomorphism. Then $R/\ker(\alpha)$ is isomorphic to $\alpha(R)$.*

Proof. Let $K = \ker(\alpha)$. Define $\beta : R/K \to \alpha(R)$ via $\beta(a + K) = \alpha(a)$. From the proof of Theorem 4.18, we see that β is an isomorphism of additive groups. Thus, it remains only to check that β respects multiplication. Take any $a, b \in R$. Then

$$\beta((a + K)(b + K)) = \beta(ab + K) = \alpha(ab) = \alpha(a)\alpha(b) = \beta(a + K)\beta(b + K),$$

as required. □

Whenever we are asked to show that a ring modulo an ideal is isomorphic to some other ring, it is usually a good indication that we should employ the First Isomorphism Theorem.

Example 9.24. We already know that for any $n \geq 2$, the additive groups $\mathbb{Z}/n\mathbb{Z}$ and $\mathbb{Z}_n$ are isomorphic. Indeed, in Example 4.21, we showed that $\alpha : \mathbb{Z} \to \mathbb{Z}_n$, given by $\alpha(a) = [a]$, is an onto group homomorphism with kernel $n\mathbb{Z}$. But, in fact, we also have $\alpha(ab) = [ab] = [a][b] = \alpha(a)\alpha(b)$, for all $a, b \in \mathbb{Z}$. Thus, α is actually an onto ring homomorphism, and we now know that $\mathbb{Z}/n\mathbb{Z}$ and $\mathbb{Z}_n$ are isomorphic rings.

Example 9.25. Let us show that $\mathbb{R}[x]/(x)$ is isomorphic to $\mathbb{R}$. To this end, let us define $\alpha : \mathbb{R}[x] \to \mathbb{R}$ via $\alpha(f(x)) = f(0)$. Now, if $f(x), g(x) \in \mathbb{R}[x]$, then

$$\alpha(f(x) + g(x)) = f(0) + g(0) = \alpha(f(x)) + \alpha(g(x)).$$

Furthermore,

$$\alpha(f(x)g(x)) = f(0)g(0) = \alpha(f(x))\alpha(g(x)).$$

Thus, α is a homomorphism. Also, if $r \in \mathbb{R}$, then simply regarding r as a constant polynomial, we have $\alpha(r) = r$; hence, α is onto. The kernel of α is the set of all polynomials $f(x)$ satisfying $f(0) = 0$. But $f(0)$ is the constant term of the polynomial. Thus, $\ker(\alpha)$ is the set of all polynomials with zero constant term, that is, the set of all polynomials that are multiples of x. Now apply Theorem 9.17.

A couple of rather interesting consequences follow.

Corollary 9.2. *Let R be a ring with identity of characteristic n. If $n = 0$, then R has a subring isomorphic to $\mathbb{Z}$. If $n \geq 2$, then R has a subring isomorphic to $\mathbb{Z}_n$.*

Proof. By Theorem 9.11, there is a homomorphism $\alpha : \mathbb{Z} \to R$ with kernel $n\mathbb{Z}$. Now, Theorem 9.10 says that $\alpha(\mathbb{Z})$ is a subring of R, and Theorem 9.17 tells us that this subring is isomorphic to $\mathbb{Z}/n\mathbb{Z}$. If $n = 0$, then $n\mathbb{Z} = \{0\}$, and there is nothing to do. Otherwise, we use Example 9.24. □

Corollary 9.3. *Let F be a field. If F has characteristic 0, then F has a subfield isomorphic to $\mathbb{Q}$. If F has prime characteristic p, then F has a subfield isomorphic to $\mathbb{Z}_p$.*

Proof. If $\mathrm{char}\, F = p > 0$, then we use the preceding corollary. If $\mathrm{char}\, F = 0$, then we note that F has a subring isomorphic to $\mathbb{Z}$. By Theorem 9.15, F also has a subfield isomorphic to the field of fractions of $\mathbb{Z}$, namely, $\mathbb{Q}$. □

The subfield discussed in Corollary 9.3 (either $\mathbb{Q}$ or $\mathbb{Z}_p$) is the smallest subfield of F, and it is called the **prime subfield**.

Theorem 9.18 (Second Isomorphism Theorem for Rings). *Let R be a ring with ideals I and J. Then $I/(I \cap J)$ is isomorphic to $(I + J)/J$.*

Proof. Define $\alpha : I \to (I + J)/J$ via $\alpha(i) = i + J$, for all $i \in I$. Consulting the proof of Theorem 4.19, we see that α is an onto homomorphism of additive groups with kernel $I \cap J$. In view of the First Isomorphism Theorem for Rings, it suffices to show that α respects multiplication. But for any $i_1, i_2 \in I$, we have $\alpha(i_1 i_2) = i_1 i_2 + J = (i_1 + J)(i_2 + J) = \alpha(i_1)\alpha(i_2)$. The proof is complete. □

Example 9.26. Let $R = \mathbb{Z}$, $I = (4)$ and $J = (6)$. Then the preceding theorem tells us that $(4)/((4) \cap (6))$ is isomorphic to $((4) + (6))/(6)$. That is, $(4)/(12)$ is isomorphic to $(2)/(6)$.

Theorem 9.19 (Third Isomorphism Theorem for Rings). *Let R be a ring, and let I and J be ideals of R with $I \subseteq J$. Then $(R/I)/(J/I)$ is isomorphic to R/J.*

Proof. Define $\alpha : R/I \to R/J$ via $\alpha(a + I) = a + J$, for all $a \in R$. The proof of Theorem 4.20 shows us that α is an onto additive group homomorphism with kernel J/I. It remains only to show that α respects multiplication, for then we can apply Theorem 9.17. Take $a, b \in R$. Then

$$\alpha((a + I)(b + I)) = \alpha(ab + I) = ab + J = (a + J)(b + J) = \alpha(a + I)\alpha(b + I).$$

We are done. $\square$

Exercises

9.41. Let R and S be rings. Show that $(R \oplus S)/(R \oplus \{0\})$ is isomorphic to S.

9.42. Let m and n be positive integers, both greater than 1. Show that the rings $(\mathbb{Z} \oplus \mathbb{Z})/((m) \oplus (n))$ and $\mathbb{Z}_m \oplus \mathbb{Z}_n$ are isomorphic.

9.43. Let I be the set of all polynomials $f(x) \in \mathbb{Z}[x]$ such that the constant term of $f(x)$ is a multiple of 5. Show that $\mathbb{Z}[x]/I$ is isomorphic to $\mathbb{Z}_5$.

9.44. Let I be the set of all matrices in $M_2(\mathbb{Z})$ in which every entry is even. Show that $M_2(\mathbb{Z})/I$ is isomorphic to $M_2(\mathbb{Z}_2)$.

9.45. Show that the rings $(3\mathbb{Z}/60\mathbb{Z})/(12\mathbb{Z}/60\mathbb{Z})$ and $3\mathbb{Z}/12\mathbb{Z}$ are isomorphic. Then show that both are isomorphic to $\mathbb{Z}_4$.

9.46. Let I and J be ideals in a ring R such that $I + J = R$. Show that $R/(I \cap J)$ is isomorphic to $(R/I) \oplus (R/J)$.

9.6 Prime and Maximal Ideals

We conclude this chapter by discussing two special sorts of ideals.

Definition 9.9. Let R be a ring. An ideal M of R is said to be **maximal** if

1. $M \neq R$; and
2. if I is an ideal of R containing M, then $I = M$ or $I = R$.

Example 9.27. Let $R = \mathbb{Z}$ and let n be a nonnegative integer. Then we claim that (n) is a maximal ideal of R if and only if n is prime. Indeed, (0) is certainly not maximal, as $(0) \subsetneq (2) \subsetneq R$. Also, (1) is not maximal, since $(1) = R$. If n is composite, say $n = kl$, with $1 < k, l < n$, then we note that $(n) \subsetneq (k) \subsetneq R$, so (n) is not maximal. Finally, let n be prime. Suppose that I is an ideal of R with $(n) \subsetneq I \subsetneq R$. Take $a \in I \backslash (n)$. Since a is not divisible by n, and n is prime, we know that $(a, n) = 1$. Thus, by Corollary 2.1, we can find integers u and v such that $au + nv = 1$. But as $a, n \in I$, this implies that $1 \in I$, hence $I = R$, giving us a contradiction. (As we shall see shortly, there is another way to prove this.)

Example 9.28. In any field, the ideal $\{0\}$ is maximal! Remember, by Corollary 9.1, a field only has two ideals.

In a commutative ring with identity, there is a nice test for maximality of ideals.

Theorem 9.20. *Let R be a commutative ring with identity, and M an ideal of R. Then M is maximal if and only if R/M is a field.*

Proof. Suppose that M is a maximal ideal. By Theorem 9.7, R/M is a commutative ring with identity. Furthermore, as $M \neq R$, we know that R/M consists of more than one additive left coset. But the only ring in which $0 = 1$ is the ring consisting only of zero; thus, $0 + M \neq 1 + M$. It remains to show that every nonzero element of R/M has an inverse. Let $a + M \neq 0 + M$. Now, define $I = \{m + ra : m \in M, r \in R\}$. We claim that I is an ideal of R. Taking $r = 0$, we note that $m \in I$ for all $m \in M$; thus, $M \subseteq I$ and, in particular, $0 \in I$. If $m_i \in M, r_i \in R$, then

$$(m_1 + r_1 a) - (m_2 + r_2 a) = (m_1 - m_2) + (r_1 - r_2)a \in I.$$

Also, for any $s \in R$, $s(m_1 + r_1 a) = sm_1 + sr_1 a$. As $sm_1 \in M$ and $sr_1 \in R$, we see that I has the absorption property and is, therefore, an ideal. But we noted above that $M \subseteq I$. Furthermore, $a = 0 + 1a \in I \backslash M$. By the maximality of M, we have $I = R$. In particular, $1 \in I$, so there exist $m \in M$ and $r \in R$ such that $m + ra = 1$. But then $(r + M)(a + M) = 1 - m + M = 1 + M$, since $m \in M$. That is, $r + M$ is the inverse of $a + M$, and R/M is a field.

Conversely, suppose that R/M is a field. We must show that M is maximal. If $M = R$, then R/M consists only of a single additive left coset, contradicting the fact that a field must have a distinct 0 and 1. Thus, $M \neq R$. Suppose that I is an ideal of R with $M \subsetneq I \subseteq R$. Take $a \in I \backslash M$. Now, $a + M \neq 0 + M$, so $a + M$ has an inverse, say $b + M$. Then $(a + M)(b + M) = 1 + M$; in other words, $1 - ab \in M \subseteq I$. But also $a \in I$, which means that $ab \in I$ by absorption, and therefore $1 = (1 - ab) + ab \in I$. By Theorem 9.2, $I = R$, giving us a contradiction and completing the proof. $\square$

Example 9.29. This gives us another way to deal with Example 9.27. If $n = 0$, then we note that $\mathbb{Z}/\{0\}$ is simply $\mathbb{Z}$, which is not a field. Thus, (0) is not maximal. If $n = 1$, then observe that $\mathbb{Z}/\mathbb{Z}$ is the ring with one element, which is not a field; hence, (1) is not maximal. For any $n \geq 2$, we see from Example 9.24 that $\mathbb{Z}/n\mathbb{Z}$ is isomorphic to $\mathbb{Z}_n$. But by Theorem 8.11, $\mathbb{Z}_n$ is a field if and only if n is prime. Thus, (n) is maximal if and only if n is prime.

Example 9.30. By Example 9.25, $\mathbb{R}[x]/(x)$ is isomorphic to $\mathbb{R}$, which is a field. Thus, (x) is a maximal ideal of $\mathbb{R}[x]$.

Example 9.31. In the same manner as Example 9.30, we see that $\mathbb{Z}[x]/(x)$ is isomorphic to $\mathbb{Z}$. But $\mathbb{Z}$ is not a field, and hence (x) is not maximal. In fact, we can see this by noting that (x) is properly contained in the ideal M consisting of those polynomials whose constant terms are multiples of 5. We can use Theorem 9.20 to show that M is maximal. See Exercise 9.43.

It is worth mentioning that Theorem 9.20 only applies when R is a commutative ring with identity. For instance, the ideal containing only the zero matrix is maximal in $M_2(\mathbb{R})$! See Exercise 9.54.

Definition 9.10. Let R be a commutative ring and P an ideal of R. Then we say that P is a **prime ideal**[2] if

1. $P \neq R$; and
2. if $a, b \in R$ and $ab \in P$, then either $a \in P$ or $b \in P$.

Example 9.32. In any integral domain, $\{0\}$ is a prime ideal. If $ab = 0$, then $a = 0$ or $b = 0$.

Example 9.33. Let us consider $R = \mathbb{Z}$. By the preceding example, $\{0\}$ is prime, so we know immediately that maximal and prime are not the same thing. Of course, $(1) = R$, so (1) is not prime. Suppose that $n \geq 2$. If n is composite, say $n = kl$ with $1 < k, l < n$, we see that $kl \in (n)$ but neither k nor l lies in (n). Thus, (n) is not prime. But if n is prime, then (n) is a prime ideal. Indeed, if $ab \in (n)$, then $n | ab$. Thus, by Theorem 2.7, $n|a$ or $n|b$, and hence a or b is in (n).

Once again, there is another way to handle this last example.

Theorem 9.21. *Let R be a commutative ring with identity and P an ideal. Then P is prime if and only if R/P is an integral domain.*

Proof. Suppose that P is prime. Since R is a commutative ring with identity, so is R/P, by Theorem 9.7. Also, as $P \neq R$, R/P has more than one element, and therefore $0 + P \neq 1 + P$. Thus, it remains to show that R/P has no zero divisors. Suppose that $(a + P)(b + P) = 0 + P$. Then $ab \in P$, and hence $a \in P$ or $b \in P$. That is, $a + P = 0 + P$ or $b + P = 0 + P$, and R/P is an integral domain.

Conversely, let R/P be an integral domain. As R/P cannot be the ring with one element, $P \neq R$. Suppose that $ab \in P$. Then $(a + P)(b + P) = 0 + P$. Since R/P has no zero divisors, $a + P = 0 + P$ or $b + P = 0 + P$. That is, $a \in P$ or $b \in P$, and P is prime. $\qquad\square$

Example 9.34. Let us look at Example 9.33 again. We know that $\mathbb{Z}/(0)$ is just $\mathbb{Z}$, which is an integral domain, and hence (0) is a prime ideal. If $n \geq 2$, then $\mathbb{Z}/(n)$ is isomorphic to $\mathbb{Z}_n$, by Example 9.24, and Theorem 8.11 tells us that this is an integral domain if and only if n is prime. Thus, for a nonnegative integer n, (n) is a prime ideal of $\mathbb{Z}$ if and only if n is 0 or prime.

Example 9.35. Refer to Example 9.31. We see that (x) is a prime ideal in $\mathbb{Z}[x]$, since $\mathbb{Z}[x]/(x)$ is isomorphic to $\mathbb{Z}$.

Example 9.36. Naturally, (x) is prime in $\mathbb{R}[x]$, because we saw in Example 9.30 that $\mathbb{R}[x]/(x)$ is isomorphic to $\mathbb{R}$, which is a field, hence an integral domain.

---

[2]Please note that for noncommutative rings, the definition of a prime ideal is different. We will only concern ourselves with prime ideals in commutative rings in this book.

Of course, this last example can be generalized.

Theorem 9.22. *Let R be a commutative ring with identity. Then every maximal ideal of R is also a prime ideal.*

Proof. Use the last two theorems and the fact that every field is an integral domain.

$\square$

As we have already seen, not every prime ideal is maximal. Also, this last theorem only applies to commutative rings with identity. In some commutative rings without an identity, it is possible to find maximal ideals that are not prime, and Exercise 9.51 asks for an example of this phenomenon.

Exercises

9.47. Let R be the ring from Exercise 8.17. Is the ideal (2) prime? Is it maximal?

9.48. Find all prime ideals in each of the following rings.

1. $\mathbb{Z}_{10}$
2. $\mathbb{Z}_{50}$

9.49. Let R be a finite commutative ring with identity. Show that every prime ideal of R is maximal.

9.50. Find every maximal ideal of $\mathbb{Z}_7 \oplus \mathbb{Z}_7$.

9.51. Find an example of a commutative ring having an ideal that is maximal but not prime.

9.52. Suppose that R is a commutative ring with identity in which the elements of R that are not units form an ideal. Show that this ideal is the unique maximal ideal of R.

9.53. Show that every field has the property described in the preceding exercise. Also show that $\mathbb{Z}_{p^n}$ has this property, for every prime p and positive integer n.

9.54. Show that the ideal containing only the zero matrix is maximal in $M_2(\mathbb{R})$.

9.55. Let R be a commutative ring with identity having a prime ideal I. Find a prime ideal in $R \oplus R$.

9.56. Let $R \neq \{0\}$ be a commutative ring with identity. Suppose that every proper ideal of R is prime. Show that R is an integral domain, and then use this information to show that R is, in fact, a field.

Chapter 10
Special Types of Domains

In this chapter, we begin with a specific and rather familiar sort of integral domain, and then generalize slightly in each section. First, we define a polynomial ring over a field, and show that we have a division algorithm in such a ring. As a result, this polynomial ring is a special type of ring called a Euclidean domain.

Subsequently, we demonstrate that Euclidean domains are principal ideal domains; that is, every ideal is principal. Finally, we prove that principal ideal domains are examples of unique factorization domains, in which we have something similar to the Fundamental Theorem of Arithmetic.

10.1 Polynomial Rings

We are certainly familiar with polynomials having real coefficients. There is no reason why we cannot consider coefficients in other rings.

Definition 10.1. Let R be a ring. Then a **polynomial** with coefficients in R is a formal expression

$$a_0 + a_1 x + a_2 x^2 + \cdots + a_n x^n,$$

where $a_i \in R$ and n is a nonnegative integer. Suppose that $b_0 + b_1 x + \cdots + b_m x^m$ is also a polynomial with coefficients in R. Without loss of generality, let us say that $n \le m$. Then these polynomials are equal if and only if $a_i = b_i$ for all $i \le n$ and $b_i = 0$ for all $i > n$. The set of all polynomials with coefficients in R is denoted $R[x]$.

Example 10.1. Let $R = \mathbb{Z}_5$. Then (inserting congruence class brackets for clarity), an example of a polynomial in $R[x]$ would be $f(x) = [3] + [2]x + [4]x^2$. As part of the above definition, we observe that $f(x) = g(x)$, where $g(x) = [3] + [2]x + [4]x^2 + [0]x^3$.

© Springer International Publishing AG, part of Springer Nature 2018
G. T. Lee, *Abstract Algebra*, Springer Undergraduate Mathematics Series,
https://doi.org/10.1007/978-3-319-77649-1_10

Note that the x in a polynomial is not an element of R. It is simply a placeholder in the expression of the polynomial. We could, equally well, define the polynomials in terms of sequences of elements of R (with only finitely many terms different from zero). But nobody thinks of polynomials in that way.

Definition 10.2. Let R be a ring and let $f(x) = a_0 + a_1 x + \cdots + a_n x^n \in R[x]$. Further suppose that $a_m \neq 0$ but $a_k = 0$ for all $k > m$. Then the **degree** of $f(x)$ is m, and we write $\deg(f(x)) = m$. The **leading term** of $f(x)$ is $a_m x^m$, and the **leading coefficient** is a_m. Note that the **zero polynomial**, 0, has no degree, leading term or leading coefficient. A **constant polynomial** has degree 0 (or is the zero polynomial). If R has an identity, then $f(x)$ is **monic** if its leading coefficient is 1.

Example 10.2. In $\mathbb{Q}[x]$, let $f(x) = 3 + 7x - 15x^2 + 0x^3 + 2x^4 + 0x^5$. Then $\deg(f(x)) = 4$, the leading term is $2x^4$ and the leading coefficient is 2. This polynomial is not monic.

We wish to make $R[x]$ into a ring, and so we need addition and multiplication operations. These will be exactly the same as for real polynomials. Let $f(x) = a_0 + a_1 x + \cdots + a_n x^n$ and $g(x) = b_0 + b_1 x + \cdots + b_m x^m$. Adding in terms with zero coefficients if necessary, we may assume that $m = n$. Then

$$f(x) + g(x) = (a_0 + b_0) + (a_1 + b_1)x + \cdots + (a_n + b_n)x^n.$$

Similarly,

$$f(x)g(x) = c_0 + c_1 x + c_2 x^2 + \cdots + c_{m+n} x^{m+n},$$

where

$$c_i = a_0 b_i + a_1 b_{i-1} + a_2 b_{i-2} + \cdots + a_i b_0.$$

Here, we take $a_j = 0$ if $j > n$ and $b_j = 0$ if $j > m$.

Example 10.3. In $\mathbb{Z}_7[x]$, let $f(x) = 5 + 2x + 6x^2$ and $g(x) = 3 + x + 4x^2 + 5x^3$. Then $f(x) + g(x) = 1 + 3x + 3x^2 + 5x^3$ and $f(x)g(x) = 1 + 4x + 5x^2 + 4x^3 + 6x^4 + 2x^5$.

Theorem 10.1. *If R is a ring, then so is $R[x]$.*

Proof. Let us show that $R[x]$ is an abelian group under addition. Clearly the sum of two polynomials is a polynomial. Let $f(x) = a_0 + \cdots + a_n x^n$, $g(x) = b_0 + \cdots + b_m x^m$ and $h(x) = c_0 + \cdots + c_k x^k$. Then the coefficient of x^i in $f(x) + g(x)$ is $a_i + b_i$, and similarly for $g(x) + f(x)$ (adding in terms with zero coefficients if necessary). Thus, addition is commutative. In the same way, because the addition of coefficients is associative, addition in $R[x]$ is associative. The zero polynomial is the additive identity, and $-f(x) = -a_0 - \cdots - a_n x^n$. Therefore, $R[x]$ is an abelian group under addition.

Evidently, the product of two polynomials is a polynomial. Let us check a distributive law. The coefficient of x^i in $f(x)(g(x) + h(x))$ is

$$a_0(b_i + c_i) + a_1(b_{i-1} + c_{i-1}) + \cdots + a_i(b_0 + c_0).$$

But this is $(a_0 b_i + \cdots + a_i b_0) + (a_0 c_i + \cdots + a_i c_0)$, which is the coefficient of x^i in $f(x)g(x) + f(x)h(x)$. The other distributive law is proved similarly. Finally, we must check that multiplication is associative. But by repeated application of the distributive laws, we see that we may reduce to proving that $(a_u x^u b_v x^v) c_w x^w = a_u x^u (b_v x^v c_w x^w)$, for all $a_u, b_v, c_w \in R$ and all $u, v, w \geq 0$. However, both sides of this equation are equal to $a_u b_v c_w x^{u+v+w}$, and the proof is complete. $\square$

Corollary 10.1. *Let R be a ring. Then*

1. *if R has an identity, then so does $R[x]$; and*
2. *if R is commutative, then so is $R[x]$.*

Proof. (1) The constant polynomial 1 is the identity.

(2) Repeatedly applying the distributive laws, we see that we need only check that $a_i x^i$ and $b_j x^j$ commute, where $a_i, b_j \in R$ and $i, j \geq 0$. But $a_i x^i b_j x^j = a_i b_j x^{i+j}$ and $b_j x^j a_i x^i = b_j a_i x^{i+j}$. Since R is commutative, these are equal. $\square$

When our ring is an integral domain, degrees of polynomials behave in a way we would expect.

Theorem 10.2. *Let R be an integral domain, and let $f(x)$ and $g(x)$ be nonzero polynomials in $R[x]$, of degree m and n respectively. Then*

1. $\deg(f(x) + g(x))$ *is at most the larger of m and n (or $f(x) + g(x) = 0$); and*
2. $\deg(f(x)g(x)) = m + n$.

Proof. (1) This is clear from the definition of polynomial addition.

(2) Let $f(x) = a_0 + \cdots + a_m x^m$ and $g(x) = b_0 + \cdots + b_n x^n$. Then we see from the definition of polynomial multiplication that the only term of highest degree in $f(x)g(x)$ is $a_m b_n x^{m+n}$. Furthermore, $a_m \neq 0 \neq b_n$ and, since R is an integral domain, $a_m b_n \neq 0$. Thus, $\deg(f(x)g(x)) = m + n$. $\square$

Note that the second part of the theorem fails if R is not an integral domain. For instance, in $\mathbb{Z}_6[x]$, we have $(2 + 3x)(1 + 2x) = 2 + x$, which does not have degree 2.

Corollary 10.2. *If R is an integral domain, then so is $R[x]$.*

Proof. By Corollary 10.1, $R[x]$ is a commutative ring with identity. Furthermore, $1 \neq 0$. By the preceding theorem, the product of nonzero polynomials cannot be the zero polynomial. $\square$

Why are we so interested in polynomial rings? We now know that if F is a field, then $F[x]$ is an integral domain. But it has another attractive property. Indeed, we have an analogue of the division algorithm with which we are familiar for the integers. Readers who have seen polynomial long division for real polynomials will find the procedure very similar.

Theorem 10.3. (Division Algorithm for Polynomials). *Let F be a field, and let $f(x), g(x) \in F[x]$, with $g(x) \neq 0$. Then there exist unique $q(x), r(x) \in F[x]$ such that*

$$f(x) = g(x)q(x) + r(x),$$

with either $r(x) = 0$ or $\deg(r(x)) < \deg(g(x))$.

Proof. Let us verify the existence of $q(x)$ and $r(x)$. If $f(x) = 0$, there is nothing to do; indeed, we let $q(x) = r(x) = 0$. Therefore, assume that $f(x)$ is not the zero polynomial. We proceed by strong induction on $\deg(f(x))$. Suppose that $\deg(f(x)) = 0$. If $\deg(g(x)) > 0$, then use $q(x) = 0$ and $r(x) = f(x)$. On the other hand, if $\deg(g(x)) = 0$, then $g(x) = b$ is a nonzero constant in F. As F is a field, we have $b^{-1} \in F$, and we can use $q(x) = b^{-1}f(x)$ and $r(x) = 0$.

Thus, suppose that $\deg(f(x)) = n > 0$ and that our result holds for polynomials of smaller degree. Let us write $f(x) = a_0 + a_1 x + \cdots + a_n x^n$. Also suppose that $\deg(g(x)) = m$, and write $g(x) = b_0 + b_1 x + \cdots + b_m x^m$. If $n < m$, then we can use $q(x) = 0$ and $r(x) = f(x)$. Otherwise, notice that in $f(x) - g(x)b_m^{-1}a_n x^{n-m}$, no term of degree greater than n appears, and the coefficient of x^n is $a_n - b_m b_m^{-1} a_n = 0$; thus, either $f(x) - g(x)b_m^{-1}a_n x^{n-m}$ is the zero polynomial, or it has degree strictly smaller than $f(x)$. By our inductive hypothesis, there exist $q(x), r(x) \in F[x]$ such that $f(x) - g(x)b_m^{-1}a_n x^{n-m} = g(x)q(x) + r(x)$, with $r(x) = 0$ or $\deg(r(x)) < \deg(g(x))$. But then $f(x) = g(x)(q(x) + b_m^{-1}a_n x^{n-m}) + r(x)$, as required.

Now for uniqueness. Suppose that $f(x) = g(x)q(x) + r(x) = g(x)q_1(x) + r_1(x)$, with $q(x), q_1(x), r(x), r_1(x) \in F[x]$ and each of $r(x)$ and $r_1(x)$ either is 0 or has degree smaller than that of $g(x)$. Then $g(x)(q(x) - q_1(x)) = r_1(x) - r(x)$. Suppose that $q(x) \neq q_1(x)$. By Theorem 10.2, $\deg(g(x)(q(x) - q_1(x))) \geq \deg(g(x))$, but $r_1(x) - r(x)$ cannot possibly have a degree that large. Thus, $q(x) = q_1(x)$ and hence $r(x) = r_1(x)$.   □

The proof also shows us how to construct $q(x)$ and $r(x)$. We look only at the leading terms of $f(x)$ and $g(x)$ (say, respectively, $a_n x^n$ and $b_m x^m$). Assuming that $n \geq m$, we subtract $b_m^{-1}a_n x^{n-m}g(x)$ from $f(x)$ and obtain either the zero polynomial or a polynomial of degree smaller than $\deg(f(x))$. Then repeat.

Example 10.4. Let us apply the division algorithm in $\mathbb{Q}[x]$ with $f(x) = 8x^4 - 4x^3 + 2x^2 + x + 1$ and $g(x) = 2x^2 + 3x + 7$. We take $2^{-1} \cdot 8x^{4-2} = 4x^2$, multiply by $g(x)$ and subtract from $f(x)$.

$$
\begin{array}{r}
4x^2 \\
2x^2 + 3x + 7 \overline{)\; 8x^4 - 4x^3 + 2x^2 + x + 1} \\
-8x^4 - 12x^3 - 28x^2 \\
\hline
-16x^3 - 26x^2 + x
\end{array}
$$

Next, take $2^{-1}(-16)x^{3-2} = -8x$, multiply by $g(x)$ and subtract.

$$4x^2 \quad - 8x$$
$$2x^2 + 3x + 7{\overline{\smash{\big)}\,8x^4 \;\; - 4x^3 \;\; + 2x^2 \quad + x + 1}}$$
$$\underline{- 8x^4 - 12x^3 - 28x^2}$$
$$- 16x^3 - 26x^2 \quad + x$$
$$\underline{16x^3 + 24x^2 + 56x}$$
$$- 2x^2 + 57x + 1$$

Finally, take $2^{-1}(-2)x^{2-2} = -1$, multiply by $g(x)$ and subtract.

$$4x^2 \quad - 8x - 1$$
$$2x^2 + 3x + 7{\overline{\smash{\big)}\,8x^4 \;\; - 4x^3 \;\; + 2x^2 \quad + x + 1}}$$
$$\underline{- 8x^4 - 12x^3 - 28x^2}$$
$$- 16x^3 - 26x^2 \quad + x$$
$$\underline{16x^3 + 24x^2 + 56x}$$
$$- 2x^2 + 57x + 1$$
$$\underline{2x^2 \;\; + 3x + 7}$$
$$60x + 8$$

We now have a remainder with degree smaller than $\deg(g(x))$, so we are done. Indeed, $f(x) = g(x)(4x^2 - 8x - 1) + (60x + 8)$.

Note that it is not sufficient to work in $R[x]$, where R is an integral domain. By Corollary 10.2, $R[x]$ is also an integral domain, but we cannot implement the division algorithm if we are unable to take the inverse of the leading coefficient of $g(x)$. Indeed, if we worked in $\mathbb{Z}[x]$, we would be immediately stymied if we tried to perform the division algorithm using $f(x) = 2x^2 + 3x + 5$ and $g(x) = 3x + 7$.

In fact, a polynomial ring over a field is a nice example of a special type of integral domain that we can now discuss.

Exercises

10.1. In $\mathbb{Z}_{11}[x]$, let $f(x) = 2x^3 + 4x^2 + 2x + 5$ and $g(x) = 2x^4 + 5x^3 + 7x + 1$. Find $f(x) - g(x)$ and $f(x)g(x)$.

10.2. Let $f(x) = 3x^5 + x^4 + x^3 + 3x^2 + 2x + 4$ and $g(x) = 2x^3 + 3x^2 + x + 1$ be polynomials in $\mathbb{Z}_5[x]$. Find $q(x), r(x) \in \mathbb{Z}_5[x]$, with $\deg(r(x)) < 3$, such that $f(x) = g(x)q(x) + r(x)$.

10.3. Let $f(x) = 3x^5 + 6x^4 + x^3 + 3x^2 + 2x + 4$ and $g(x) = 2x^3 + 3x^2 + x + 1$ be polynomials in $\mathbb{Z}_7[x]$. Find $q(x), r(x) \in \mathbb{Z}_7[x]$, with $\deg(r(x)) < 3$, such that $f(x) = g(x)q(x) + r(x)$.

10.4. Let R be an integral domain. Show that the units of $R[x]$ are precisely the constant polynomials a, where $a \in U(R)$.

10.5. If F is a field, is $F[x]$ a field?

10.6. Show that $2x + 1$ is a unit in $\mathbb{Z}_4[x]$. Then, for any prime p, find a unit in $\mathbb{Z}_{p^2}[x]$ that is not a constant polynomial.

10.7. For any ring R, show that R and $R[x]$ have the same characteristic.

10.8. If R and S are isomorphic rings, show that $R[x]$ and $S[x]$ are also isomorphic.

10.9. Let S be a subring of R. Show that $S[x]$ is a subring of $R[x]$. In particular, if S is an ideal of R, show that $S[x]$ is an ideal of $R[x]$.

10.10. Let R be a commutative ring with identity and P a prime ideal of R. Show that $P[x]$ is a prime ideal of $R[x]$.

10.2   Euclidean Domains

A Euclidean domain is an integral domain having an additional property.

Definition 10.3. Let R be an integral domain. Then a **Euclidean function** is a function ε from the set of nonzero elements of R to the nonnegative integers such that, for all nonzero $a, b \in R$, we have

1. $\varepsilon(a) \le \varepsilon(ab)$; and
2. there exist $q, r \in R$ such that $a = bq + r$, and either $r = 0$ or $\varepsilon(r) < \varepsilon(b)$.

Definition 10.4. A **Euclidean domain** is an integral domain having a Euclidean function.

We have already seen several examples of Euclidean domains.

Example 10.5. The integers form a Euclidean domain. We already know that $\mathbb{Z}$ is an integral domain. Define $\varepsilon(a) = |a|$. If a and b are nonzero integers, then $|ab| = |a||b| \ge |a|$. Furthermore, by the division algorithm, there exist $q, r \in \mathbb{Z}$ such that $a = |b|q + r$, with $0 \le r < |b|$. If $b > 0$, we are done. Otherwise, simply note that $a = b(-q) + r$.

Example 10.6. Any field is a Euclidean domain. See Exercise 10.12.

Example 10.7. If F is a field, then $F[x]$ is a Euclidean domain. Indeed, Corollary 10.2 tells us that it is an integral domain. For any $0 \ne f(x) \in F[x]$, let $\varepsilon(f(x)) = \deg(f(x))$. If $0 \ne g(x) \in F[x]$, then by Theorem 10.2, $\deg(f(x)g(x)) = \deg(f(x)) + \deg(g(x)) \ge \deg(f(x))$. The division algorithm for polynomials completes the proof.

Let us construct a new Euclidean domain.

Example 10.8. Let $R = \{a + bi : a, b \in \mathbb{Z}\}$. We call this the ring of **Gaussian integers**. By Exercises 8.17 and 8.27, R is a subring of $F = \{a + bi : a, b \in \mathbb{Q}\}$ which, in turn, is a subfield of $\mathbb{C}$. We claim that R is a Euclidean domain. It is surely an integral domain, since it is a unital subring of a field and therefore has no zero divisors. It remains to construct a Euclidean function.

Define $\varepsilon : F \to \mathbb{Q}$ via $\varepsilon(a + bi) = a^2 + b^2$. In particular, if $0 \neq a + bi \in R$, then $\varepsilon(a + bi) \in \mathbb{N}$. If $a, b, c, d \in \mathbb{Q}$, then

$$
\begin{aligned}
\varepsilon((a + bi)(c + di)) &= \varepsilon((ac - bd) + (ad + bc)i) \\
&= (ac - bd)^2 + (ad + bc)^2 \\
&= a^2c^2 + b^2d^2 + a^2d^2 + b^2c^2 \\
&= (a^2 + b^2)(c^2 + d^2) \\
&= \varepsilon(a + bi)\varepsilon(c + di).
\end{aligned}
$$

In particular, if $a + bi$ and $c + di$ are nonzero elements of R, then

$$
\varepsilon(a + bi) \leq \varepsilon((a + bi)(c + di)).
$$

Take any nonzero $u, v \in R$. Then as F is a field, $uv^{-1} \in F$. Let us write $uv^{-1} = s + ti$, with $s, t \in \mathbb{Q}$. Choose integers m and n such that $|s - m| \leq \frac{1}{2}$ and $|t - n| \leq \frac{1}{2}$. Then

$$
\begin{aligned}
u - v(m + ni) &= u - v((s + ti) + ((m - s) + (n - t)i)) \\
&= u - v(uv^{-1}) + v((s - m) + (t - n)i) \\
&= v((s - m) + (t - n)i).
\end{aligned}
$$

Now,

$$
\varepsilon((s - m) + (t - n)i) = (s - m)^2 + (t - n)^2 \leq \frac{1}{2}.
$$

Therefore,

$$
\varepsilon(u - v(m + n)i) = \varepsilon(v)\varepsilon((s - m) + (t - n)i) < \varepsilon(v).
$$

Letting $q = m + ni$ and $r = u - v(m + ni)$, we have $u = vq + r$ and we are done.

What is so special about Euclidean domains? Let us begin with some definitions.

Definition 10.5. Let R be a commutative ring with identity. If $a, b \in R$, then we say that a **divides** b, and write $a|b$, if there exists a $c \in R$ such that $b = ac$.

Of course, this agrees with our definition of divisibility in $\mathbb{Z}$. We are very much interested in extending the notion of a greatest common divisor as well. For an arbitrary ring, this is problematic, as there is no particular notion of ordering. But for a Euclidean domain, we have ε!

Definition 10.6. Let R be a Euclidean domain, and let $a, b \in R$, not both zero. Then a nonzero element d of R is said to be a **greatest common divisor** (or **gcd**) if

1. $d|a$ and $d|b$; and
2. whenever c is an element of R satisfying $c|a$ and $c|b$, we have $\varepsilon(c) \leq \varepsilon(d)$.

Certainly a gcd must exist. Indeed, 1 is a common divisor of any two elements, so the set of common divisors is not empty. Furthermore, by definition of a Euclidean function, if $c|a$ and $a \neq 0$, then $\varepsilon(c) \leq \varepsilon(a)$. Thus, there is an upper bound on the ε values of the common divisors, so we can select one having the largest possible value.

Notice that we called d "a gcd", not "the gcd". Indeed, this definition does not produce a unique gcd. In particular, in $\mathbb{Z}$, we see that both 5 and -5 would meet the description of "a gcd" of 10 and 35. However, when we say "the gcd", we will still mean the positive one; that is, $(10, 35) = 5$, not -5.

Similarly, if F is a field, suppose that $d(x)$ is a gcd of $f(x)$ and $g(x)$. If u is a nonzero element of F, we see immediately that $ud(x)$ also divides both $f(x)$ and $g(x)$, and that $\deg(ud(x)) = \deg(d(x))$. Thus, $ud(x)$ is also a gcd. But again, we can choose a specific gcd here.

Definition 10.7. Let F be a field and let $f(x)$ and $g(x)$ be polynomials in $F[x]$, not both the zero polynomial. By **the gcd** of $f(x)$ and $g(x)$ we mean a monic gcd. When we write $(f(x), g(x))$, we mean specifically this monic gcd.

For more general Euclidean domains, we cannot easily single out a particular gcd in this manner. But we will see that, in fact, the gcds are all related to each other in a nice way. While proving this, we can produce some other interesting results. For instance, the Euclidean domain is so named because there is a Euclidean algorithm just like in $\mathbb{Z}$.

Theorem 10.4 (Euclidean Algorithm for Euclidean Domains). *Let R be a Euclidean domain. Take $a, b \in R$ with $b \neq 0$. If $b|a$, then b is a gcd of a and b. Otherwise, apply the division algorithm repeatedly. To wit, write*

$$a = bq_1 + r_1$$
$$b = r_1q_2 + r_2$$
$$r_1 = r_2q_3 + r_3$$
$$\vdots$$
$$r_{k-2} = r_{k-1}q_k + r_k$$
$$r_{k-1} = r_kq_{k+1} + 0,$$

where all $q_i, r_i \in R$ and $r_i \neq 0$, with $\varepsilon(r_1) < \varepsilon(b)$ and $\varepsilon(r_j) < \varepsilon(r_{j-1})$ for all $j \geq 2$. Then r_k is a gcd of a and b.

Proof. If $b|a$, then b is a common divisor of a and b. Also, if $c|b$, then $\varepsilon(c) \leq \varepsilon(b)$, and so b is a gcd. Assume that b does not divide a, and we perform the division algorithm repeatedly, as indicated.

Note, first of all, that this process must end, as the $\varepsilon(r_i)$ are strictly decreasing integers and cannot be negative. Suppose that $c|a$ and $c|b$, say $a = ca_1$ and $b = ca_2$, with $a_i \in R$. Then $r_1 = c(a_1 - a_2 q_1)$, and hence $c|r_1$. Similarly, any common divisor of b and r_1 must also divide a. Thus, the set of common divisors of a and b is precisely the same as the set of common divisors of b and r_1. In particular, they have the same set of gcds.

By the same argument, the gcds of b and r_1 are the same as those of r_1 and r_2. We then repeat this and find that the gcds of a and b are the same as the gcds of r_k and 0. But as everything divides 0, we are looking only for the largest value of ε among the divisors of r_k. However, as if $u|v$ and $v \neq 0$, then $\varepsilon(u) \leq \varepsilon(v)$, we see that r_k is a gcd of r_k and 0, as required. □

Corollary 10.3. *Let R be a Euclidean domain. Take $a, b \in R$ with $b \neq 0$. Let d be the gcd of a and b found in the preceding theorem. Then there exist $u, v \in R$ such that $d = au + bv$.*

Proof. If $b|a$, then $d = b = a(0) + b(1)$. Assume otherwise. We have $d = r_k = r_{k-2} + r_{k-1}(-q_k)$, a multiple of r_{k-2} plus a multiple of r_{k-1}. But the preceding equation is $r_{k-3} = r_{k-2}q_{k-1} + r_{k-1}$. Thus,

$$d = r_{k-2} + (r_{k-3} + r_{k-2}(-q_{k-1}))(-q_k).$$

We have written d as a multiple of r_{k-3} plus a multiple of r_{k-2}. Now move backwards through the equations, and we will eventually write d as a multiple of a plus a multiple of b. □

Example 10.9. Let us apply the Euclidean algorithm and its corollary in $\mathbb{Z}_7[x]$, starting with $f(x) = 2x^3 + 4x^2 + x + 1$ and $g(x) = 6x^3 + 4x^2 + 4x + 5$. We write

$$2x^3 + 4x^2 + x + 1 = (6x^3 + 4x^2 + 4x + 5)(5) + (5x^2 + 2x + 4)$$
$$6x^3 + 4x^2 + 4x + 5 = (5x^2 + 2x + 4)(4x + 2) + (5x + 4)$$
$$5x^2 + 2x + 4 = (5x + 4)(x + 1) + 0.$$

Thus, $5x + 4$ is a gcd of $f(x)$ and $g(x)$. Now let us apply the method discussed in the proof of the preceding corollary. We have

$$5x + 4 = g(x) - (5x^2 + 2x + 4)(4x + 2)$$
$$= g(x) - (f(x) - g(x)(5))(4x + 2)$$
$$= f(x)(3x + 5) + g(x)(6x + 4).$$

If we want to use $(f(x), g(x))$, we must make it monic. Now, $5^{-1} = 3$, and therefore $(f(x), g(x)) = 3(5x + 4) = x + 5$. Then we get

$$x + 5 = 3(5x + 4) = f(x)(2x + 1) + g(x)(4x + 5).$$

Corollary 10.4. *Let R be a Euclidean domain, and let $a, b \in R$ with $b \neq 0$. Let d be the gcd of a and b found in Theorem 10.4. Then if $c \in R$ is a divisor of both a and b, then $c \mid d$.*

Proof. We have $d = au + bv$, for some $u, v \in R$. If $c \mid a$ and $c \mid b$, then $c \mid d$. □

Let us now discuss how the gcds of two elements of a Euclidean domain relate to each other.

Definition 10.8. Let R be a commutative ring with identity. If $a, b \in R$, then we say that a and b are **associates** if there exists a unit u of R such that $b = au$.

Note that if $b = au$, where u is a unit, then $a = bu^{-1}$. Thus, if a is an associate of b, then b is an associate of a.

Example 10.10. In $\mathbb{Z}$, the only units are 1 and -1, so the only associates of a are a and $-a$.

Example 10.11. Let F be a field. The units in $F[x]$ are the nonzero constants. (See Exercise 10.4.) Thus, the associates of $f(x)$ are of the form $af(x)$, where $0 \neq a \in F$.

Lemma 10.1. *Let R be an integral domain. Then a and b are associates in R if and only if $a \mid b$ and $b \mid a$.*

Proof. If a and b are associates, the fact that $a \mid b$ and $b \mid a$ follows from the definition. Suppose that $a \mid b$ and $b \mid a$, say $b = ar$ and $a = bs$, with $r, s \in R$. Then $a = ars$. If $a = 0$, then $b = 0$, so $a = b \cdot 1$. Otherwise, by cancellation, $rs = 1$, and hence r is a unit. □

Theorem 10.5. *Let R be a Euclidean domain. Take $a, b \in R$, not both 0. Let d be any gcd of a and b. Then $c \in R$ is a gcd of a and b if and only if c and d are associates.*

Proof. Suppose that c is a gcd of a and b. Let g be the gcd of a and b found in Theorem 10.4. By Corollary 10.4, c and d divide g. Applying the division algorithm, we have $c = gq + r$, where $q, r \in R$ and either $r = 0$ or $\varepsilon(r) < \varepsilon(g)$. Suppose the latter. Now, $c \mid g$, and therefore $c \mid r = c - gq$. But then $\varepsilon(c) \leq \varepsilon(r)$. However, if c and g are both gcds, we must have $\varepsilon(c) = \varepsilon(g)$, giving us a contradiction. Therefore, $r = 0$ and $g \mid c$. By the preceding lemma, g and c are associates, say $c = gu$, with u a unit in R. By the same argument, $d = gv$, where v is a unit in R. Then $c = duv^{-1}$, where uv^{-1} is a unit in R, and hence c and d are associates.

Conversely, let c and d be associates. Then since $d \mid a$ and $d \mid b$, we have $c \mid a$ and $c \mid b$ as well. Furthermore, since $c \mid d$ and $d \mid c$ we can only have $\varepsilon(c) = \varepsilon(d)$. Therefore, c is a gcd of a and b. □

We can now feel better about Definition 10.7, where we referred to "the" monic gcd of $f(x)$ and $g(x)$ in $F[x]$. As any two gcds are associates, and the only units are nonzero elements of F, there can only be one gcd that is monic.

Time to tidy up! We can strengthen Corollary 10.3. It actually applies to any gcd, not just the one found in Theorem 10.4.

Theorem 10.6. *Let R be a Euclidean domain. Take $a, b \in R$, not both 0. Let d be a gcd of a and b. Then there exist $u, v \in R$ such that $d = au + bv$.*

Proof. Without loss of generality, assume that $b \neq 0$, and calculate the gcd g of a and b from Theorem 10.4. Then by Corollary 10.3, $g = au + bv$, for some $u, v \in R$. But by Theorem 10.5, $d = gw$, for some unit w of R. Thus, $d = auw + bvw$. □

We conclude by strengthening Corollary 10.4.

Theorem 10.7. *Let R be a Euclidean domain. Take $a, b \in R$, not both 0. Then the following are equivalent for an element d of R:*

1. *d is a gcd of a and b; and*
2. *$d|a$, $d|b$, and if $c|a$ and $c|b$, then $c|d$.*

Proof. Suppose (1) holds. Without loss of generality, assume that $b \neq 0$. By definition, $d|a$ and $d|b$. Suppose that $c|a$ and $c|b$. If g is the gcd of a and b found in Theorem 10.4, then by Corollary 10.4, $c|g$. But Theorem 10.5 tells us that $g|d$. Thus, $c|d$.

Conversely, suppose that (2) holds. Then d is a common divisor of a and b. Suppose that c is another common divisor of a and b. Then by assumption, $c|d$. But this means that $\varepsilon(c) \leq \varepsilon(d)$; hence, d is a gcd. □

A nice feature of Theorem 10.7 is that it shows that gcds in a Euclidean domain do not depend upon the particular Euclidean function that is used.

Exercises

10.11. In an integral domain, if a and ab are associates, show that $a = 0$ or b is a unit.

10.12. Show that every field is a Euclidean domain.

10.13. Let R be a Euclidean domain. Let n be the smallest value of $\varepsilon(s)$, for all $0 \neq s \in R$. Show that for each $0 \neq a \in R$ we have $\varepsilon(a) = n$ if and only if a is a unit.

10.14. Find all units in the ring of Gaussian integers.

10.15. In $\mathbb{Q}[x]$, let $f(x) = 3x^4 + 7x^3 + 13x^2 + 7x + 6$ and $g(x) = 2x^4 + 7x^3 + 13x^2 + 11x + 3$. Find $(f(x), g(x))$.

10.16. In $\mathbb{Z}_5[x]$, let $f(x) = 3x^4 + 3x^3 + x + 1$ and $g(x) = 2x^3 + 4x^2 + x + 1$. Find $(f(x), g(x))$.

10.17. Taking $f(x)$ and $g(x)$ as in Exercise 10.15, find $u(x), v(x) \in \mathbb{Q}[x]$ such that $(f(x), g(x)) = f(x)u(x) + g(x)v(x)$.

10.18. Taking $f(x)$ and $g(x)$ as in Exercise 10.16, find $u(x), v(x) \in \mathbb{Z}_5[x]$ such that $(f(x), g(x)) = f(x)u(x) + g(x)v(x)$.

10.19. Find a gcd for $5 + 7i$ and $1 + 3i$ in the ring of Gaussian integers.

10.20. Let R be a Euclidean domain having the following additional property: for every $a, b \in R$ such that a, b and $a + b$ are all nonzero, $\varepsilon(a + b)$ is no bigger than the larger of $\varepsilon(a)$ and $\varepsilon(b)$. (For example, if F is a field, the degree function on $F[x] \setminus \{0\}$ has this property.) Show that in the second part of the definition of a Euclidean function, the elements q and r are uniquely determined.

10.3 Principal Ideal Domains

Let us discuss another sort of integral domain with a nice property.

Definition 10.9. A **principal ideal domain** (or **PID**) is an integral domain in which every ideal is principal.

A field F is an obvious example of a PID; indeed, its only ideals are (0) and $F = (1)$. But we can obtain others through the following theorem.

Theorem 10.8. *Every Euclidean domain is a PID.*

Proof. Let R be a Euclidean domain with Euclidean function ε, and I an ideal of R. If $I = \{0\}$, then $I = (0)$, and there is nothing to do. Assume that $I \neq \{0\}$. Among the nonzero elements of I, choose b so that $\varepsilon(b)$ is as small as possible. (Since ε takes on values that are nonnegative integers, there must be a smallest such value.) We claim that $I = (b)$. Take $a \in I$. As ε is a Euclidean function, we have $a = bq + r$, where $q, r \in R$ and either $r = 0$ or $\varepsilon(r) < \varepsilon(b)$. If $r = 0$, then $b|a$, as required. Otherwise, we note that $a, b \in I$, and since I is an ideal, $r = a - bq \in I$. But by the minimality of $\varepsilon(b)$, this is impossible. $\qquad\square$

Example 10.12. Since $\mathbb{Z}$ is a Euclidean domain, it is a PID.

Example 10.13. Let F be a field. Since $F[x]$ is a Euclidean domain, it is a PID.

Proving that an integral domain is not a Euclidean domain can be a bit tricky; it is often simpler to show that is not a PID, from which it follows that it is not a Euclidean domain.

Example 10.14. We claim that $\mathbb{Z}[x]$ is not a PID, and hence not a Euclidean domain. To prove this, consider the set I of all $f(x) \in \mathbb{Z}[x]$ whose constant terms are divisible by 5. We saw in Exercise 9.2 that I is an ideal. But it is not principal. Indeed, suppose that $I = (f(x))$. Then as the constant polynomial 5 is in I, we see that $f(x)|5$. In view of Theorem 10.2, $f(x)$ is a constant polynomial. As it divides 5, the constant must be in $\{\pm 1, \pm 5\}$. However, $(1) = (-1) = \mathbb{Z}[x]$, whereas $(5) = (-5) = 5\mathbb{Z}[x]$, which does not include x. But $x \in I$, and therefore $5\mathbb{Z}[x] \neq I$.

We might, at this point, ask if every PID is a Euclidean domain. The answer is no, but this is not obvious. Theodore S. Motzkin showed that there is a subring of the complex numbers that is a PID but not a Euclidean domain. We will not use this fact, but the interested reader can find an accessible proof in the paper of Wilson [1].

Let us explore a couple of other properties of PIDs. The following theorem shows that a PID has the **ascending chain condition**.

Theorem 10.9. *Let R be a PID. Suppose that R has ideals I_k, $k \in \mathbb{N}$, such that $I_1 \subseteq I_2 \subseteq I_3 \subseteq \cdots$. Then there exists a positive integer n such that $I_k = I_n$ for all $k \geq n$.*

Proof. Let $I = \bigcup_{k=1}^{\infty} I_k$. We claim that I is an ideal. Certainly $0 \in I_1 \subseteq I$. If $a, b \in I$, then there exist positive integers k and l such that $a \in I_k$ and $b \in I_l$. Let m be the larger of k and l. Then $a, b \in I_m$, and hence $a - b \in I_m \subseteq I$. Similarly, if $a \in I$, say $a \in I_k$, and $r \in R$, then $ra \in I_k \subseteq I$. Thus, I is an ideal. As R is a PID, we must have $I = (c)$ for some $c \in I$. But then $c \in I_n$, for some positive integer n. It now follows that $I = (c) \subseteq I_n$. That is, $I = I_n$, and hence $I_k = I_n$ for all $k \geq n$. $\square$

We are familiar with the notion of a prime positive integer. Let us extend the idea.

Definition 10.10. Let R be an integral domain. Then an element p of R is **prime** if it is not zero, not a unit, and if $p|ab$, with $a, b \in R$, then $p|a$ or $p|b$.

We observe that the definition of a prime positive integer that we introduced in Chapter 2 is different. However, Theorem 2.7 assures us that the definitions are equivalent, for positive integers. Of course, the positive integers do not form a ring, so in $\mathbb{Z}$, we see that the primes are $\pm 2, \pm 3, \pm 5, \ldots$. (Note that 1 and -1 are units, so we exclude them.)

We have an easy lemma.

Lemma 10.2. *Let R be an integral domain, and take $0 \neq p \in R$. Then p is prime if and only if (p) is a prime ideal.*

Proof. Let p be prime. If $(p) = R$, then there exists an $r \in R$ such that $rp = 1$; hence, p is a unit. But primes cannot be units, so this is impossible. If $ab \in (p)$, then $p|ab$, and hence $p|a$ or $p|b$. Thus, $a \in (p)$ or $b \in (p)$, and (p) is a prime ideal. Conversely, suppose that (p) is a prime ideal and $p|ab$. Then $ab \in (p)$, and hence $a \in (p)$ or $b \in (p)$. That is, $p|a$ or $p|b$. Furthermore, if p is a unit, then by Theorem 9.2, $(p) = R$, which contradicts the assumption that (p) is a prime ideal. Thus, p is prime. $\square$

Definition 10.11. Let R be an integral domain, and take $p \in R$. We say that p is **irreducible** if it is not zero, not a unit, and if $p = ab$, with $a, b \in R$, then either a or b must be a unit.

This is, essentially, the definition we used for a prime positive integer. As we noted above, in the integers, these concepts are equivalent. What is the general situation?

Theorem 10.10. *Let R be an integral domain. Then every prime in R is irreducible.*

Proof. Let p be a prime, and suppose that $p = ab$, with $a, b \in R$. Then $p|ab$, so $p|a$ or $p|b$. Without loss of generality, say $p|a$. But $a|p$ as well. By Lemma 10.1, p and a are associates. Thus, by Exercise 10.11, b is a unit, as required. $\qquad\square$

Unfortunately, the converse is not true in general.

Example 10.15. Let $R = \{a + b\sqrt{5}i : a, b \in \mathbb{Z}\}$. It is easy to check that R is a unital subring of $\mathbb{C}$, and hence an integral domain. We can define a function called a **norm** on R via $N(a + b\sqrt{5}i) = a^2 + 5b^2$. If $u, v \in R$, then $N(uv) = N(u)N(v)$. (This is the same calculation as in Example 10.8.) We claim that 3 is irreducible in R. If $3 = uv$, then $9 = N(3) = N(u)N(v)$. Noting that the norms of elements of R are nonnegative integers, we can only have $N(u) = N(v) = 3$ or, without loss of generality, $N(u) = 1$ and $N(v) = 9$. But the equation $a^2 + 5b^2 = 3$ has no solution in the integers, so $N(u) = N(v) = 3$ is impossible. Also, the only solutions to $a^2 + 5b^2 = 1$ are $a \in \{1, -1\}$ and $b = 0$. However, 1 and -1 are units in R. Also, 3 is clearly not a unit, and the claim is proved. Nevertheless, 3 is not prime. To see this, we note that $(2 + \sqrt{5}i)(2 - \sqrt{5}i) = 9$. Of course, $3|9$, but 3 does not divide $2 + \sqrt{5}i$ or $2 - \sqrt{5}i$.

The good news, however, is that in a PID, primeness and irreducibility are equivalent.

Theorem 10.11. *Let R be a PID and $p \in R$. Then p is prime if and only if p is irreducible.*

Proof. In view of Theorem 10.10, we only need to show the converse. Let p be irreducible, and let $I = (p)$. We claim that I is a maximal ideal of R. If not, suppose that J is an ideal of R with $I \subsetneq J \subsetneq R$. Since R is a PID, we have $J = (a)$, for some $a \in J$. Now, $p \in I \subseteq J$, so $p = ab$, for some $b \in R$. As p is irreducible, either a or b is a unit. If a is a unit, then by Theorem 9.2, $J = R$, which is not permitted. Therefore, b is a unit. But then $a = pb^{-1} \in I$. Thus, $J \subseteq I$, which is also not allowed. On the other hand, if $I = R$, then p is a unit, which is impossible. Our claim is proved.

By Theorem 9.22, a maximal ideal is necessarily prime. Lemma 10.2 completes the proof. $\qquad\square$

Exercises

10.21. With R as in Example 10.15, show that $1 + 2\sqrt{5}i$ is irreducible, but not prime.

10.22. Let S be $\{a + b\sqrt{3}i \; : \; a, b \in \mathbb{Z}\}$, a subring of $\mathbb{C}$. Show that $1 + \sqrt{3}i$ is irreducible, but not prime.

10.23. Show that R and S from the preceding two exercises are not PIDs.

10.24. Let R be an integral domain. Show that an associate of an irreducible element is irreducible, and an associate of a prime element is prime.

10.25. If R is a Euclidean domain, does it follow that $R[x]$ is a Euclidean domain? Prove that it does, or give an explicit counterexample.

10.26. Let R be a PID. Show that every proper ideal of R is a subset of a maximal ideal of R.

10.27. Let R be an integral domain and p a prime in R. If $p | a_1 a_2 \cdots a_n$, with $a_i \in R$, show that some a_i is divisible by p.

10.28. Let R be a PID and $0 \neq a \in R$. Show that a is irreducible if and only if (a) is a maximal ideal.

10.29. Let R be an integral domain, but not a field. Show that there exist infinitely many ideals $I_1, I_2, \ldots$ of R such that I_{n+1} is a proper subset of I_n for all n.

10.30. Let R be an integral domain. If $R[x]$ is a PID, show that R is a field.

10.4 Unique Factorization Domains

We now reach our main conclusion, which is that every PID has an analogue of the Fundamental Theorem of Arithmetic.

Definition 10.12. Let R be an integral domain. We say that R is a **unique factorization domain** (or **UFD**) if

1. every nonzero, nonunit element of R can be written as a product of one or more irreducibles; and
2. the product is unique up to order and associates; that is, if $p_1 p_2 \cdots p_k = q_1 q_2 \cdots q_l$, for some irreducibles p_i and q_i, then $k = l$ and, after rearranging, each p_i is an associate of q_i.

Theorem 10.12. *Every PID is a UFD.*

Proof. Let R be a PID. We shall prove that R satisfies the first part of the definition of a UFD. Take any nonzero nonunit $a_1 \in R$, and suppose that a_1 is not a product of irreducibles. If a_1 is irreducible then we have an immediate contradiction. Therefore, we may write $a_1 = a_2 b_2$, where a_2 and b_2 are nonunits in R. If a_2 and b_2 are both products of irreducibles then again, we have a contradiction, as a_1 is then a product of irreducibles. Without loss of generality, let us say that a_2 is not a product of irreducibles. In particular, it is not irreducible, so write $a_2 = a_3 b_3$, where a_3 and b_3 are nonunits, and so forth. Then we have $a_{i+1}|a_i$ for all positive integers i. Furthermore, as b_{i+1} is not a unit, we see that a_i and a_{i+1} are not associates. By Lemma 10.1, a_i does not divide a_{i+1}. In particular, $a_i \in (a_{i+1})$, but $a_{i+1} \notin (a_i)$, so $(a_i) \subsetneq (a_{i+1})$ for all positive integers i. But this contradicts Theorem 10.9, and we see that each nonzero nonunit is a product of irreducibles.

Now let us verify the uniqueness. Suppose that $p_1 \cdots p_k = q_1 \cdots q_l$, where the p_i and q_i are irreducible, and $k \leq l$. Then $p_1|q_1 \cdots q_l$. By Theorem 10.11, p_1 is prime. Thus, p_1 divides one of the terms in the product. After rearranging, we may assume that $p_1|q_1$. Let us write $q_1 = p_1 u_1$, with $u_1 \in R$. As q_1 is irreducible and p_1 is not a unit, we see that u_1 is a unit, and hence p_1 and q_1 are associates. Thus, $p_1 p_2 \cdots p_k = u_1 p_1 q_2 \cdots q_l$. Cancelling, we have $p_2 \cdots p_k = u_1 q_2 \cdots q_l$. Now, $p_2|u_1 q_2 \cdots q_l$, and since p_2 is prime, it divides a term in the product. Since a divisor of a unit is a unit, we cannot have $p_2|u_1$, and therefore $p_2|q_i$, for some $i \geq 2$. Rearranging, we have $p_2|q_2$. Just as before, we see that $q_2 = p_2 u_2$, for some unit u_2. Repeating, we find that p_i and q_i are associates, $1 \leq i \leq k$. If $k = l$, we are done. Otherwise, we have $1 = u_1 \cdots u_k q_{k+1} \cdots q_l$. But nonunits cannot divide 1, so we have a contradiction. □

Our examples of UFDs will largely be PIDs.

Example 10.16. As we already knew from the Fundamental Theorem of Arithmetic, $\mathbb{Z}$ is a UFD.

Example 10.17. For any field F, $F[x]$ is a UFD.

There are also UFDs that are not PIDs. In fact, $\mathbb{Z}[x]$ is such a ring. We opt to postpone the proof of this until Section 11.2.

What sort of integral domains are not UFDs? Either of the two conditions could fail. Let us first consider one where nonzero nonunit elements are not necessarily products of irreducibles.

Example 10.18. Let R be the subset of $\mathbb{Q}[x]$ consisting of all polynomials with an integer constant term. It is easy to see that R is a unital subring of $\mathbb{Q}[x]$. As $\mathbb{Q}[x]$ is an integral domain, so is R. We claim that the only units of R are the constant polynomials 1 and -1. Indeed, a unit is necessarily a unit in $\mathbb{Q}[x]$ as well. By Exercise 10.4, our unit is a nonzero constant a. But as the constant term of an element of R must be an integer, we see that if $af(x) = 1$, then a can only be ± 1, proving the claim. In particular, x is a nonzero nonunit. If we write $x = p_1(x) \cdots p_k(x)$, a product of irreducibles, then all but one of the $p_i(x)$ (say $p_1(x)$) are integers and $p_1(x) = qx$, for some $0 \neq q \in \mathbb{Q}$. But qx is not irreducible; indeed, $qx = 2\left(\frac{q}{2}x\right)$, and neither 2 nor $\frac{q}{2}x$ is a unit. Thus, x is not a product of irreducibles, and R is not a UFD.

Even if every nonzero nonunit is a product of irreducibles, this product may not be unique.

Example 10.19. Consider the ring $R = \{a + b\sqrt{5}i : a, b \in \mathbb{Z}\}$ from Example 10.15. We noted in that example that 3 is irreducible. Applying a similar argument, we can see that $2 + \sqrt{5}i$ and $2 - \sqrt{5}i$ are irreducible. (We do have to check that they are not units, but if $uv = 1$, then $N(u)N(v) = N(1) = 1$, and as we noted in Example 10.15, this must mean that $u = v = \pm 1$.) Thus, we can write $9 = 3 \cdot 3 = (2 + \sqrt{5}i)(2 - \sqrt{5}i)$, giving two different products of irreducibles. As the only units are ± 1, we see that 3 is not an associate of $2 + \sqrt{5}i$ or $2 - \sqrt{5}i$. Therefore, our factorization is not unique.

We close with a few remarks concerning divisibility in a UFD.

Theorem 10.13. *Let R be a UFD, and let a and b be nonzero nonunit elements of R. Then there exist irreducibles $p_1, \ldots, p_k$, none of which are associates, such that $a = u p_1^{m_1} \cdots p_k^{m_k}$ and $b = v p_1^{n_1} \cdots p_k^{n_k}$, for some units $u, v \in R$ and some nonnegative integers m_i and n_i. Furthermore, $a | b$ if and only if $m_i \leq n_i$ for all i.*

Proof. Write each of a and b as a product of irreducibles. List all of the irreducibles that appear, and if some are associates, say $q_1, q_2, \ldots$, then delete all but one. Let $p_1, \ldots, p_k$ be the irreducibles that remain. Then a can be written as a product of irreducibles, each of which is an associate of some p_i, and so can be written as a product of a unit and p_i. Gathering the units together, we obtain our expression for a, and similarly for b.

If $m_i \leq n_i$ for all i, then we see that $b = avu^{-1}p_1^{n_1-m_1} \cdots p_k^{n_k-m_k}$; hence, $a | b$. Conversely, without loss of generality, suppose that $m_1 > n_1$. If $a | b$, then write $b = ac$. As R is an integral domain, we can use cancellation, and obtain

$$p_1^{m_1-n_1} p_2^{m_2} \cdots p_k^{m_k} c = u^{-1} v p_2^{n_2} \cdots p_k^{n_k}.$$

Here, c is either a unit or a product of irreducibles. By unique factorization, p_1 must be an associate of one of $u^{-1}v p_2, p_3, \ldots, p_k$. But by our choice of the p_i, this is impossible. $\square$

A UFD does not necessarily have anything comparable to a Euclidean function, so we cannot order elements in any logical way. However, we can obtain the equivalent form of a gcd given in Theorem 10.7.

Theorem 10.14. *Let R be a UFD. Take any nonzero nonunits $a, b \in R$, and write them in the form $a = u p_1^{m_1} \cdots p_k^{m_k}$, $b = v p_1^{n_1} \cdots p_k^{n_k}$, as in Theorem 10.13. Let $d = p_1^{l_1} \cdots p_k^{l_k}$, where l_i is the smaller of m_i and n_i, for all i. Then $d | a$, $d | b$, and if $c | a$ and $c | b$, then $c | d$.*

Proof. Theorem 10.13 tells us that $d|a$ and $d|b$. Suppose that $c|a$ and $c|b$. If c is a unit, then surely $c|d$. Suppose it is not. Then write $a = cr$, with $r \in R$. Now c can be written as a product of irreducibles, and r is a unit or a product of irreducibles. By unique factorization, all of these irreducibles must be associates of the p_i. Using Theorem 10.13 again, we can write $c = w p_1^{j_1} \cdots p_k^{j_k}$, where w is a unit and $j_k \leq m_k$, for all k. By the same argument, as $c|b$, we have $j_k \leq n_k$, and hence $j_k \leq l_k$, for all k. Therefore, $c|d$.   □

Exercises

10.31. Show that $1 + i$ is prime in the ring R of Gaussian integers.

10.32. In the ring of Gaussian integers, which of the numbers 3, 5 and 7 are irreducible?

10.33. Must a unital subring of a UFD be a UFD? Prove that it must, or give an explicit counterexample.

10.34. Let R be a UFD. Suppose that a and b are nonzero nonunit elements of R. If d_1 and d_2 are gcds of a and b (in the sense discussed in the second part of Theorem 10.7), show that d_1 and d_2 are associates.

10.35. Let $R = \{a + b\sqrt{6}i : a, b \in \mathbb{Z}\}$. Find $a, b, c, d \in R$ such that $10 = ab = cd$, but a, b, c and d are all irreducible and neither of $\{a, b\}$ is an associate of either of $\{c, d\}$. Conclude that R is not a UFD.

10.36. Let R be a UFD, and let p be an irreducible element of R. If a and b are nonzero nonunits of R, and $p|ab$, writing both a and b as products of irreducibles, show that p is an associate of at least one of the irreducibles appearing in at least one of these products.

10.37. Show that every irreducible in a UFD is prime.

10.38. Let R be a UFD. Suppose that there exist $a_1, a_2, \ldots \in R$ such that $(a_1) \subseteq (a_2) \subseteq \cdots$. Show that there exists an i such that $(a_i) = (a_{i+1})$.

Reference

1. Wilson, J.C.: A principal ideal ring that is not a Euclidean ring. Math. Mag. **46**, 34–38 (1973)

Part IV
Fields and Polynomials

Chapter 11
Irreducible Polynomials

Let $F[x]$ be the polynomial ring over a field F. If $f(x) \in F[x]$, we can now discuss some conditions under which $f(x)$ is irreducible.

11.1 Irreducibility and Roots

For any field F, we recall that the polynomial ring $F[x]$ is a UFD (see Example 10.17). Also, by Exercise 10.4, the units in $F[x]$ are the nonzero elements of F. Thus, every polynomial of degree greater than 0 is a product of one or more irreducibles. Here, a polynomial $f(x)$ of degree greater than 0 is irreducible over F if, whenever $f(x) = g(x)h(x)$ for some $g(x), h(x) \in F[x]$, either $g(x)$ or $h(x)$ is an element of F. Otherwise, $f(x)$ is reducible. Note that irreducibility depends very much upon the particular field.

Example 11.1. The polynomial $x^2 - 2$ is irreducible over $\mathbb{Q}$, but reducible over $\mathbb{R}$, since $x^2 - 2 = (x - \sqrt{2})(x + \sqrt{2})$.

Let $f(x) = a_0 + a_1 x + \cdots + a_n x^n \in F[x]$. If $r \in F$, we can **evaluate** $f(x)$ at r, and obtain

$$f(r) = a_0 + a_1 r + a_2 r^2 + \cdots + a_n r^n.$$

In this way, we obtain a function (not a homomorphism!) $\alpha : F \to F$ given by $\alpha(r) = f(r)$. In dealing with polynomials in $\mathbb{R}[x]$, we are accustomed to identifying the polynomial $f(x)$ with this function α. But over a more general field, we cannot do this. Indeed, two different polynomials can induce the same function.

Example 11.2. In $\mathbb{Z}_5[x]$, the polynomials $f(x) = x^3 + x + 1$ and $g(x) = x^5 + x^3 + 1$ induce the same function. That is, $f(r) = g(r)$ for all $r \in \mathbb{Z}_5$. (There are only five elements in $\mathbb{Z}_5$, so this is easily checked.)

It is worth mentioning that we do obtain a homomorphism if we fix an element r of the field and consider evaluating polynomials at r.

© Springer International Publishing AG, part of Springer Nature 2018
G. T. Lee, *Abstract Algebra*, Springer Undergraduate Mathematics Series,
https://doi.org/10.1007/978-3-319-77649-1_11

Lemma 11.1. *Let R be a commutative ring and fix $r \in R$. Then the function $\alpha :$ $R[x] \to R$ given by $\alpha(f(x)) = f(r)$ is a homomorphism.*

Proof. Let $f(x) = a_0 + \cdots + a_n x^n$ and $g(x) = b_0 + \cdots + b_n x^n$ be arbitrary polynomials in $R[x]$ (adding in terms with coefficient zero if necessary). Then

$$
\begin{aligned}
\alpha(f(x) + g(x)) &= a_0 + b_0 + a_1 r + b_1 r + \cdots + a_n r^n + b_n r^n \\
&= (a_0 + \cdots + a_n r^n) + (b_0 + \cdots + b_n r^n) \\
&= \alpha(f(x)) + \alpha(g(x)).
\end{aligned}
$$

Also, writing $f(x)g(x) = c_0 + \cdots + c_{2n} x^{2n}$, where $c_i = a_0 b_i + \cdots + a_i b_0$, we have

$$
\alpha(f(x)g(x)) = c_0 + \cdots + c_{2n} r^{2n},
$$

whereas

$$
\alpha(f(x))\alpha(g(x)) = (a_0 + \cdots + a_n r^n)(b_0 + \cdots + b_n r^n).
$$

But for any i,

$$
a_0 b_i r^i + a_1 r b_{i-1} r^{i-1} + \cdots + a_i r^i b_0 = c_i r^i,
$$

and so $\alpha(f(x)g(x)) = \alpha(f(x))\alpha(g(x))$. □

We can now use the division algorithm to write a polynomial over a field F as a multiple of $x - a$, for any $a \in F$, plus a constant.

Theorem 11.1 (Remainder Theorem). *Let F be a field and $f(x) \in F[x]$. Take any $a \in F$. Then there exists a $q(x) \in F[x]$ such that*

$$
f(x) = (x - a)q(x) + f(a).
$$

Proof. By the division algorithm for polynomials, $f(x) = (x - a)q(x) + r(x)$, where $q(x), r(x) \in F[x]$, and either $r(x)$ is the zero polynomial, or $\deg(r(x)) < \deg(x - a) = 1$. That is, $r(x)$ is some constant, $b \in F$. By the preceding lemma,

$$
f(a) = (a - a)q(a) + b = b.
$$ □

It is crucial for us to know if a polynomial has any roots.

Definition 11.1. Let F be a field and $f(x) \in F[x]$. If $a \in F$, then we say that a is a **root** of $f(x)$ if $f(a) = 0$.

Example 11.3. The polynomial $x^2 - 2$ has no roots in $\mathbb{Q}$. However, if we regard it as a polynomial over $\mathbb{R}$, we see that $\sqrt{2}$ and $-\sqrt{2}$ are roots.

Recall that if $f(x), g(x) \in F[x]$, we say that $f(x)$ divides $g(x)$, and write $f(x)|g(x)$, if there exists an $h(x) \in F[x]$ such that $g(x) = f(x)h(x)$.

Theorem 11.2 (Factor Theorem). *Let F be a field and $f(x) \in F[x]$. Take any $a \in F$. Then a is a root of $f(x)$ if and only if $(x - a)|f(x)$.*

Proof Suppose that a is a root of $f(x)$. By the Remainder Theorem, we have $f(x) = (x - a)q(x)$, and hence $(x - a)|f(x)$. Conversely, suppose that $(x - a)|f(x)$. Then $f(x) = (x - a)g(x)$, for some $g(x) \in F[x]$. In this case, $f(a) = (a - a)g(a) = 0$, and hence a is a root. $\square$

Example 11.4. In $\mathbb{Z}_7[x]$, let $f(x) = 3x^3 + 5x^2 + 4x + 4$. We note that 2 is a root. Thus, $x - 2$ (in other words, $x + 5$) must divide $f(x)$. In fact, $f(x) = (x - 2)(3x^2 + 4x + 5)$.

Corollary 11.1. *Let F be a field and $f(x) \in F[x]$. If $\deg(f(x)) > 1$ and $f(x)$ has a root in F, then $f(x)$ is reducible over F.*

Proof. Let a be a root of $f(x)$. By the Factor Theorem, $f(x) = (x - a)g(x)$, for some $g(x) \in F[x]$. Since $\deg(f(x)) > 1$, we note that $g(x)$ is not a constant. Thus, $f(x)$ is reducible. $\square$

The converse is false!

Example 11.5. In $\mathbb{R}[x]$, let $f(x) = x^4 + 2x^2 + 1$. For any $a \in \mathbb{R}$, we have $f(a) \geq 1$; thus, $f(x)$ has no real roots. However, $f(x) = (x^2 + 1)^2$. Thus, $f(x)$ is reducible.

However, for polynomials of degree 2 and 3, the converse does hold.

Corollary 11.2. *Let F be a field and $f(x) \in F[x]$. Then*

1. *if $\deg(f(x)) = 1$, then $f(x)$ is irreducible over F; and*
2. *if $f(x)$ has degree 2 or 3, then $f(x)$ is irreducible over F if and only if it has no roots in F.*

Proof. (1) If $f(x) = g(x)h(x)$, then by Theorem 10.2, either $g(x)$ or $h(x)$ has degree 0.

(2) If $f(x)$ is irreducible, then the preceding corollary tells us that $f(x)$ has no roots. Suppose that $f(x)$ is reducible, say $f(x) = g(x)h(x)$ for some nonconstant polynomials $g(x)$ and $h(x)$ in $F[x]$. As the sum of their degrees is 2 or 3, either $g(x)$ or $h(x)$ must have degree 1. Without loss of generality, say $g(x) = ax + b$, with $a, b \in F$ and $a \neq 0$. But then notice that $f(-a^{-1}b) = (a(-a^{-1}b) + b)h(-a^{-1}b) = 0$. Thus, $-a^{-1}b$ is a root of $f(x)$. $\square$

We can also put a limit on the number of roots of a polynomial.

Corollary 11.3. *Let F be a field and $f(x) \in F[x]$ a nonzero polynomial. If $f(x)$ has degree n, then $f(x)$ has at most n roots in F.*

Proof. We proceed by induction on n. If $n = 0$, then $f(x)$ is a nonzero constant polynomial, which clearly has no roots. Assume that our result is true for n, and let $\deg(f(x)) = n + 1$. If $f(x)$ has no roots, then we are done. Otherwise, let a be a root. By Theorem 11.2, $f(x) = (x - a)g(x)$, for some $g(x) \in F[x]$. Furthermore, by Theorem 10.2, $\deg(g(x)) = n$. Thus, our inductive hypothesis tells us that $g(x)$ has at most n roots. Let b be any root of $f(x)$. Then $0 = f(b) = (b - a)g(b)$. Therefore, either $b - a = 0$ (and $b = a$) or $g(b) = 0$ (and b is among the at most n roots of $g(x)$). Thus, $f(x)$ has at most $n + 1$ roots, as required. $\square$

Exercises

11.1. Are the following polynomials irreducible in $\mathbb{Z}_7[x]$?

1. $x^3 + 5x^2 + 4x + 3$
2. $x^3 + x^2 + 1$
3. $x^4 + x^2 + 2$

11.2. Write each of the following as products of irreducibles in $\mathbb{Z}_5[x]$.

1. $x^3 + 3x^2 + 3x + 2$
2. $x^3 + 2x^2 + 4x + 2$
3. $x^4 + 2x^3 + 4x + 3$

11.3. Find every irreducible polynomial of degree 3 over $\mathbb{Z}_2$.

11.4. If we divide $3x^{59} + 4x^{16} + 2$ by $x + 5$ in $\mathbb{Z}_7[x]$, what is the remainder? (The answer must be in $\{0, 1, \ldots, 6\}$.)

11.5. Let F be an infinite field. If $f(x), g(x) \in F[x]$, and $f(a) = g(a)$ for all $a \in F$, show that $f(x) = g(x)$.

11.6. Let p be a prime. Find infinitely many polynomials $f_1(x), f_2(x), \ldots$ in $\mathbb{Z}_p[x]$ such that $f_i(a) = 0$ for all $a \in \mathbb{Z}_p$ and all positive integers i.

11.7. Is Lemma 11.1 still true for noncommutative rings?

11.8. Let R be an integral domain. Show that $U(R)$ has at most n elements of order n, for every positive integer n. Also give an example of a commutative ring R with identity which is not an integral domain for which this is not true.

11.9. Let p be a prime number. Show that the following are equivalent:

1. $x^2 + 1$ is reducible in $\mathbb{Z}_p[x]$; and
2. there exist nonnegative integers m and n such that $p = m + n$ and $p|(mn - 1)$.

11.10. Show that Theorems 11.1 and 11.2 remain true if F is replaced with an integral domain.

11.2 Irreducibility over the Rationals

If we have a polynomial $f(x) \in \mathbb{Q}[x]$, then by multiplying by a suitable positive integer, we obtain a polynomial in $\mathbb{Z}[x]$. It is often simpler to start with a polynomial with integer coefficients.

As we noted in the preceding section, a polynomial of degree greater than 1 in $\mathbb{Q}[x]$ is necessarily reducible if it has a root. Of course, there are infinitely many possible roots, so testing them all is impossible. However, we can narrow the possible roots down to a finite set of rational numbers.

Theorem 11.3 (Rational Roots Theorem). *Let* $f(x) = a_0 + a_1 x + \cdots + a_n x^n \in \mathbb{Z}[x]$, *with* $a_n \neq 0$. *Suppose that* $q \in \mathbb{Q}$ *is a root of* $f(x)$. *If* $q = \frac{r}{s}$, *with* $r, s \in \mathbb{Z}$ *and* $(r, s) = 1$, *then* $r \mid a_0$ *and* $s \mid a_n$.

Proof. We have

$$0 = f\left(\frac{r}{s}\right) = a_0 + \frac{a_1 r}{s} + \cdots + \frac{a_{n-1} r^{n-1}}{s^{n-1}} + \frac{a_n r^n}{s^n}.$$

Multiplying through by s^n, we obtain

$$a_0 s^n + a_1 r s^{n-1} + \cdots + a_{n-1} r^{n-1} s + a_n r^n = 0.$$

As s divides every term except $a_n r^n$, it also divides $a_n r^n$. Since $(r, s) = 1$, Corollary 2.2 tells us that $s \mid a_n$. Similarly, r divides every term except $a_0 s^n$, so it also divides $a_0 s^n$. Since $(r, s) = 1$, we see that $r \mid a_0$. $\qquad\square$

Example 11.6. Let $f(x) = 3x^3 + 2x^2 - 2x - 8$. In view of the Rational Roots Theorem, the only possible rational roots of $f(x)$ are $\pm 1, \pm 2, \pm 4, \pm 8, \pm\frac{1}{3}, \pm\frac{2}{3}, \pm\frac{4}{3}$ and $\pm\frac{8}{3}$. Trying them all, we see that the only rational root of $f(x)$ is $\frac{4}{3}$.

Of course, polynomials can be reducible without having roots. If we wish to restrict our attention to polynomials in $\mathbb{Z}[x]$, we must be sure that it makes sense to do so. At first blush, it seems conceivable that we could have a polynomial in $\mathbb{Z}[x]$ that factors into a product of polynomials of lower degree in $\mathbb{Q}[x]$, but not in $\mathbb{Z}[x]$. In fact, this does not happen. Let us see why.

Definition 11.2. If $f(x)$ is a nonzero polynomial in $\mathbb{Z}[x]$, then the **content** of $f(x)$ is the largest positive integer that divides every coefficient of $f(x)$. We say that $f(x)$ is **primitive** if its content is 1.

Example 11.7. The polynomial $6x^3 - 15x^2 + 81x - 12$ has content 3, whereas $5x^2 + 14x - 2$ is primitive.

We can now present a famous result due to Carl F. Gauss.

Lemma 11.2 (Gauss's Lemma). *The product of two primitive polynomials in* $\mathbb{Z}[x]$ *is also primitive.*

Proof. Let $f(x) = a_0 + \cdots + a_n x^n$ and $g(x) = b_0 + \cdots + b_m x^m$ be primitive. Suppose that $f(x)g(x)$ is not primitive. Let p be a prime dividing the content of $f(x)g(x)$. As p cannot divide all of the coefficients of $f(x)$, let i be the smallest nonnegative integer such that p does not divide a_i. Similarly, let j be the smallest nonnegative integer such that b_j is not divisible by p. Then the coefficient of x^{i+j} in $f(x)g(x)$ is

$$a_0 b_{i+j} + a_1 b_{i+j-1} + \cdots + a_{i-1} b_{j+1} + a_i b_j + a_{i+1} b_{j-1} + \cdots + a_{i+j-1} b_1 + a_{i+j} b_0,$$

where we add terms with coefficient zero if necessary. Now, this coefficient must be divisible by p. Also, p divides a_k, $0 \le k < i$, and p divides b_l, $0 \le l < j$. Thus, every term in the sum is divisible by p except $a_i b_j$, which means that $p | a_i b_j$ as well. But this contradicts Theorem 2.7. □

As a consequence, we can see that if a polynomial in $\mathbb{Z}[x]$ is reducible in $\mathbb{Q}[x]$, then it is reducible in $\mathbb{Z}[x]$ as well.

Theorem 11.4. *Let $f(x)$ be a polynomial in $\mathbb{Z}[x]$, and suppose that $f(x) = g(x)h(x)$, where $g(x), h(x) \in \mathbb{Q}[x]$. Then there is a positive rational number q such that $qg(x)$ and $\frac{1}{q}h(x)$ lie in $\mathbb{Z}[x]$.*

Proof. Assume, first of all, that $f(x)$ is primitive. Choose positive integers a and b such that $ag(x), bh(x) \in \mathbb{Z}[x]$. Then $abf(x) = (ag(x))(bh(x))$.

Let c be the content of $ag(x)$ and d the content of $bh(x)$. Then $\frac{a}{c}g(x), \frac{b}{d}h(x) \in \mathbb{Z}[x]$, and both are primitive polynomials. By Gauss's lemma, their product, $\frac{ab}{cd}f(x)$, is also primitive. Thus, the content of $abf(x)$ is cd. But as $f(x)$ is primitive, the content of $abf(x)$ is also ab. Thus, $ab = cd$, and hence letting $q = \frac{a}{c}$, we see that $\frac{b}{d} = \frac{1}{q}$.

Suppose that $f(x)$ is not primitive. If it is the zero polynomial, then either $g(x)$ or $h(x)$ must be as well. Without loss of generality, say that $h(x)$ is the zero polynomial. Then let q be a positive integer such that $qg(x) \in \mathbb{Z}[x]$. On the other hand, if $f(x)$ is not the zero polynomial, then let k be its content. Writing $f(x) = kf_1(x)$, with $f_1(x) \in \mathbb{Z}[x]$, we have $f_1(x) = \left(\frac{1}{k}g(x)\right)h(x)$. By the argument above, there exists a positive rational number q such that $\frac{q}{k}g(x), \frac{1}{q}h(x) \in \mathbb{Z}[x]$. But then $qg(x), \frac{1}{q}h(x) \in \mathbb{Z}[x]$ as well. □

Example 11.8. The polynomial $f(x) = 3x^3 + 2x^2 - 2x - 8$ has $\frac{4}{3}$ as a rational root. Thus, by Theorem 11.2, $g(x) = x - \frac{4}{3}$ is a divisor of $f(x)$ in $\mathbb{Q}[x]$. Performing polynomial long division, we see that $f(x) = g(x)h(x)$, where $h(x) = 3x^2 + 6x + 6$. Using $q = 3$ in the above theorem, we find that $f(x) = (3x - 4)(x^2 + 2x + 2)$, and we have a factorization in $\mathbb{Z}[x]$.

Even if a polynomial has coefficients in $\mathbb{Z}$, it can still be difficult to tell if it is irreducible over $\mathbb{Q}$. One nice result that can be rather helpful is attributed to F. Gotthold M. Eisenstein, although a proof was first published by Theodor Schönemann.

Theorem 11.5 (Eisenstein's Criterion). *Let* $f(x) = a_0 + a_1 + \cdots + a_n x^n \in \mathbb{Z}[x]$, *with* $n \geq 1$ *and* $a_n \neq 0$. *Suppose that there exists a prime* p *such that* $p|a_i$, $0 \leq i < n$, *but* $p \nmid a_n$ *and* $p^2 \nmid a_0$. *Then* $f(x)$ *is irreducible in* $\mathbb{Q}[x]$.

Proof. If $f(x)$ is reducible, then by Theorem 11.4, there exist nonconstant polynomials $g(x) = b_0 + \cdots + b_l x^l$ and $h(x) = c_0 + \cdots + c_m x^m$ in $\mathbb{Z}[x]$, with $b_l \neq 0 \neq c_m$ and $f(x) = g(x)h(x)$. Now, p divides $a_0 = b_0 c_0$, but p^2 does not. Thus, p divides exactly one of $\{b_0, c_0\}$. Without loss of generality, say $p|b_0$. But p does not divide $a_n = b_l c_m$. Thus, p divides neither b_l nor c_m. Let i be the smallest positive integer such that $p \nmid b_i$. Then

$$a_i = b_0 c_i + b_1 c_{i-1} + \cdots + b_{i-1} c_1 + b_i c_0.$$

Now, $p|b_j$, $0 \leq j < i$. Furthermore, as $i \leq l < n$, we know that $p|a_i$. Thus, $p|b_i c_0$. But p divides neither b_i nor c_0, and we have a contradiction. $\qquad\square$

Example 11.9. The polynomial $13x^3 - 42x^2 + 81x - 15$ is irreducible over $\mathbb{Q}$, using Eisenstein's criterion with $p = 3$.

Example 11.10. For any positive integer n and any prime p, we observe that $x^n - p$ is irreducible over $\mathbb{Q}$.

Note that if F is a subfield of K, and $f(x)$ is a reducible polynomial in $F[x]$, then it is also necessarily reducible in $K[x]$ (just using the same factorization). Of course, the fact that it is reducible in $K[x]$ does not imply that it is reducible in $F[x]$, as we illustrated in Example 11.3.

But the relationship between $\mathbb{Z}[x]$ and $\mathbb{Q}[x]$ is backwards. Indeed, we have seen that if a polynomial in $\mathbb{Z}[x]$ is reducible in $\mathbb{Q}[x]$, then it is also reducible in $\mathbb{Z}[x]$. The other direction does not work!

Example 11.11. Let $f(x) = 2x - 6$. Then by Corollary 11.2, $f(x)$ is irreducible in $\mathbb{Q}[x]$. But $f(x)$ is reducible in $\mathbb{Z}[x]$; indeed, $f(x) = 2(x - 3)$, and neither 2 nor $x - 3$ is a unit in $\mathbb{Z}[x]$.

The problem, then, is that the nonzero constants are not necessarily units in $\mathbb{Z}[x]$, and this affects irreducibility.

Lemma 11.3. *Let* $f(x) \in \mathbb{Z}[x]$. *Then* $f(x)$ *is irreducible in* $\mathbb{Z}[x]$ *if and only if either*

1. $f(x)$ *is a (positive or negative) prime in* $\mathbb{Z}$; *or*
2. $f(x)$ *is a primitive polynomial that is irreducible in* $\mathbb{Q}[x]$.

Proof. Note that a unit in $\mathbb{Z}[x]$ is also a unit in $\mathbb{Q}[x]$, and hence a constant. But the only constants having inverses in $\mathbb{Z}[x]$ are ± 1, so those are the only units.

Suppose that $f(x)$ is a constant $c \in \mathbb{Z}$. If c is prime, then its only factorizations are $1 \cdot c$ and $(-1)(-c)$, so $f(x)$ is irreducible. Otherwise, c has some other factorization, and $f(x)$ is not irreducible.

So, let $\deg(f(x)) \geq 1$. Suppose that $f(x)$ is irreducible in $\mathbb{Z}[x]$. If $f(x)$ has content $d > 1$, then $f(x)$ has a factorization $d\left(\frac{1}{d}f(x)\right)$, and therefore $f(x)$ is reducible. So, we may assume that $f(x)$ is primitive. If it is reducible in $\mathbb{Q}[x]$, then by Theorem 11.4, it is reducible in $\mathbb{Z}[x]$ as well. Conversely, assume that $f(x)$ is irreducible in $\mathbb{Q}[x]$ and primitive. If $f(x) = g(x)h(x)$, with $g(x), h(x) \in \mathbb{Z}[x]$, then we have a factorization in $\mathbb{Q}[x]$ as well, which would make $f(x)$ reducible over $\mathbb{Q}$, unless either $g(x)$ or $h(x)$ is a constant. Without loss of generality, let $g(x) = e \neq 0$. If $e = \pm 1$, then $g(x)$ is a unit in $\mathbb{Z}[x]$. If not, then $f(x)$ has content $|e|$ times the content of $h(x)$, contradicting the assumption that $f(x)$ is primitive. Thus, $f(x)$ is irreducible in $\mathbb{Z}[x]$ in this case.   □

Let us now present the counterexample promised in Section 10.4. We already know that $\mathbb{Z}[x]$ is not a PID. But we have the following.

Theorem 11.6. *The ring $\mathbb{Z}[x]$ is a UFD.*

Proof. Let $f(x) \in \mathbb{Z}[x]$ be a nonzero nonunit. We will show that $f(x)$ is a product of irreducibles. First, suppose that $\deg(f(x)) = n \geq 1$. We claim that $f(x)$ is a product of polynomials in $\mathbb{Z}[x]$ that are irreducible in $\mathbb{Q}[x]$. Our proof is by strong induction on n. If $n = 1$, then $f(x)$ is irreducible in $\mathbb{Q}[x]$ and there is nothing to do. Let $n \geq 2$, and suppose that our claim holds for polynomials of smaller degree. If $f(x)$ is irreducible in $\mathbb{Q}[x]$, then again, there is nothing to do. Otherwise, we know that $f(x) = g(x)h(x)$, where $g(x)$ and $h(x)$ are polynomials of degree less than n in $\mathbb{Q}[x]$. By Theorem 11.4, we may choose $g(x)$ and $h(x)$ to be in $\mathbb{Z}[x]$. Then our inductive hypothesis tells us that $g(x)$ and $h(x)$ are products of polynomials in $\mathbb{Z}[x]$ that are irreducible in $\mathbb{Q}[x]$, and hence, so is $f(x)$, proving the claim.

If $f(x) = f_1(x) \cdots f_k(x)$, where each $f_i(x)$ is irreducible in $\mathbb{Q}[x]$, then let c_i be the content of $f_i(x)$. We now have

$$f(x) = (c_1 \cdots c_k)\left(\frac{1}{c_1}f_1(x)\right) \cdots \left(\frac{1}{c_k}f_k(x)\right),$$

where each $\frac{1}{c_i}f_i(x)$ is irreducible in $\mathbb{Z}[x]$, by the preceding lemma.

Thus, bringing the $\deg(f(x)) = 0$ possibility back into consideration, we see that $f(x)$ is either an integer not in $\{0, \pm 1\}$, or a nonzero integer multiplied by a product of irreducibles in $\mathbb{Z}[x]$.

It remains only to consider the case of an integer. But the Fundamental Theorem of Arithmetic tells us that any integer not in $\{0, \pm 1\}$ is a product of (positive or negative) primes, which are certainly irreducible in $\mathbb{Z}[x]$. We do still have to deal with $f(x) = (-1)g_1(x) \cdots g_k(x)$, where each $g_i(x)$ is irreducible, but then this is $(-g_1(x))g_2(x) \cdots g_k(x)$, and $-g_1(x)$ is irreducible as well.

Let us verify the uniqueness. Suppose that

$$f(x) = p_1 \cdots p_k g_1(x) \cdots g_l(x) = q_1 \cdots q_m h_1(x) \cdots h_n(x),$$

where each p_i and q_i is a (positive or negative) prime in $\mathbb{Z}$, and each $g_i(x)$ and $h_i(x)$ is a primitive polynomial which is irreducible in $\mathbb{Q}[x]$. (We allow the possibility that k, l, m or n may be zero.) By Gauss's lemma, the product of primitive polynomials is primitive. Thus, the content of $f(x)$ is $|p_1 \cdots p_k| = |q_1 \cdots q_l|$. By the Fundamental Theorem of Arithmetic, $k = l$ and after rearranging, each $p_i = \pm q_i$. Cancelling, we have $g_1(x) \cdots g_m(x) = \pm h_1(x) \cdots h_n(x)$. But these are products of irreducible polynomials in $\mathbb{Q}[x]$. As $\mathbb{Q}[x]$ is a UFD, $m = n$ and, after rearranging, each $g_i(x) = q_i h_i(x)$, for some $q_i \in \mathbb{Q}$. Write $q_i = \frac{r_i}{s_i}$, with $r_i, s_i \in \mathbb{Z}$ and $s_i \neq 0$. Then $s_i g_i(x) = r_i h_i(x)$. As $g_i(x)$ and $h_i(x)$ are primitive, looking at the content of each side of the equation, we have $|s_i| = |r_i|$, and hence $q_i \in \{1, -1\}$. We are done.   □

Exercises

11.11. Find all rational roots of each of the following polynomials.

1. $x^3 - 7x^2 + 5x + 2$
2. $6x^4 - x^3 + 4x^2 - x - 2$

11.12. Are the following polynomials irreducible over $\mathbb{Q}$?

1. $3x^4 + 15x^3 - 25x^2 + 45x + 10$
2. $2x^3 + 5x^2 + x + 7$
3. $x^{14} - 75$

11.13. Write each of the following polynomials as a product of irreducibles in $\mathbb{Q}[x]$.

1. $x^4 - 10x^3 + 35x^2 - 48x + 18$
2. $x^4 + 2x^3 + x^2 + 3x + 2$

11.14. Write each of the following polynomials as a product of irreducibles in $\mathbb{Z}[x]$.

1. $6x^4 + 84x^3 - 126x$
2. $6x^4 - 3x^3 + 18x^2 - 3x - 3$

11.15. Let F be a field, $a \in F$ and $f(x) \in F[x]$. Show that $f(x)$ is irreducible if and only if $f(x + a)$ is irreducible.

11.16. Modify Eisenstein's criterion as follows, namely, insist that $p|a_i, 1 \leq i \leq n$, but $p \nmid a_0$ and $p^2 \nmid a_n$. Show that the result still holds.

11.17. Is $7x^6 + 21x^5 - 49x^3 + 14x^2 + 7x + 2$ reducible or irreducible over $\mathbb{Q}$?

11.18. Let R be a Euclidean domain. If $f(x) \in R[x]$ is a nonzero polynomial, let us say that it is primitive if the only common divisors of its coefficients are the units of R. Show that Gauss's lemma holds in $R[x]$.

11.3   Irreducibility over the Real and Complex Numbers

While the real numbers may seem like a more natural field with which to work, the complex numbers have a more attractive algebraic structure. Indeed, we wish to consider the complex numbers, because there are nonconstant real polynomials, such as $x^2 + 1$, having no real roots. The complex numbers do not have this problem. Indeed, this is a famous result known as the Fundamental Theorem of Algebra. There are many different proofs of this theorem. Curiously, to the best of the author's knowledge, all of these proofs require results from outside of algebra. A proof that is mostly algebraic can be found in the advanced textbook of Dummit and Foote [1]. (Sadly, the algebra involved is somewhat beyond the scope of this course.)

Theorem 11.7 (Fundamental Theorem of Algebra). *Let $f(x)$ be a nonconstant polynomial in $\mathbb{C}[x]$. Then $f(x)$ has a root in $\mathbb{C}$.*

We say that the field of complex numbers is **algebraically closed**.

Corollary 11.4. *If $f(x) \in \mathbb{C}[x]$, then $f(x)$ is irreducible if and only if $\deg(f(x)) = 1$.*

Proof. Combine Theorem 11.7 with Corollaries 11.1 and 11.2.   □

Corollary 11.5. *Let $f(x) \in \mathbb{C}[x]$ be a nonconstant polynomial. Then there exist $a, c_1, c_2, \ldots, c_n \in \mathbb{C}$ such that $f(x) = a(x - c_1)(x - c_2) \cdots (x - c_n)$.*

Proof. We proceed by induction on $\deg(f(x)) = n$. If $n = 1$, then $f(x) = ax + b = a(x - (-a^{-1}b))$, for some $a, b \in \mathbb{C}$, with $a \neq 0$. Suppose that the result is true for n, and let $\deg(f(x)) = n + 1$. By Theorem 11.7, $f(x)$ has a root, $c_{n+1} \in \mathbb{C}$. But then Theorem 11.2 tells us that $f(x) = g(x)(x - c_{n+1})$, where $\deg(g(x)) = n$, by Theorem 10.2. Now apply our inductive hypothesis to $g(x)$.   □

Thus, complex polynomials behave as nicely as we could possibly wish. What about real polynomials? The situation there is slightly more complicated.

Lemma 11.4. *Let $f(x) \in \mathbb{R}[x]$. If $c, d \in \mathbb{R}$, and $c + di$ is a complex root of $f(x)$, then so is $c - di$.*

Proof. Write $\overline{c + di} = c - di$. Let $f(x) = a_0 + \cdots + a_n x^n$, $a_i \in \mathbb{R}$. Then if $z = c + di$, we have

$$f(\bar{z}) = a_0 + a_1\bar{z} + a_2(\bar{z})^2 + \cdots + a_n(\bar{z})^n.$$

But by Example 9.13, the function mapping z to $\bar{z}$ is a homomorphism. Thus,

$$f(\bar{z}) = a_0 + a_1\bar{z} + a_2\bar{z}^2 + \cdots + a_n\bar{z}^n$$
$$= \overline{a_0} + \overline{a_1}\bar{z} + \overline{a_2}\bar{z}^2 + \cdots + \overline{a_n}\bar{z}^n$$
$$= \overline{a_0 + a_1 z + a_2 z^2 + \cdots + a_n z^n}$$
$$= \overline{f(z)} = \bar{0} = 0,$$

making using of the fact that each $\overline{a_i} = a_i$, since $a_i \in \mathbb{R}$.   □

We can use this to classify the irreducible real polynomials.

Theorem 11.8. *Let $f(x) \in \mathbb{R}[x]$. Then $f(x)$ is irreducible over $\mathbb{R}$ if and only if either*

1. $\deg(f(x)) = 1$; *or*
2. $f(x) = ax^2 + bx + c$, *where $a \neq 0$ and $b^2 < 4ac$.*

Proof. Since $\mathbb{R}$ is a field, constant polynomials are either 0 or a unit, and therefore need not be considered. If $\deg(f(x)) = 1$, then Corollary 11.2 tells us that $f(x)$ is indeed irreducible. Therefore, let $f(x)$ have degree at least 2. Suppose that $f(x)$ is irreducible.

By Theorem 11.7, $f(x)$ has a root $z = a + bi \in \mathbb{C}$. If $z \in \mathbb{R}$, then by Corollary 11.1, we have a contradiction. Assume otherwise. By Lemma 11.4, $a - bi$ is also a root. Expressing $f(x)$ as in Corollary 11.5, we see that $(x - (a + bi))(x - (a - bi)) | f(x)$ in $\mathbb{C}[x]$. But $(x - (a + bi))(x - (a - bi)) = x^2 - 2ax + (a^2 + b^2) \in \mathbb{R}[x]$. Thus, applying the division algorithm, we see that there exist $q(x), r(x) \in \mathbb{R}[x]$ such that $f(x) = (x^2 - 2ax + (a^2 + b^2))q(x) + r(x)$, and $r(x) = 0$ or $\deg(r(x)) < 2$. By the uniqueness of the division algorithm in $\mathbb{C}[x]$, we must have $r(x) = 0$ and $x^2 - 2ax + (a^2 + b^2)$ divides $f(x)$ in $\mathbb{R}[x]$. In particular, if $\deg(f(x)) > 2$, then $f(x)$ must be reducible.

Thus, we may assume that $f(x) = ax^2 + bx + c$, with $a \neq 0$. By Corollary 11.2, such a polynomial is irreducible over $\mathbb{R}$ if and only if it has no roots in $\mathbb{R}$. But the quadratic formula tells us that this happens if and only if $b^2 - 4ac < 0$. We are done.   □

We can use this to recover a well-known fact from calculus.

Corollary 11.6. *Let $f(x) \in \mathbb{R}[x]$ be a polynomial of odd degree. Then $f(x)$ has a real root.*

Proof. We know that $\mathbb{R}[x]$ is a UFD. Thus, write $f(x)$ as a product of irreducible polynomials. By the preceding theorem, each such irreducible has degree 1 or 2. Since $f(x)$ has odd degree, at least one of these irreducible polynomials has degree 1. Therefore, there exist $a, b \in \mathbb{R}$, with $a \neq 0$, such that $ax + b$ divides $f(x)$. But then $-a^{-1}b$ is a root of $f(x)$.   □

Exercises

11.19. Given that a is a root of $f(x)$, find all complex roots of $f(x)$.

1. $f(x) = x^3 - 11x^2 + 41x - 91, a = 2 + 3i$
2. $f(x) = x^4 + x^2 - 2x + 6, a = 1 - i$

11.20. Given that a is a root of $f(x)$, find all complex roots of $f(x)$.

1. $f(x) = x^3 + x^2 - 5x - 21, a = 3$
2. $f(x) = x^4 - 6x^3 + 33x^2 - 84x + 136, a = 1 + 4i$

11.21. Write each of the following polynomials as a product of irreducibles in $\mathbb{Q}[x]$, $\mathbb{R}[x]$ and $\mathbb{C}[x]$.

1. $x^4 - 10$
2. $x^3 + x^2 + 5x - 22$

11.22. Write each of the following polynomials as a product of irreducibles in $\mathbb{Q}[x]$, $\mathbb{R}[x]$ and $\mathbb{C}[x]$.

1. $x^3 + 12$
2. $x^4 + 4x^2 + 4$

11.23. Find a nonzero polynomial in $\mathbb{R}[x]$ having $2 - 5i, 4 + i$ and 6 as roots.

11.24. Let $f(x)$ and $g(x)$ be nonzero polynomials in $\mathbb{Q}[x]$. Consider the gcds of $f(x)$ and $g(x)$ in $\mathbb{Q}[x]$, $\mathbb{R}[x]$ and $\mathbb{C}[x]$. Must these gcds be the same, or can they be different?

11.4   Irreducibility over Finite Fields

When our field is finite, we have the luxury of taking a brute force approach to factoring polynomials. That is, we can simply list all of the polynomials of suitable degrees, and see if the products work. Of course, we can save ourselves some effort by narrowing the possibilities first.

Example 11.12. Let $f(x) = x^4 + x^3 + x^2 + x + 1 \in \mathbb{Z}_2[x]$. We claim that $f(x)$ is irreducible over $\mathbb{Z}_2$. If not, there are two possibilities. First, $f(x)$ could be a product of a degree 1 polynomial and a degree 3 polynomial. But if it has a degree 1 polynomial as a factor, then it has a root. There are only two possible roots in $\mathbb{Z}_2$, namely, 0 and 1, and neither works. Second, $f(x)$ could be a product of two polynomials of degree 2. Now, the only possible coefficients are 0 and 1. Furthermore, the leading coefficients and the constant terms must multiply to give 1. Thus, the only possible factors are $x^2 + 1$ and $x^2 + x + 1$. But $x^2 + 1$ has 1 as a root, and $f(x)$ does not, so only $x^2 + x + 1$ remains. However, $(x^2 + x + 1)^2 = x^4 + x^2 + 1 \neq f(x)$. Thus, $f(x)$ is indeed irreducible.

Example 11.13. Let $f(x) = 3x^5 + x^4 + 4x^3 + 4x^2 + 3x + 2 \in \mathbb{Z}_5[x]$. We would like to write $f(x)$ as a product of irreducibles. The first thing we should do is check for roots. We run through the five elements of $\mathbb{Z}_5$ and find that 3 is a root. Thus, $x - 3$ (or, equivalently, $x + 2$) divides $f(x)$. Performing polynomial long division, we find that $f(x) = (x + 2)(3x^4 + 4x^2 + x + 1)$. Let $g(x) = 3x^4 + 4x^2 + x + 1$. Evaluating $g(x)$ at each element of $\mathbb{Z}_5$, we see that $g(x)$ has no roots. Thus, if it is to be factored, it must be as a product of two polynomials of degree 2. Up to a unit in $\mathbb{Z}_5$, these factors would have to be $x^2 + ax + b$ and $3x^2 + cx + d$, for some $a, b, c, d \in \mathbb{Z}_5$. Furthermore, looking at the constant terms, we have $bd = 1$. Thus, once b is decided, $d = b^{-1}$. Looking at the coefficients of x^3, we have $3a + c = 0$. Thus, once a is decided, we have $c = 2a$. Trying the various possibilities for a and b, we have $g(x) = (x^2 + 2x + 3)(3x^2 + 4x + 2)$. Since $g(x)$ has no roots, this cannot be factored any further. Thus, $f(x) = (x + 2)(x^2 + 2x + 3)(3x^2 + 4x + 2)$ is a product of irreducibles in $\mathbb{Z}_5[x]$.

Our ability to handle polynomials over finite fields can be helpful when we consider polynomials in $\mathbb{Q}[x]$.

Theorem 11.9. *Let $f(x) = a_0 + a_1x + \cdots + a_nx^n \in \mathbb{Z}[x]$. Let p be a prime such that $p \nmid a_n$. Reducing all of the coefficients modulo p, if $[a_0] + [a_1]x + \cdots + [a_n]x^n$ is irreducible in $\mathbb{Z}_p[x]$, then $f(x)$ is irreducible in $\mathbb{Q}[x]$.*

Proof. Suppose $f(x)$ is reducible in $\mathbb{Q}[x]$. Then by Theorem 11.4, we must have $f(x) = g(x)h(x)$, where $g(x) = b_0 + \cdots + b_kx^k$ and $h(x) = c_0 + \cdots + c_mx^m$ are polynomials in $\mathbb{Z}[x]$, with $k, m > 0$ and $b_k \neq 0 \neq c_m$. Now, we have

$$a_i = b_0c_i + b_1c_{i-1} + \cdots + b_ic_0,$$

for each i. By Example 9.12, the function from $\mathbb{Z}$ to $\mathbb{Z}_p$ sending d to $[d]$ is a ring homomorphism. Thus,

$$[a_i] = [b_0][c_i] + [b_1][c_{i-1}] + \cdots + [b_i][c_0].$$

It now follows that

$$[a_0] + \cdots + [a_n]x^n = ([b_0] + \cdots + [b_k]x^k)([c_0] + \cdots + [c_m]x^m).$$

That is, $[a_0] + \cdots + [a_n]x^n$ is reducible, unless one of the factors is a constant polynomial. But as $p \nmid a_n$, we see that the degree of $[a_0] + \cdots + [a_n]x^n$ is $n = k + m$. The only way the product will have the correct degree is if $[b_k] \neq [0] \neq [c_m]$. This contradiction completes the proof. $\qquad\square$

Note that the condition that p does not divide the leading coefficient is important. Indeed, $3x^2 + x - 4$ is reducible in $\mathbb{Q}[x]$, as it is $(x - 1)(3x + 4)$. But if we tried to use $p = 3$, we would obtain $x + 2 \in \mathbb{Z}_3[x]$, which is certainly irreducible. Also, the converse of the theorem is not true. For instance, $x^2 + 1$ is irreducible over $\mathbb{Q}$ (or, for that matter, $\mathbb{R}$), but in $\mathbb{Z}_5[x]$, we have $x^2 + 1 = (x + 2)(x + 3)$.

Example 11.14. We claim that $15x^4 - 29x^3 + 13x^2 + 33x - 201$ is irreducible over $\mathbb{Q}[x]$. Use Theorem 11.9 with $p = 2$. In $\mathbb{Z}_2[x]$, we obtain $x^4 + x^3 + x^2 + x + 1$ which, by Example 11.12, is irreducible.

Sometimes, we might have to try more than one prime.

Example 11.15. Let $f(x) = 5x^3 + 3x^2 + x + 1$. If we use $p = 2$, we obtain $x^3 + x^2 + x + 1 \in \mathbb{Z}_2[x]$. But this polynomial has 1 as a root, so it is reducible. No help here! Let us try $p = 3$. Then we get $2x^3 + x + 1 \in \mathbb{Z}_3[x]$. By Corollary 11.2, it is irreducible if it has no roots. But trying 0, 1 and 2, we see that it has no roots in $\mathbb{Z}_3$. Thus, $f(x)$ is irreducible in $\mathbb{Q}[x]$.

Exercises

11.25. Are the following polynomials reducible or irreducible over the rationals?

1. $f(x) = x^3 + 5x^2 + 2x + 16$
2. $f(x) = 22x^4 - 9x^3 + 16x^2 + 18x + 20$

11.26. Are the following polynomials reducible or irreducible over the rationals?

1. $f(x) = 9x^4 - 15x^3 + 8x^2 - 6x + 25$
2. $f(x) = 2x^4 + 11x^3 + 16x^2 + 5x + 6$

11.27. Let F be a finite field with n elements. How many monic irreducible polynomials of degree 2 are there in $F[x]$?

11.28. Write each of the following polynomials as a product of irreducibles in $\mathbb{Z}_{11}[x]$.

1. $2x^3 + 3x^2 + 9x + 10$
2. $x^4 + 4x^3 + 5x^2 + x + 7$

11.29. Let p be an odd prime. Show that $x^4 + 1$ is reducible over $\mathbb{Z}_p$ in each of the following cases:

1. there exists an $a \in \mathbb{Z}_p$ such that $a^2 = p - 1$;
2. there exists an $a \in \mathbb{Z}_p$ such that $a^2 = p - 2$; or
3. there exists an $a \in \mathbb{Z}_p$ such that $a^2 = 2$.

11.30. Show that $x^4 + 1$ is irreducible in $\mathbb{Q}[x]$ but reducible in $\mathbb{Z}_p[x]$ for every prime p. (Thus, the converse of Theorem 11.9 is wildly false!)

Reference

1. Dummit, D.S., Foote, R.M.: Abstract Algebra, 3rd edn. Wiley, Hoboken (2004)

Chapter 12
Vector Spaces and Field Extensions

We begin this chapter with some basic facts about vector spaces. These will be familiar (at least in the case of real vector spaces) to those readers who have studied linear algebra. We then focus our attention on the particular case of a field extension. A number of properties of field extensions are discussed.

Let F be a field and $f(x) \in F[x]$ a nonconstant polynomial. We demonstrate how to create a field extension in which $f(x)$ splits into a product of polynomials of degree 1. This leads to a classification of all finite fields.

12.1   Vector Spaces

We begin with the definition of a vector space. In most linear algebra courses, vector spaces are defined over $\mathbb{R}$ or, occasionally, $\mathbb{C}$. But we can do the same thing over any field.

If F is a field and V is a set, then a **scalar multiplication** on V is a function from $F \times V$ to V. If $a \in F, v \in V$, then we write av for the image of (a, v) under such a function.

Definition 12.1. Let F be a field. Then a **vector space** over F is a set V having an addition operation and a scalar multiplication such that

1. V is an abelian group under addition;
2. $av \in V$ for all $a \in F$ and all $v \in V$;
3. $(a + b)v = av + bv$ for all $a, b \in F$ and all $v \in V$;
4. $a(u + v) = au + av$ for all $a \in F$ and all $u, v \in V$;
5. $a(bv) = (ab)v$ for all $a, b \in F$ and all $v \in V$; and
6. $1v = v$ for all $v \in V$.

© Springer International Publishing AG, part of Springer Nature 2018
G. T. Lee, *Abstract Algebra*, Springer Undergraduate Mathematics Series,
https://doi.org/10.1007/978-3-319-77649-1_12

Of course, condition (2) is redundant, given the definition of a scalar multiplication, but we include it, because it must be checked.

Certainly the most familiar vector space over $\mathbb{R}$ is $\mathbb{R}^n$. We can generalize this.

Example 12.1. Let F be a field. For any positive integer n, let $F^n = \underbrace{F \oplus F \oplus \cdots \oplus F}_{n \text{ times}}$.

Then F^n is a vector over F with the usual addition operation and scalar multiplication $a(b_1, \ldots, b_n) = (ab_1, \ldots, ab_n)$, for any $a, b_1, \ldots, b_n \in F$.

Example 12.2. Let F be any field. Then $F[x]$ is a vector space over F with the usual polynomial addition and $a(b_0 + b_1 x + \cdots + b_n x^n) = ab_0 + ab_1 x + \cdots + ab_n x^n$, for any $a, b_0, \ldots, b_n \in F$.

Example 12.3. Let m and n be any positive integers, and let V be the set of $m \times n$ matrices with entries in a field F. Then V is a vector space over F using matrix addition and scalar multiplication.

The least exciting example of a vector space is the following.

Example 12.4. Let F be any field and V the trivial additive group, $\{0\}$. Then V is a vector space using the only available addition and scalar multiplication options, $0 + 0 = 0$ and $a0 = 0$, for all $a \in F$.

The most important example for our purposes is the following.

Definition 12.2. If F and K are fields, with F a subfield of K, then we say that K is an **extension field** of F.

Example 12.5. Any extension field K of F is a vector space over F, using the addition operation in K and multiplication in K as the scalar multiplication. (All the properties are immediate, except that $1v = v$ for all $v \in K$. To be sure of that, we must know that the identity of F is the identity of K. But this follows from Theorem 8.12.) For example, $\mathbb{R}$ and $\mathbb{C}$ are vector spaces over $\mathbb{Q}$.

Let us mention a few basic properties of vector spaces.

Theorem 12.1. *Let V be a vector space over F. Then*

1. *$a0 = 0$ for all $a \in F$;*
2. *$0v = 0$ for all $v \in V$; and*
3. *$(-1)v = -v$ for all $v \in V$.*

Proof. (1) Note that $a0 = a(0 + 0) = a0 + a0$. Adding $-(a0)$ to both sides, we see that $a0 = 0$.

(2) We have $0v = (0 + 0)v = 0v + 0v$. Adding $-0v$ to both sides, we obtain the desired conclusion.

(3) Observe that $v + (-1)v = 1v + (-1)v = (1 - 1)v = 0v = 0$. Thus, $(-1)v = -v$. $\qquad \square$

We do have to be a bit careful about which 0 we are using. For example, when we write $0v = 0$ in the theorem above, the first 0 is in F and the second is in V.

Definition 12.3. Let V be a vector space over a field F. Then a subset W of V is said to be a **subspace** of V if it is a vector space over F using the same addition and scalar multiplication.

Example 12.6. If F is a subfield of K, and K is a subfield of L, then L is a vector space over F having K and F as subspaces.

Example 12.7. Regarding $\mathbb{R}[x]$ as a vector space over $\mathbb{Q}$, we note that $\mathbb{Q}[x]$ is a subspace.

There is a simple test for a subspace.

Theorem 12.2. *Let F be a field and V a vector space over F. Then a subset W of V is a subspace if and only if*

1. $0 \in W$;
2. $w_1 + w_2 \in W$ for all $w_1, w_2 \in W$ *(closure under addition); and*
3. $aw \in W$ for all $a \in F$ and $w \in W$ *(closure under scalar multiplication).*

Proof. If W is a subspace then, in particular, it is an additive subgroup, so (1) and (2) hold. Part (3) is one of the conditions for a vector space. Conversely, suppose that (1), (2) and (3) hold. Noting that (3) tells us that $-w = (-1)w \in W$, for all $w \in W$, we see from Theorem 3.10 that W is an additive subgroup of V. We are given closure under scalar multiplication. The remaining vector space properties hold in V, and therefore in any subset of V. $\qquad\square$

Note that in the preceding theorem, condition (1) could be replaced with the condition that W is not the empty set, for if $w \in W$, then $-w \in W$, and therefore $0 = w + (-w) \in W$.

Example 12.8. Let $V = \mathbb{R}^4$, which is a vector space over $\mathbb{R}$. We claim that $W = \{(a, b, 2a - b + 3c, c) : a, b, c \in \mathbb{R}\}$ is a subspace of V. Letting $a = b = c = 0$, we see that $(0, 0, 0, 0) \in W$. To check closure under addition, take $a_i, b_i, c_i \in \mathbb{R}$. Then

$$(a_1, b_1, 2a_1 - b_1 + 3c_1, c_1) + (a_2, b_2, 2a_2 - b_2 + 3c_2, c_2)$$
$$= (a_1 + a_2, b_1 + b_2, 2(a_1 + a_2) - (b_1 + b_2) + 3(c_1 + c_2), c_1 + c_2) \in W.$$

Similarly, if $a \in \mathbb{R}$, then

$$a(a_1, b_1, 2a_1 - b_1 + 3c_1, c_1) = (aa_1, ab_1, 2aa_1 - ab_1 + 3ac_1, ac_1) \in W.$$

Thus, we have closure under scalar multiplication, and the claim is proved.

Exercises

12.1. Let F be a field and n a positive integer. If V is the set of all polynomials of degree n in $F[x]$, together with the zero polynomial, is V a subspace of $F[x]$?

12.2. Let F be a field and n a positive integer. If V is the set of all polynomials of degree at most n in $F[x]$, together with the zero polynomial, show that V is a subspace of $F[x]$.

12.3. Let V be a vector space having subspaces U and W. Show that $U \cap W$ is a subspace of V. Extend this to the intersection of an arbitrary collection of subspaces.

12.4. Let V be a vector space having subspaces U and W. Show that $U + W$ (regarding U and W as additive subgroups of V) is a subspace of V.

12.5. Let V and W be vector spaces over a field F. A function $\alpha : V \to W$ is said to be a linear transformation if $\alpha(v_1 + v_2) = \alpha(v_1) + \alpha(v_2)$ and $\alpha(av_1) = a\alpha(v_1)$ for all $a \in F$, $v_1, v_2 \in V$. If U is a subspace of V, show that $\alpha(U)$ is a subspace of W.

12.6. Let F, V, W and α be as in the preceding exercise. Show that the kernel of α (regarding α as a homomorphism of additive groups) is a subspace of V. Further show that α is one-to-one if and only if the kernel is $\{0\}$.

12.7. Let $F = \mathbb{Z}_{11}$ and $V = F^3$. If $W = \{(a, b, c) \in V : 2a + 3b + 7c = 0\}$, is W a subspace of V?

12.8. Let F be a field with vector spaces V and W. Let $U = V \times W$ be the direct product of the additive groups V and W. Define a scalar multiplication on U via $a(v, w) = (av, aw)$ for all $a \in F$, $v \in V$ and $w \in W$. Is U a vector space over F?

12.9. Let F be a field of characteristic 3 and V a vector space over F. Show that $v + v + v = 0$ for all $v \in F$.

12.10. Suppose that V is a vector space over an infinite field F. Show that V is not the union of a finite number of proper subspaces.

12.2   Basis and Dimension

In order to define a basis for a vector space, we must first discuss linear combinations of vectors.

Definition 12.4. Let V be a vector space over a field F. If $v_1, v_2, \ldots, v_k \in V$, then a vector $v \in V$ is said to be a **linear combination** of the v_i if $v = a_1 v_1 + \cdots + a_k v_k$, for some $a_i \in F$.

Example 12.9. Let $F = \mathbb{Q}$ and $V = F^3$. If $v_1 = (2, -3, 7)$ and $v_2 = (4, 0, 1)$, then $(24, -6, 19)$ is a linear combination of v_1 and v_2, since $(24, -6, 19) = 2v_1 + 5v_2$.

Definition 12.5. Let F be a field and V a vector space over F. Let $v_1, v_2, \ldots, v_k \in V$. We say that the v_i are **linearly dependent** if there exist $a_1, \ldots, a_k \in F$, not all zero, such that $a_1 v_1 + \cdots + a_k v_k = 0$. Otherwise, the v_i are **linearly independent**.

Example 12.10. Let $F = \mathbb{Z}_5$ and $V = F^3$. The vectors $(2, 1, 3)$, $(1, 3, 0)$ and $(2, 1, 4)$ are linearly dependent, since $3(2, 1, 3) + (1, 3, 0) + 4(2, 1, 4) = (0, 0, 0)$. On the other hand, $(1, 0, 4)$, $(3, 2, 1)$ and $(2, 0, 2)$ are linearly independent. Indeed, if $a_1(1, 0, 4) + a_2(3, 2, 1) + a_3(2, 0, 2) = (0, 0, 0)$, then looking at the middle entry, we see immediately that $a_2 = 0$. Then $a_1 + 2a_3 = 4a_1 + 2a_3 = 0$. This yields $3a_1 = 0$, and hence $a_1 = 0$ and, finally, $a_3 = 0$.

Here is a handy test for linear dependence.

Theorem 12.3. *Let V be a vector space over a field F and $v_1, \ldots, v_k \in V$. Then the v_i are linearly dependent if and only if either*

1. $v_1 = 0$; *or*
2. *there exists an $m \geq 2$ such that v_m is a linear combination of $v_1, \ldots, v_{m-1}$.*

Proof. Suppose that the v_i are linearly dependent. Choose $a_i \in F$, not all zero, such that $a_1 v_1 + \cdots + a_k v_k = 0$. Let m be the largest positive integer such that $a_m \neq 0$. Then $a_1 v_1 + \cdots + a_m v_m = 0$. If $m = 1$, then $a_1 v_1 = 0$, with $a_1 \neq 0$. Thus, $v_1 = a_1^{-1} 0 = 0$, giving case (1). If $m > 1$, then $v_m = -a_m^{-1} a_1 v_1 - \cdots - a_m^{-1} a_{m-1} v_{m-1}$, and so v_m is a linear combination of $v_1, \ldots, v_{m-1}$, which proves case (2).

Conversely, suppose that (1) or (2) is satisfied. If $v_1 = 0$, then $1 v_1 + 0 v_2 + \cdots + 0 v_k = 0$, meaning that the v_i are linearly dependent. If $v_m = b_1 v_1 + \cdots + b_{m-1} v_{m-1}$, for some $b_i \in F$, then

$$b_1 v_1 + \cdots + b_{m-1} v_{m-1} - 1 v_m + 0 v_{m+1} + \cdots + 0 v_k = 0.$$

Again, the v_i are linearly dependent. □

Linear independence is most useful when combined with another property.

Definition 12.6. Let V be a vector space over a field F, and let $v_1, \ldots, v_k \in V$. Then we say that the v_i **span** V if every $v \in V$ is a linear combination of the v_i.

Example 12.11. Regarding $\mathbb{C}$ as a vector space over $\mathbb{R}$, we note that 1 and i span $\mathbb{C}$, as $a + bi = a1 + bi$.

Example 12.12. Let $F = \mathbb{R}$ and $V = \mathbb{R}^3$. Then the vectors $(1, 0, 0)$, $(0, 1, 0)$ and $(0, 0, 1)$ span V, since $(a, b, c) = a(1, 0, 0) + b(0, 1, 0) + c(0, 0, 1)$.

The following lemma describes a very nice relationship between linear independence and spanning.

Lemma 12.1. *Let V be a vector space over a field F. Suppose that $v_1, \ldots, v_k$ span V. If $w_1, \ldots, w_l \in V$, and $l > k$, then the w_i are linearly dependent.*

Proof. Since the v_i span V, we know that w_1 is a linear combination of the v_i. Let us say that $w_1 = a_1 v_1 + \cdots + a_k v_k$, with $a_i \in F$. If all of the a_i are zero, then w_1 is the zero vector. Thus, by Theorem 12.3, we are done. Therefore, we may assume that some a_i is nonzero. Without loss of generality, say $a_1 \neq 0$. We now observe that $w_1, v_2, v_3, \ldots, v_k$ span V. Indeed, if $v \in V$, then $v = b_1 v_1 + \cdots + b_k v_k$, for some $b_i \in F$. But

$$v_1 = a_1^{-1} w_1 - a_1^{-1} a_2 v_2 - \cdots - a_1^{-1} a_k v_k.$$

Thus,

$$v = b_1 a_1^{-1} w_1 + (b_2 - b_1 a_1^{-1} a_2) v_2 + \cdots + (b_k - b_1 a_1^{-1} a_k) v_k,$$

proving the claim.

Now consider w_2. It is a linear combination of $w_1, v_2, v_3, \ldots, v_k$. Let us say that $w_2 = c_1 w_1 + c_2 v_2 + c_3 v_3 + \cdots + c_k v_k$, with $c_i \in F$. If $c_i = 0$ for all $i \geq 2$, then w_2 is a linear combination of w_1, proving that the w_i are linearly dependent. Thus, we may assume that there exists an $i \geq 2$ with $c_i \neq 0$. Without loss of generality, say $c_2 \neq 0$. But then v_2 is a linear combination of $w_1, w_2, v_3, v_4, \ldots, v_k$. And just as before, we now deduce that $w_1, w_2, v_3, v_4, \ldots, v_k$ span V.

Repeat this argument. We will conclude either that the w_i are linearly dependent or, eventually, that $w_1, \ldots, w_k$ span V. But then w_{k+1} is a linear combination of $w_1, \ldots, w_k$. By Theorem 12.3, the w_i are linearly dependent. $\square$

What we really need is a basis for a vector space.

Definition 12.7. Let V be a vector space over a field F. We say that $v_1, \ldots, v_k \in V$ form a **basis** for V if they are linearly independent and span V.

Example 12.13. Regarding $\mathbb{C}$ as a vector space over $\mathbb{R}$, we can see that 1 and i form a basis for $\mathbb{C}$.

Example 12.14. For any field F and any positive integer n, the vectors

$$(1, 0, 0, \ldots, 0), (0, 1, 0, 0, \ldots, 0), \ldots, (0, 0, \ldots, 0, 1)$$

form a basis for F^n.

Example 12.15. Let F be any field and $V = F[x]$. Then V has no finite basis. Indeed, if $v_1, \ldots, v_k \in V$, then any linear combination of these vectors must have degree no larger than the maximum of the degrees of the v_i. On the other hand, for any positive integer n, let W be the set of all polynomials having degree at most n (including the zero polynomial). By Exercise 12.2, W is a subspace of V, and the polynomials $1, x, x^2, \ldots, x^n$ form a basis.

Theorem 12.4. *Let V be a vector space over a field F. If $v_1, \ldots, v_k$ form a basis for V, then every element of V can be written uniquely in the form $a_1 v_1 + \cdots + a_k v_k$, with $a_i \in F$.*

Proof. Since a basis spans the space, only the uniqueness needs to be proved. Suppose that

$$a_1 v_1 + \cdots + a_k v_k = b_1 v_1 + \cdots + b_k v_k,$$

with $a_i, b_i \in F$. Then

$$(a_1 - b_1)v_1 + \cdots + (a_k - b_k)v_k = 0.$$

By linear independence, $a_i = b_i$ for all i.   □

Bases are not unique. For instance, $(1, 0)$ and $(0, 1)$ form a basis for $\mathbb{R}^2$ over $\mathbb{R}$, but so do $(1, 3)$ and $(5, 2)$. However, any two bases for a vector space must have the same number of vectors.

Theorem 12.5. *Let V be a vector space over a field F. If $v_1, \ldots, v_k$ and $w_1, \ldots, w_l$ are bases for V, then $k = l$.*

Proof. Suppose the theorem is false. Without loss of generality, say $k < l$. Then $v_1, \ldots, v_k$ span V. Since $k < l$, Lemma 12.1 tells us that $w_1, \ldots, w_l$ are linearly dependent. We have a contradiction.   □

Definition 12.8. Let V be a vector space over a field F. If $v_1, \ldots, v_k$ is a basis for V, then we say that V has **dimension** k, and write $\dim V = k$ (or $\dim_F V = k$, if the field is unclear from the context). We also stipulate that $\dim\{0\} = 0$. In either of these cases, V is **finite-dimensional**. If V has no finite basis, then V is **infinite-dimensional**.

Example 12.16. For any field F and positive integer n, $\dim F^n = n$. See Example 12.14.

Example 12.17. The dimension of $\mathbb{C}$ over $\mathbb{R}$ is 2. See Example 12.13.

Example 12.18. If F is any field, then $F[x]$ is infinite-dimensional. The vector space consisting of the polynomials of degree at most n over F, including the zero polynomial, has dimension $n + 1$. See Example 12.15.

In a finite-dimensional space, we can discard vectors from a spanning set to obtain a basis, or add vectors to a linearly independent set to obtain a basis.

Theorem 12.6. *Let V be any vector space over a field F, with $V \neq \{0\}$. Take $v_1, \ldots, v_k \in V$. Then*

1. *if $v_1, \ldots, v_k$ span V, then some subset of $\{v_1, \ldots, v_k\}$ is a basis for V; and*
2. *if $v_1, \ldots, v_k$ are linearly independent, and $\dim V = n < \infty$, then there exist $v_{k+1}, \ldots, v_n \in V$ such that $v_1, \ldots, v_n$ form a basis for V.*

Proof. (1) We proceed by induction on k. If $k = 1$, then since v_1 spans V, and $V \neq \{0\}$, we see that $v_1 \neq 0$, and hence v_1 is linearly independent. (If $av_1 = 0$, and $0 \neq a \in F$, then $0 = a^{-1}av_1 = v_1$.) Thus, v_1 is a basis. Suppose the result is true for k, and let $v_1, \ldots, v_{k+1}$ span V. If they are linearly independent, there is nothing to do. Otherwise, refer to Theorem 12.3. If $v_1 = 0$, then $v_2, \ldots, v_{k+1}$ span V as well. By our inductive hypothesis we are done. Otherwise, some v_l is a linear combination of $v_1, \ldots, v_{l-1}$. Without loss of generality, say $l = k+1$. Write $v_{k+1} = a_1 v_1 + \cdots + a_k v_k$. If $v \in V$, we know that

$$v = b_1 v_1 + \cdots + b_{k+1} v_{k+1},$$

for some $b_i \in F$. But then

$$v = (b_1 + a_1 b_{k+1})v_1 + \cdots + (b_k + a_k b_{k+1})v_k.$$

Thus, $v_1, \ldots, v_k$ span V. Our inductive hypothesis completes the proof.

(2) If $v_1, \ldots, v_k$ span V, there is nothing to do. Otherwise, find $v_{k+1} \in V$ which is not a linear combination of $v_1, \ldots, v_k$. Suppose that $v_1, \ldots, v_{k+1}$ are linearly dependent. Then $a_1 v_1 + \cdots + a_{k+1} v_{k+1} = 0$, for some $a_i \in F$. If $a_{k+1} = 0$, then $v_1, \ldots, v_k$ are linearly dependent, which is not the case. Otherwise,

$$v_{k+1} = -a_{k+1}^{-1} a_1 v_1 - \cdots - a_{k+1}^{-1} a_k v_k;$$

that is, v_{k+1} is a linear combination of $v_1, \ldots, v_k$, giving us a contradiction. Therefore, $v_1, \ldots, v_{k+1}$ is a linearly independent set. Now repeat. This process must stop, because Lemma 12.1 tells us that V cannot have a linearly independent set with more than n vectors. $\qquad\square$

Example 12.19. Let $F = \mathbb{Q}$ and $V = \mathbb{Q}^3$. The vectors $(3, -7, 0)$ and $(1, 2, 0)$ are easily seen to be linearly independent. Furthermore, $(2, 5, 8)$ is not a linearly combination of these two vectors. Thus, since $\dim V = 3$, we see that the vectors $(3, -7, 0), (1, 2, 0), (2, 5, 8)$ form a basis for V.

Exercises

12.11. Let $F = \mathbb{R}$ and $V = \mathbb{R}^3$. Are the following sets of vectors in V linearly dependent or independent over F?

1. $(1, 3, 5), (2, 1, 4), (7, 11, 23)$
2. $(1, 3, 4), (2, 2, 1), (3, 6, 3)$

12.12. Let $F = \mathbb{Z}_7$ and $V = M_2(F)$. Are the following sets of vectors in V linearly dependent or independent over F?

1. $\begin{pmatrix} 2 & 3 \\ 4 & 5 \end{pmatrix}, \begin{pmatrix} 6 & 0 \\ 0 & 3 \end{pmatrix}, \begin{pmatrix} 4 & 0 \\ 3 & 2 \end{pmatrix}$

2. $\begin{pmatrix} 1 & 2 \\ 3 & 4 \end{pmatrix}, \begin{pmatrix} 2 & 3 \\ 4 & 5 \end{pmatrix}, \begin{pmatrix} 3 & 4 \\ 5 & 6 \end{pmatrix}$

12.13. Do the following vectors span $\mathbb{Q}^3$ (as a vector space over $\mathbb{Q}$)?

1. $(1, 0, 2), (2, 5, 3), (3, 5, 5)$
2. $(1, 0, 2), (2, 3, 5), (0, 0, 4)$

12.14. Do the following matrices span $M_2(\mathbb{Z}_5)$ (as a vector space over $\mathbb{Z}_5$), namely
$\begin{pmatrix} 1 & 1 \\ 0 & 0 \end{pmatrix}, \begin{pmatrix} 0 & 1 \\ 1 & 0 \end{pmatrix}, \begin{pmatrix} 0 & 0 \\ 1 & 1 \end{pmatrix}, \begin{pmatrix} 1 & 0 \\ 0 & 1 \end{pmatrix}$?

12.15. Let $V = M_2(\mathbb{C})$. Find the dimension of V as a vector space over $\mathbb{C}$, and as a vector space over $\mathbb{R}$.

12.16. Let $F = \mathbb{Z}_7$ and $V = \{(a, b, c) \in F^3 : c = 3a + 5b\}$. Find the dimension of V over F.

12.17. Let F be a field and V a finite-dimensional vector space. If W is a subspace of V, show that $\dim W \leq \dim V$, with equality if and only if $W = V$. (Do not assume, to begin with, that W is finite-dimensional.)

12.18. Suppose that a vector space V with dimension n has subspaces U and W with dimensions m and k, respectively. If $m + k > n$, show that $U \cap W \neq \{0\}$.

12.19. Let F, V, W and α be as in Exercise 12.5. Suppose that $v_1, \ldots, v_n \in V$ are linearly independent and α is one-to-one. Show that $\alpha(v_1), \ldots, \alpha(v_n)$ are linearly independent.

12.20. Let F be a field and V a finite-dimensional vector space over F. Say $\dim V = n \in \mathbb{N}$. Show that there exists a bijective linear transformation (see Exercise 12.5 for the definition) $\alpha : V \to F^n$.

12.3 Field Extensions

Let us now focus on our main vector space of interest: the field extension.

Definition 12.9. Let K be a field extension of F. Then the **degree** of the extension is the dimension of K over F. We write $[K : F] = \dim_F K$. The extension is **finite** if $[K : F] < \infty$ and, in particular, **quadratic** if $[K : F] = 2$.

Example 12.20. As we observed in Example 12.17, $\mathbb{C}$ is a quadratic extension of $\mathbb{R}$.

Example 12.21. Let $K = \{a + b\sqrt[3]{2} + c\sqrt[3]{4} : a, b, c \in \mathbb{Q}\}$. We claim that K is a subfield of $\mathbb{R}$ and, therefore, an extension field of $\mathbb{Q}$. All of the properties of a subfield are easy to verify except, perhaps, that nonzero elements have inverses. Take $0 \neq a + b\sqrt[3]{2} + c\sqrt[3]{4} \in K$. Then notice that $(a + bx + cx^2, x^3 - 2)$ divides $x^3 - 2$. But, by Example 11.10, $x^3 - 2$ is irreducible over $\mathbb{Q}$. Thus, the gcd can only be a constant polynomial (in fact, 1, since we assume it to be monic). As $\mathbb{Q}[x]$ is a Euclidean domain, Theorem 10.6 guarantees that we can write

$$1 = u(x)(x^3 - 2) + v(x)(a + bx + c^2),$$

for some $u(x), v(x) \in \mathbb{Q}[x]$. But then $1 = v(\sqrt[3]{2})(a + b\sqrt[3]{2} + c\sqrt[3]{4})$. As it is easy to see that $v(\sqrt[3]{2}) \in K$, we have an inverse for $a + b\sqrt[3]{2} + c\sqrt[3]{4}$ in K, as claimed.

In fact, $[K : \mathbb{Q}] = 3$. To see this, we observe that $\{1, \sqrt[3]{2}, \sqrt[3]{4}\}$ is a basis for K over $\mathbb{Q}$. Clearly, these numbers span K. If they are linearly dependent, then there are rational numbers a, b, c, not all zero, such that $\sqrt[3]{2}$ is a root of $a + bx + cx^2$. But again, $1 = (a + bx + cx^2, x^3 - 2)$, and we write $1 = u(x)(a + bx + cx^2) + v(x)(x^3 - 2)$, for some $u(x), v(x) \in \mathbb{Q}[x]$. Evaluating at $\sqrt[3]{2}$, we obtain $1 = 0$, giving a contradiction and establishing that we have a basis.

We are, in fact, engaging in a small abuse of notation here. If K is an extension field of F then, of course, F is also an additive subgroup of K. We could also use the notation $[K : F]$ to mean the index of F in K as additive subgroup. This is not the same as the degree of the extension! For the remainder of the book, when we write $[K : F]$, we will mean the degree of the extension.

In the particular case of a finite field, we can illustrate the difference. By Lagrange's theorem, the index of the additive groups would be $\frac{|K|}{|F|}$. However, the degree is calculated as follows.

Theorem 12.7. *Let K be a field extension of F, such that K is a finite field. Then $[K : F] = \log_{|F|} |K|$.*

Proof. First, we note that K must be finite-dimensional over F. Indeed, the elements of K must span K, and by Theorem 12.6, we can obtain a finite basis. Let $[K : F] = n$, and suppose that $\{v_1, \ldots, v_n\}$ is a basis for K over F. By Theorem 12.4, the elements of K are uniquely of the form $a_1 v_1 + \cdots + a_n v_n$, with $a_i \in F$. As there are $|F|$ choices for each a_i, the total number of elements of K is $|F|^n$. Taking the base $|F|$ logarithm, we obtain our result. $\qquad\square$

Degrees of extensions behave in a nice way.

Theorem 12.8. *Let K be a finite extension of F and L a finite extension of K. Then $[L : F] = [L : K][K : F]$.*

Proof. Let $\{v_1, \ldots, v_n\}$ be a basis for K over F, and let $\{w_1, \ldots, w_m\}$ be a basis for L over K. We claim that $\{v_i w_j : 1 \le i \le n, 1 \le j \le m\}$ is a basis for L over F. This will complete the proof.

Take any $l \in L$. Then $l = a_1 w_1 + \cdots + a_m w_m$, for some $a_i \in K$. But $a_i = b_{i1} v_1 + \cdots + b_{in} v_n$, for some $b_{ij} \in F$. Thus,

$$l = b_{11} v_1 w_1 + b_{12} v_2 w_1 + \cdots + b_{1n} v_n w_1 + \cdots + b_{m1} v_1 w_m + \cdots + b_{mn} v_n w_m.$$

That is, the $v_i w_j$ span L over F. Suppose that they are linearly dependent. Then there exist $b_{ij} \in F$, not all zero, such that

$$0 = b_{11}v_1w_1 + \cdots + b_{1n}v_nw_1 + \cdots + b_{m1}v_1w_m + \cdots + b_{mn}v_nw_m$$
$$= (b_{11}v_1 + \cdots + b_{1n}v_n)w_1 + \cdots + (b_{m1}v_1 + \cdots + b_{mn}v_n)w_m.$$

As each $b_{i1}v_1 + \cdots + b_{in}v_n \in K$, and the w_i are linearly independent over K, we have $b_{i1}v_1 + \cdots + b_{in}v_n = 0$, for all i. But the $b_{ij} \in F$, and the v_j are linearly independent over F. Thus, all of the b_{ij} are zero. The proof is complete. □

Example 12.22. Let $K = \{a+b\sqrt{2} : a, b \in \mathbb{Q}\}$. By Example 8.30, K is an extension field of $\mathbb{Q}$. Clearly, 1 and $\sqrt{2}$ span K over $\mathbb{Q}$. If they were linearly dependent, then $\sqrt{2}$ would lie in $\mathbb{Q}$, which is not the case. Thus, $[K : \mathbb{Q}] = 2$. Let $L = \{c + d\sqrt{3} : c, d \in K\}$. We claim that L is a subfield of $\mathbb{R}$ and, hence, an extension of K. All of the subfield properties are easy to check except perhaps the existence of inverses. Let $0 \neq c + d\sqrt{3} \in L$. Then $(c + d\sqrt{3})(c - d\sqrt{3}) = c^2 - 3d^2$. Suppose that this is 0. Then $c - d\sqrt{3} = 0$. If $d = 0$, then so is c, giving us a contradiction. Otherwise, $\sqrt{3} = cd^{-1} \in K$. Thus, we can write $a + b\sqrt{2} = \sqrt{3}$, with $a, b \in \mathbb{Q}$. Then $a^2 + 2b^2 + 2ab\sqrt{2} = 3$. If $b = 0$, then $\sqrt{3} = a \in \mathbb{Q}$, which is not true. If $a = 0$, then $\sqrt{\frac{3}{2}} = b \in \mathbb{Q}$. But then $2x^2 - 3$ has a rational root which, by Theorem 11.5, is not the case. Thus, $ab \neq 0$, and $\sqrt{2} \in \mathbb{Q}$, giving us a contradiction. Therefore, $(c + d\sqrt{3})^{-1} = \frac{c-d\sqrt{3}}{c^2-3d^2} \in L$. Now, 1 and $\sqrt{3}$ span L over K. If they were linearly dependent, then we would have $\sqrt{3} \in K$ which, as we have just seen, is not the case. Therefore, $[L : K] = 2$. By the theorem above, $[L : \mathbb{Q}] = [L : K][K : \mathbb{Q}] = 4$.

One particular type of extension is especially important.

Definition 12.10. Let K be a field extension of F. If $a \in K$, then we write $F(a)$ for the intersection of all subfields of K containing F and a. We say that K is a **simple extension** of F if $K = F(a)$ for some $a \in K$.

By Exercise 8.33, the intersection of some set of fields is a field. Thus, $F(a)$ is always a field. Indeed, it is the smallest subfield of K containing F and a.

Example 12.23. By Example 8.30, $\{a + b\sqrt{2} : a, b \in \mathbb{Q}\}$ is a subfield of $\mathbb{R}$. Thus, since any field including $\mathbb{Q}$ and $\sqrt{2}$ would surely contain this field, it is $\mathbb{Q}(\sqrt{2})$.

Example 12.24. In a similar manner, we note that $\mathbb{Q}(\sqrt[3]{2})$ would have to contain $\sqrt[3]{2}$ and $(\sqrt[3]{2})^2$. Example 12.21 shows us that $\mathbb{Q}(\sqrt[3]{2}) = \{a + b\sqrt[3]{2} + c\sqrt[3]{4} : a, b, c \in \mathbb{Q}\}$.

Let us concentrate on simple extensions. In fact, we need to break them down into two types, depending upon one specific property of the element a.

Definition 12.11. Let K be a field extension of F and $a \in K$. We say that a is **algebraic** over F if there exists a nonzero polynomial $f(x) \in F[x]$ such that $f(a) = 0$. Otherwise, a is **transcendental** over F.

Example 12.25. The number $\sqrt[3]{2}$ is algebraic over $\mathbb{Q}$, since it is a root of $x^3 - 2 \in \mathbb{Q}[x]$.

Example 12.26. The number $\sqrt{2} + \sqrt{3}$ is algebraic over $\mathbb{Q}$, since it is a root of $x^4 - 10x^2 + 1$.

Finding examples of real numbers that are transcendental over $\mathbb{Q}$ is a bit tricky. As it happens, the constants e and π are both transcendental. (This is a difficult result. For a proof, see the advanced monograph of Baker [1].) Of course, the underlying field is important! If we let $F = \mathbb{Q}(\pi^2)$, then π is algebraic over F, as π is a root of $x^2 - \pi^2 \in F[x]$.

We are primarily interested in algebraic elements. However, we can mention one important fact about transcendental elements. If F is a field, then $F[x]$ is an integral domain, and so we can consider its field of fractions. Denote this field of fractions by $F(x)$.

Theorem 12.9. *Let K be an extension field of F, and let $a \in K$ be transcendental over F. Then $F(a)$ is isomorphic to $F(x)$. In particular, $F(a)$ is of infinite degree over F.*

Proof. Define $\alpha : F[x] \to K$ via $\alpha(f(x)) = f(a)$. By Lemma 11.1, α is a homomorphism. If $f(x) \in \ker(\alpha)$, then $f(a) = 0$. Since a is transcendental, $f(x)$ is the zero polynomial. Thus, α is one-to-one, and $F[x]$ is isomorphic to $\alpha(F[x])$. Also, $f(a) \in F(a)$ for all $f(x) \in F[x]$; thus, $\alpha(F[x])$ is a subring of $F(a)$. By Theorem 9.15, there is a subfield L of $F(a)$ such that L is isomorphic to $F(x)$ and contains $\alpha(F[x])$. Clearly $\alpha(b) = b$ for all $b \in F$ and $\alpha(x) = a$; thus, $\alpha(F[x])$ contains both F and a. But $F(a)$ is the smallest subfield of K containing both F and a; thus, $F(a) = L$.

Suppose that $[F(a) : F] = n < \infty$. Then according to Lemma 12.1, the elements $1, a, a^2, \ldots, a^n$ must be linearly dependent over F. But then there exist $c_i \in F$, not all zero, such that a is a root of $c_0 + c_1 x + \cdots + c_n x^n$. That is, a is algebraic, giving us a contradiction. $\qquad\square$

Now suppose that a is algebraic over F. We know that it satisfies a nonzero polynomial in $F[x]$. But one particular such polynomial is key.

Definition 12.12. Let K be an extension field of F and let $a \in K$ be algebraic over F. Then the **minimal polynomial** of a over F is the monic irreducible polynomial $m(x) \in F[x]$ such that $m(a) = 0$.

Example 12.27. The minimal polynomial of $\sqrt[3]{2}$ over $\mathbb{Q}$ is $x^3 - 2$. Indeed, $\sqrt[3]{2}$ is a root, and the polynomial is irreducible by Example 11.10.

Example 12.28. The minimal polynomial of $\sqrt{2} + \sqrt{3}$ over $\mathbb{Q}$ is $x^4 - 10x^2 + 1$. As we noted in Example 12.26, $\sqrt{2} + \sqrt{3}$ is a root. Suppose it were reducible over $\mathbb{Q}$. The Rational Roots Theorem shows us that it has no roots in $\mathbb{Q}$. Thus, it would have to factor as a product of two polynomials of degree 2. By Theorem 11.4, these polynomials may be assumed to be in $\mathbb{Z}[x]$. Looking at the coefficients, we see immediately that (up to multiplying both factors by -1) the only possibilities are $(x^2 + ax + 1)(x^2 - ax + 1)$ and $(x^2 + ax - 1)(x^2 - ax - 1)$, for some $a \in \mathbb{Z}$. But then $2 - a^2 = -10$ or $-2 - a^2 = -10$. Neither of these has a solution in $\mathbb{Z}$.

We were a bit bold in our definition of the minimal polynomial. Indeed, we assumed that such a polynomial exists, and that there is only one. Fortunately, our presumptuousness was justified; in fact, we can say more.

Theorem 12.10. *Let K be an extension field of F, and let $a \in K$ be algebraic over F. Then*

1. *the minimal polynomial $m(x)$ of a over F exists, and is the unique monic polynomial of smallest degree in $F[x]$ of which a is a root; and*
2. *if $f(x) \in F[x]$, then $f(a) = 0$ if and only if $m(x)|f(x)$.*

Proof. Let $I = \{f(x) \in F[x] : f(a) = 0\}$. We claim that I is an ideal of $F[x]$. Surely $0 \in I$. If $f(x), g(x) \in I$, then $f(a) - g(a) = 0$, and hence $f(x) - g(x) \in I$. Also, if $h(x) \in F[x]$, then $f(a)h(a) = 0$, and hence $f(x)h(x) \in I$, proving the claim.

We know that $F[x]$ is a Euclidean domain and hence, by Theorem 10.8, a PID. Thus, let $I = (m(x))$. Since a is algebraic, $m(x)$ is not the zero polynomial. As $(m(x)) = (cm(x))$ if $0 \neq c \in F$, we may as well assume that $m(x)$ is monic. Now, $f(x) \in I$ if and only if $m(x)|f(x)$, as required by (2). As such, $\deg(m(x)) \leq \deg f(x)$, unless $f(x) = 0$. If $\deg(m(x)) = \deg(f(x))$, then $f(x)$ is simply $m(x)$ multiplied by an element of F. If $f(x)$ is also monic, then $f(x) = m(x)$. Thus, $m(x)$ satisfies condition (1) as well.

We must still establish that $m(x)$ is actually the minimal polynomial of a over F. To demonstrate this, we must show that $m(x)$ is irreducible. But if $m(x) = f(x)g(x)$, with $f(x), g(x) \in F[x]$, then $0 = m(a) = f(a)g(a)$. Thus, $f(a) = 0$ or $g(a) = 0$. Without loss of generality, say $f(a) = 0$. Then $m(x)|f(x)$. But also $f(x)|m(x)$. It now follows that $\deg(f(x)) = \deg(m(x))$, and hence $g(x)$ is a constant polynomial. Thus, $m(x)$ is irreducible, and hence a minimal polynomial for a. If $g(x)$ is another minimal polynomial, then $g(x) \in I$, and hence $m(x)|g(x)$. But $g(x)$ is irreducible, and therefore $g(x) = cm(x)$ for some $c \in F$. As $m(x)$ and $g(x)$ are both monic, $m(x) = g(x)$, and the proof is complete. $\square$

We can use the minimal polynomial to describe the simple extension.

Theorem 12.11. *Let L be an extension field of F, and let $a \in L$ be algebraic over F. If $m(x)$ is the minimal polynomial of a over F, let $n = \deg(m(x))$. Then*

1. *$[F(a) : F] = n$;*
2. *$\{1, a, a^2, \ldots, a^{n-1}\}$ is a basis for $F(a)$ over F; and*
3. *$F(a)$ is isomorphic to $F[x]/(m(x))$.*

Proof. Of course, (1) follows immediately from (2), so let us prove (2). Suppose that $1, a, \ldots, a^{n-1}$ are linearly dependent. Then there exist $c_0, \ldots, c_{n-1} \in F$, not all zero, such that $c_0 + c_1 a + \cdots + c_{n-1} a^{n-1} = 0$. That is, a is a root of $c_0 + c_1 x + \cdots + c_{n-1} x^{n-1}$. But this polynomial has degree smaller than that of $m(x)$, contradicting Theorem 12.10. Thus, $1, a, \ldots, a^{n-1}$ are linearly independent.

We claim that they span $F(a)$. We know that $F(a)$ is the smallest field containing F and a (and, hence, all of the a^i). Therefore, it is sufficient to show that $K =$

$\{c_0 + c_1 a + \cdots + c_{n-1} a^{n-1} : c_i \in F\}$ is a field. Clearly, it contains 1 and is closed under subtraction. To show that it is closed under multiplication, it is enough to show that $a^i \in K$ for all positive integers i. Our proof is by strong induction upon i. If $i < n$, there is nothing to do, so let $i \geq n$ and suppose it is true for smaller exponents. Writing $m(x) = b_0 + \cdots + b_{n-1} x^{n-1} + x^n$, we have $a^i = a^n a^{i-n} = (-b_0 - b_1 a - \cdots - b_{n-1} a^{n-1}) a^{i-n}$. But this is a linear combination of terms of the form a^j, with $j < i$. Thus, by our inductive hypothesis, $a^i \in K$. Finally, we must check that every nonzero element of K has an inverse in K. But a nonzero element of K has the form $f(a)$, for some $0 \neq f(x) \in F[x]$, with $\deg(f(x)) < n$. Now, $(f(x), m(x)) | m(x)$. As $m(x)$ is irreducible, $(f(x), m(x))$ is either 1 or an associate of $m(x)$. However, $\deg(f(x)) < \deg(m(x))$. Thus, $(f(x), m(x)) = 1$. By Theorem 10.6, there exist $u(x), v(x) \in F[x]$ such that $f(x)u(x) + m(x)v(x) = 1$. Since $m(a) = 0$, we have $f(a)u(a) = 1$. Furthermore, as we noted above, $u(a) \in K$, so $f(a)$ has an inverse in K. Therefore, K is a field, and (2) is proved.

(3) Define $\alpha : F[x] \to F(a)$ via $\alpha(f(x)) = f(a)$. By Lemma 11.1, α is a homomorphism. In view of (2), it is onto. The kernel is the set of all polynomials in $F[x]$ of which a is a root. By Theorem 12.10, this is $(m(x))$. Apply the First Isomorphism Theorem. □

Example 12.29. As $x^2 + 1$ is the minimal polynomial of i over $\mathbb{R}$, we see that $\mathbb{C} = \mathbb{R}(i)$ is isomorphic to $\mathbb{R}[x]/(x^2 + 1)$.

Example 12.30. As we saw in Example 12.28, the minimal polynomial of $\sqrt{2} + \sqrt{3}$ over $\mathbb{Q}$ is $x^4 - 10x^2 + 1$. Therefore, $\mathbb{Q}(\sqrt{2} + \sqrt{3})$ is isomorphic to

$$\mathbb{Q}[x]/(x^4 - 10x^2 + 1).$$

Furthermore, letting $a = \sqrt{2} + \sqrt{3}$, the elements of $\mathbb{Q}(a)$ are precisely $c_0 + c_1 a + c_2 a^2 + c_3 a^3$, with $c_i \in \mathbb{Q}$. Addition works in the obvious way. To demonstrate multiplication, let us try $(2 - 3a + 4a^2)(5 + a - 6a^2 + 2a^3)$. We get $10 - 13a + 5a^2 + 26a^3 - 30a^4 + 8a^5$. Now, $a^4 = 10a^2 - 1$ and $a^5 = 10a^3 - a$. Thus, our product is $40 - 21a - 295a^2 + 106a^3$.

Our last theorem has an interesting immediate consequence.

Corollary 12.1. *Let K be an extension field of F. If $a, b \in K$, and a and b have the same minimal polynomial over F, then $F(a)$ is isomorphic to $F(b)$.*

Proof. If $m(x)$ is the minimal polynomial, then by Theorem 12.11, both $F(a)$ and $F(b)$ are isomorphic to $F[x]/(m(x))$. □

Example 12.31. Let ω be a primitive cube root of unity in $\mathbb{C}$. (That is, $\omega^3 = 1$ but $\omega \neq 1$.) Then $\sqrt[3]{2}$ and $\omega\sqrt[3]{2}$ are both roots of $x^3 - 2 \in \mathbb{Q}[x]$. As we have observed, $x^3 - 2$ is irreducible over $\mathbb{Q}$, so it is the minimal polynomial of both $\sqrt[3]{2}$ and $\omega\sqrt[3]{2}$. Thus, $\mathbb{Q}(\sqrt[3]{2})$ is isomorphic to $\mathbb{Q}(\omega\sqrt[3]{2})$. These fields are clearly distinct, as $\mathbb{Q}(\sqrt[3]{2})$ is a subfield of $\mathbb{R}$, but $\omega\sqrt[3]{2} \notin \mathbb{R}$.

Exercises

12.21. Find the minimal polynomial of $\sqrt{5} + \sqrt{7}$ over $\mathbb{Q}$.

12.22. Find the minimal polynomial of $\sqrt[3]{3} + \sqrt[3]{9}$ over $\mathbb{Q}$.

12.23. Let K be a finite extension field of F. Show that every element of K is algebraic over F.

12.24. Let K be an extension field of F and L an extension field of K. If $a \in L$ is algebraic over F, show that $[K(a) : K] \leq [F(a) : F]$.

12.25. Suppose that we have subfields F_n of K with $F_1 \subseteq F_2 \subseteq F_3 \subseteq \cdots$. Show that $\bigcup_{n=1}^{\infty} F_n$ is a field.

12.26. For each positive integer n, let $a_n = \sqrt[2^n]{2}$. If $K = \bigcup_{n=1}^{\infty} \mathbb{Q}(a_n)$, show that K is an infinite field extension of $\mathbb{Q}$, but every element of K is algebraic over $\mathbb{Q}$.

12.27. Let K be a field extension of F. Show that for every $a \in K$, $F(a^2) \subseteq F(a)$. Also, give an explicit example illustrating that we do not have $F(a^2) = F(a)$ in general.

12.28. Let K be a field extension of F and $a \in K$. Show that a is algebraic over F if and only if a^2 is algebraic over F.

12.29. Let K be an extension field of $\mathbb{C}$. If $a \in K$ is algebraic over $\mathbb{C}$, show that $a \in \mathbb{C}$.

12.30. Let K be a finite extension of F. If R is a subring of K containing F, show that R is a field.

12.4 Splitting Fields

Let us now take a slightly different perspective from the preceding section. Given a field F, instead of looking at elements of extension fields and finding their minimal polynomials, let us instead take a nonconstant polynomial $f(x) \in F[x]$ and see if we can find a field containing F and a root of $f(x)$. For instance, suppose that we only knew about the rational numbers, and we wanted to construct a field having a root of $x^2 - 2$.

Definition 12.13. Let F be a field and let $f(x) \in F[x]$ be a nonconstant polynomial. If K is an extension field of F, then we say that $f(x)$ **splits** over K if there exist $a, a_1, \ldots, a_n \in K$ such that $f(x) = a(x - a_1) \cdots (x - a_n)$. In particular, K is a **splitting field** for $f(x)$ if $f(x)$ splits over K, and if L is any subfield of K with $F \subseteq L \subsetneq K$, then $f(x)$ does not split over L.

To put this another way, if K is an extension field of F and $b_1, \ldots, b_n \in K$, write $F(b_1, \ldots, b_n)$ for the intersection of all subfields of K containing F and all of the b_i. If $f(x) \in F[x]$ splits over K, and $a_1, \ldots, a_n$ are the roots of $f(x)$ in K, then K is a splitting field of $f(x)$ if and only if $K = F(a_1, \ldots, a_n)$.

But how to construct such a field? The following observation is helpful.

Lemma 12.2. *Every nonzero prime ideal in a PID is maximal.*

Proof. Let I be a nonzero prime ideal in a PID R. Then $I = (a)$, for some $a \in I$. By Lemma 10.2, a is prime. In particular, by Theorem 10.10, a is irreducible. Since I is prime, $I \neq R$. Suppose that J is an ideal of R with $I \subsetneq J \subsetneq R$. Let $J = (b)$. Then $a \in (b)$, so $b | a$. As a is irreducible, b is a unit or an associate of a. In the former case, $J = R$. In the latter, $a | b$, and hence $J = I$. Either way, we have a contradiction. $\square$

The next lemma is the key to our construction.

Lemma 12.3. *Let F be a field and $f(x)$ an irreducible polynomial in $F[x]$. Let $K = F[x]/(f(x))$. Then K is a field containing (an isomorphic copy of) F and a root a of $f(x)$. In fact, $K = F(a)$.*

Proof. We know that $F[x]$ is a Euclidean domain and hence, by Theorem 10.8, a PID. By Theorem 10.11, $f(x)$ is prime. Thus, by Lemma 10.2, $(f(x))$ is a prime ideal. The preceding lemma tells us that $(f(x))$ is maximal. By Theorem 9.20, K is indeed a field. Define $\alpha : F \to K$ via $\alpha(b) = b + (f(x))$. It is immediate that α is a homomorphism. If $\alpha(b) = 0$, then $b \in (f(x))$, which means that $f(x) | b$. As b is a constant, $b = 0$, and hence α is one-to-one. Thus, K contains an isomorphic copy of F, namely $\alpha(F)$. Finally, let us show that K contains a root of $f(x)$. But this root is $a = x + (f(x))$. Indeed, $f(a) = f(x) + (f(x)) = 0 + (f(x))$, as required. Clearly, $F(a)$ would have to contain $x^n + (f(x))$ for all $n \geq 0$. Thus, $K = F(a)$. $\square$

Let us combine the preceding lemma with Theorem 12.11. We see that if $f(x) \in F[x]$ is irreducible of degree n, then the field K has, as a basis over F, the terms $x^i + (f(x))$, with $0 \leq i < n$. This allows us for the first time to create finite fields other than $\mathbb{Z}_p$, where p is a prime.

Example 12.32. Suppose we wish to construct a field of order 125. In view of Theorem 12.7, we would need an extension of degree 3 of $\mathbb{Z}_5$. Consider $f(x) = x^3 + 3x^2 + x + 2 \in \mathbb{Z}_5[x]$. By Corollary 11.2, it is irreducible over $\mathbb{Z}_5$ if it has no roots in $\mathbb{Z}_5$. There are only five possible roots, and none of them work. Therefore, $f(x)$ is irreducible and $F[x]/(f(x))$ is a field of order 125. The elements are $a_0 + a_1 x + a_2 x^2 + (f(x))$, with $a_i \in \mathbb{Z}_5$. Addition works in the obvious way. As an example of multiplication, we have (letting $I = (f(x))$)

$$(2 + 4x + 3x^2 + I)(1 + 4x + I) = 2 + 2x + 4x^2 + 2x^3 + I$$
$$= 2 + 2x + 4x^2 + 2(-3x^2 - x - 2) + I$$
$$= 3 + 3x^2 + I.$$

We can now construct splitting fields.

Theorem 12.12. *Let F be a field and $f(x) \in F[x]$ a nonconstant polynomial. Then there is a splitting field of $f(x)$ over F.*

Proof. First, let us prove the existence of a field extension in which $f(x)$ splits. We proceed by induction on $n = \deg(f(x))$. If $n = 1$, then F will suffice. Assume that $n \geq 2$ and the $n - 1$ case holds. We know that $F[x]$ is a UFD. Thus, write $f(x) = g_1(x) \cdots g_k(x)$, where the $g_i(x)$ are irreducible in $F[x]$. By Lemma 12.3, there is an extension field K of F in which $g_1(x)$ has a root, a. Then by Theorem 11.2, $g_1(x) = (x - a)h_1(x)$, for some $h_1(x) \in K[x]$. Thus, in $K[x]$, we have $f(x) = (x - a)h_1(x)g_2(x) \cdots g_k(x)$. Now, $h_1(x)g_2(x) \cdots g_k(x)$ has degree $n - 1$. Thus, by our inductive hypothesis, it splits in some extension field L of K. Hence, L is an extension field of F, and $f(x)$ splits over L.

Let us write $f(x) = b(x - b_1) \cdots (x - b_n)$, with $b, b_1, \ldots, b_n \in L$. Then $F(b_1, \ldots, b_n)$ is a splitting field for $f(x)$ over F. $\qquad\square$

But we can go one step further. We want to show that splitting fields are unique up to isomorphism. (The proof is a bit technical, but the result will pay dividends when we classify the finite fields.) To this end, we need to sharpen Corollary 12.1 a bit. If $\alpha : R \to S$ is a ring homomorphism, and $f(x) = c_0 + \cdots + c_n x^n \in R[x]$, then we write $\alpha(f(x)) = \alpha(c_0) + \cdots + \alpha(c_n)x^n \in S[x]$.

Lemma 12.4. *Let $\alpha : F \to K$ be an isomorphism of fields. Let $f(x) \in F[x]$ be an irreducible polynomial. Suppose that a is a root of $f(x)$ in some extension field of F and b is a root of $\alpha(f(x))$ in some extension field of K. Then there exists an isomorphism $\beta : F(a) \to K(b)$ such that $\beta(c) = \alpha(c)$ for all $c \in F$ and $\beta(a) = b$.*

Proof. Define $\gamma : F[x] \to F(a)$ via $\gamma(g(x)) = g(a)$. By Lemma 11.1, γ is a homomorphism. By Theorem 12.10, $\ker(\gamma) = (f(x))$. (We assumed that $f(x)$ was monic in that theorem, but that is immaterial here.) In view of Theorem 12.11, γ is onto. Thus, the proof of the First Isomorphism Theorem shows us that the map $\rho : F[x]/(f(x)) \to F(a)$ given by $\rho(g(x) + (f(x))) = g(a)$ is an isomorphism. We also note that if $c \in F$, then $\rho(c + (f(x))) = c$ and $\rho(x + (f(x))) = a$. In precisely the same manner, the map $\tau : K[x]/(\alpha(f(x))) \to K(b)$ given by $\tau(h(x) + (\alpha(f(x)))) = h(b)$ is an isomorphism, $\tau(d + (\alpha(f(x)))) = d$ for all $d \in K$ and $\tau(x + (\alpha(f(x)))) = b$.

Now, the function from $F[x]$ to $K[x]$ mapping each $u(x)$ to $\alpha(u(x))$ is easily seen to be an isomorphism. Composing that with the obvious homomorphism from $K[x]$ to $K[x]/(\alpha(f(x)))$, we obtain a homomorphism from $F[x]$ onto $K[x]/(\alpha(f(x)))$ with kernel $(f(x))$. In view of the First Isomorphism Theorem, we have an isomorphism $\sigma : F[x]/(f(x)) \to K[x]/(\alpha(f(x)))$ given by

$$\sigma(u(x) + (f(x))) = \alpha(u(x)) + (\alpha(f(x))).$$

Notice that $\sigma(c + (f(x))) = \alpha(c) + (\alpha(f(x)))$ for all $c \in F$, and $\sigma(x + (f(x))) = x + (\alpha(f(x)))$.

From Theorem 9.12, we learn that $\tau\sigma\rho^{-1} : F(a) \to K(b)$ is an isomorphism. Furthermore, if $c \in F$, then

$$\tau\sigma\rho^{-1}(c) = \tau\sigma(c + (f(x))) = \tau(\alpha(c) + (\alpha(f(x)))) = \alpha(c),$$

and

$$\tau\sigma\rho^{-1}(a) = \tau\sigma(x + (f(x))) = \tau(x + (\alpha(f(x)))) = b.$$

Letting $\beta = \tau\sigma\rho^{-1}$, we are done.   □

This allows us to prove the uniqueness of splitting fields.

Theorem 12.13. *Let $\alpha : F \to K$ be a field isomorphism, and let $f(x) \in F[x]$ be a nonconstant polynomial. If L is a splitting field of $f(x)$ over F, and M is a splitting field of $\alpha(f(x))$ over K, then there is an isomorphism $\beta : L \to M$ such that β and α agree on F.*

Proof. We proceed by induction on $n = \deg(f(x))$. If $n = 1$, then we can only have $L = F$ and $M = K$. Thus, letting $\beta = \alpha$ will suffice. Assume that the result is true for polynomials of degree $n - 1$. As $f(x)$ is a product of irreducibles in $F[x]$, let us say that $f(x) = g(x)h(x)$, where $g(x)$ is irreducible and $h(x) \in F[x]$. Let a be a root of $g(x)$ in L and b a root of $\alpha(g(x))$ in M. By the preceding lemma, there is an isomorphism $\gamma : F(a) \to K(b)$ such that γ agrees with α on F and $\gamma(a) = b$. We have $f(x) = (x - a)u(x)$, for some $u(x) \in F(a)[x]$, by Theorem 11.2. Also, $\gamma(f(x)) = (x - \gamma(a))\gamma(u(x)) = (x - b)\gamma(u(x))$ in $K(b)[x]$. Now, L is a splitting field for $u(x)$ over $F(a)$ and M is a splitting field for $\gamma(u(x))$ over $K(b)$. Since $\deg(u(x)) = n - 1$, our inductive hypothesis completes the proof.   □

Corollary 12.2. *Let F be a field and $f(x) \in F[x]$ a nonconstant polynomial. Then any two splitting fields of $f(x)$ over F are isomorphic.*

Proof. In the preceding theorem, let $\alpha : F \to F$ be the identity automorphism.   □

Exercises

12.31. Construct an extension field F of $\mathbb{Z}_7$ having order 7^3. In particular, if $F = \mathbb{Z}_7(a)$, what do all of the elements of F look like? To which of these elements is $(a^2 + 5a + 4)(3a^2 + 6)$ equal?

12.32. Construct an extension field F of $\mathbb{Z}_3$ having order 81. In particular, if $F = \mathbb{Z}_3(a)$, what do all of the elements of F look like? To which of these elements is $(a^3 + 2a^2 + 2)(2a^2 + a + 1)$ equal?

12.33. Show that $\mathbb{Q}(\sqrt[3]{2}, \omega)$ is a splitting field of $x^3 - 2$ over $\mathbb{Q}$, where $\omega \in \mathbb{C}$, $\omega^3 = 1$, but $\omega \neq 1$.

12.34. Let F be a field and $f(x) \in F[x]$ a nonconstant polynomial. If K is a splitting field of $f(x)$ over F and L is any extension field of F, suppose that $\alpha : K \to L$ is a homomorphism satisfying $\alpha(c) = c$ for all $c \in F$. If $a \in K$ is a root of $f(x)$, show that $\alpha(a)$ is also a root of $f(x)$.

12.35. Find every automorphism of $\mathbb{Q}(\sqrt{2})$.

12.36. Construct a splitting field for $x^3 + 2x + 1$ over $\mathbb{Z}_3$. Show that it has degree 3 over $\mathbb{Z}_3$.

12.37. Let F be any field and $f(x) \in F[x]$ a nonconstant polynomial. If we let $g(x) = f(x + 1)$, show that $f(x)$ and $g(x)$ have the same splitting fields over F.

12.38. Let F be a field and $f(x) \in F[x]$ a polynomial with $\deg(f(x)) = n \in \mathbb{N}$. Show that $f(x)$ has a splitting field K over F with $[K : F] \leq n!$.

12.5 Applications to Finite Fields

Let us see what we can deduce about finite fields. If F is a finite field, we know that its prime subfield must be isomorphic to $\mathbb{Z}_p$, for some prime p. By Theorem 12.7, F must have order p^n, for some positive integer n. We will construct a field of order p^n and show that, up to isomorphism, there is only one such field.

The following concept looks suspiciously like calculus, but is not.

Definition 12.14. Let F be a field and $f(x) = a_0 + a_1 x + a_2 x^2 + \cdots + a_n x^n \in F[x]$. Then the **formal derivative** of $f(x)$ is $f'(x) = a_1 + 2a_2 x + \cdots + na_n x^{n-1}$.

Note that this has nothing whatsoever to do with limits, as limits do not necessarily make sense in an arbitrary field. The formula happens to agree with the one used for the derivative of real polynomials. We will also not be disturbed by the fact that the following lemma extends the similarity to calculus.

Lemma 12.5. Let F be a field, $f(x), g(x) \in F[x]$ and $a \in F$. Then

1. $(af(x))' = af'(x)$;
2. $(f(x) + g(x))' = f'(x) + g'(x)$; and
3. $(f(x)g(x))' = f'(x)g(x) + f(x)g'(x)$.

Proof. The first two parts follow immediately from the definition. The third is left as Exercise 12.40. $\qquad\square$

Definition 12.15. Let F be a field, $f(x) \in F[x]$ and $a \in F$. We say that a is a **multiple root** of $f(x)$ if $(x - a)^2 | f(x)$.

Example 12.33. In $\mathbb{Q}[x]$, 2 is a multiple root of $x^5 - 4x^4 + 7x^3 - 7x^2 - 8x + 20$, since the polynomial factors as $(x - 2)^2(x^3 + 3x + 5)$.

Theorem 12.14. Let F be a field, $f(x) \in F[x]$ and let $a \in F$ be a root of $f(x)$. Then a is a multiple root of $f(x)$ if and only if $f'(a) = 0$.

Proof. Suppose that a is a multiple root of $f(x)$, say $f(x) = (x - a)^2 g(x)$, with $g(x) \in F[x]$. Then by Lemma 12.5, $f'(x) = 2(x - a)g(x) + (x - a)^2 g'(x)$. Thus, $f'(a) = 0$. Conversely, suppose that $f'(a) = 0$. By Theorem 11.2, $f(x) = (x-a)h(x)$, for some $h(x) \in F[x]$. Thus, $f'(x) = h(x)+(x-a)h'(x)$. As $f'(a) = 0$, we have $0 = h(a) + (a - a)h'(a) = h(a)$. By Theorem 11.2, $(x - a)|h(x)$, and hence $(x - a)^2|f(x)$. $\qquad\square$

Corollary 12.3. *Let F be a field and let $f(x) \in F[x]$ be irreducible. Let K be a splitting field of $f(x)$ over F. If $f(x)$ has a multiple root in K, then $f'(x)$ is the zero polynomial.*

Proof. Let a be the multiple root. Then (multiplying $f(x)$ by a suitable element of F to make it monic), we see that $f(x)$ is the minimal polynomial of a over F. By Theorem 12.10, $f(x)|f'(x)$. But if $f'(x) \neq 0$, then $\deg(f'(x)) < \deg(f(x))$, which is impossible. Therefore, $f'(x)$ is the zero polynomial. $\qquad\square$

Definition 12.16. A field F is said to be **perfect** if no irreducible $f(x) \in F[x]$ has multiple roots in any splitting field of $f(x)$ over F.

We digress from our discussion of finite fields to mention the following.

Theorem 12.15. *Every field of characteristic zero is perfect.*

Proof. If $f(x) = a_0 + \cdots + a_n x^n$, with $a_n \neq 0$ and $n \geq 1$, then $f'(x) = a_1 + \cdots + na_n x^{n-1}$ has leading coefficient $na_n \neq 0$. Thus, $f'(x)$ is not the zero polynomial. Apply Corollary 12.3. $\qquad\square$

Actually, finite fields are perfect too! Let us see why.

Lemma 12.6. *Let F be a finite field of characteristic p. Then the function $\alpha : F \to F$ given by $\alpha(a) = a^p$ is an automorphism.*

Proof. Since F is commutative, $\alpha(ab) = (ab)^p = a^p b^p = \alpha(a)\alpha(b)$, for all $a, b \in F$. By Theorem 8.14, $\alpha(a + b) = (a + b)^p = a^p + b^p = \alpha(a) + \alpha(b)$. If $a^p = 0$, then since F is a field, $a = 0$. Thus, α is one-to-one. Since F is finite, α is onto as well. $\qquad\square$

Theorem 12.16. *Every finite field is perfect.*

Proof. Suppose that F has characteristic p. Let $f(x) \in F[x]$ be irreducible. Suppose that $f(x) = a_0 + a_1 x + \cdots + a_n x^n$. If $f(x)$ has multiple roots in a splitting field, then by Corollary 12.3, $f'(x) = 0$. Thus, $ka_k = 0$, for $1 \leq k \leq n$. If $p \nmid k$, then as $(k, p) = 1$, we may write $ku + pv = 1$, for some $u, v \in \mathbb{Z}$. Therefore, $a_k = uka_k + pva_k = 0 + 0 = 0$. Thus,

$$f(x) = a_0 + a_p x^p + a_{2p} x^{2p} + \cdots + a_{mp} x^{mp}.$$

In view of the preceding lemma, there exist $b_i \in F$ such that $b_i^p = a_{ip}$. But now Theorem 8.14 tells us that

$$(b_0 + b_1 x + b_2 x^2 + \cdots + b_m x^m)^p = b_0^p + b_1^p x^p + \cdots + b_m^p x^{mp}$$
$$= a_0 + a_p x^p + \cdots + a_{mp} x^{mp}$$
$$= f(x).$$

That is, $f(x)$ is reducible. This contradiction completes the proof. $\square$

What would an imperfect field look like? Clearly, it would have to be an infinite field of prime characteristic. Exercise 12.44 shows how to construct an imperfect field.

Back to the finite fields!

Lemma 12.7. *Let F be a field of prime characteristic p and let n be a positive integer. If $K = \{a \in F : a^{p^n} = a\}$, then K is a subfield of F.*

Proof. See Exercise 8.40.

Theorem 12.17. *Let p be a prime and n a positive integer. Then a field F has order p^n if and only if it is a splitting field of $x^{p^n} - x$ over the prime subfield, (an isomorphic copy of) $\mathbb{Z}_p$.*

Proof. Let F have order p^n. Then $U(F)$ has order $p^n - 1$. Thus, if $0 \neq a \in F$, then $a^{p^n - 1} = 1$, and hence $a^{p^n} - a = 0$. Clearly, $0^{p^n} - 0 = 0$ as well. Thus, every element of F is a root of $x^{p^n} - x$. By Corollary 11.3, $x^{p^n} - x$ can only have p^n roots. Thus, $x^{p^n} - x$ splits over F, and surely it cannot split over any smaller field, as all of the roots must be present. Therefore, F is a splitting field of $x^{p^n} - x$ over $\mathbb{Z}_p$.

Conversely, let F be a splitting field of $x^{p^n} - x$ over $\mathbb{Z}_p$. By Lemma 12.7, the roots of $x^{p^n} - x$ form a subfield K of F. Since $x^{p^n} - x$ splits over K, we must have $F = K$. Furthermore, the formal derivative of $x^{p^n} - x$ is -1, which has no roots. Therefore, by Theorem 12.14, $x^{p^n} - x$ has no multiple roots. In particular, $|F| = p^n$, as required. $\square$

Theorem 12.18. *If k is a positive integer, then there is a field of order k if and only if $k = p^n$ for some prime p and positive integer n. All fields of order p^n are isomorphic.*

Proof. By Theorem 12.7, a finite field must have order p^n. Theorem 12.12 tells us that $x^{p^n} - x$ has a splitting field over $\mathbb{Z}_p$. By Theorem 12.17, this splitting field has order p^n. But Theorem 12.17 also says that every field of order p^n is such a splitting field. By Corollary 12.2, these splitting fields are isomorphic. $\square$

The unique (up to isomorphism) field of order p^n is called the **Galois field** of order p^n.

We can also determine the subfields of a finite field. In order to do so, we will need the following theorem, which is of interest on its own.

Theorem 12.19. *Let F be a field. Then any finite subgroup G of $U(F)$ is cyclic.*

Proof. Since G is a finite abelian group, Theorem 5.3 tells us that it is a direct product of cyclic groups. If all of these cyclic groups have relatively prime orders, then by Theorem 5.4, G is cyclic, and we are done. Otherwise, we may assume that G has a subgroup $\langle a \rangle \times \langle b \rangle$, and there exists a prime p dividing the orders of a and b. By Cauchy's theorem, $\langle a \rangle$ and $\langle b \rangle$ each contain an element of order p. Thus, G has a subgroup isomorphic to $\mathbb{Z}_p \times \mathbb{Z}_p$. But every element of $\mathbb{Z}_p \times \mathbb{Z}_p$ has order 1 or p. That is, we have at least p^2 roots for the polynomial $x^p - 1 \in F[x]$, which has degree p, giving us a contradiction and completing the proof.   □

Theorem 12.20. *Let F be a field of order p^n, for some prime p and positive integer n. Then every subfield of F has order p^m, for some positive divisor m of n. Furthermore, for each positive divisor m of n, F has exactly one subfield of order p^m, namely $\{a \in F : a^{p^m} = a\}$.*

Proof. Let K be a subfield of F. Then K and F have the same prime subfield, (an isomorphic copy of) $\mathbb{Z}_p$. By Theorems 12.7 and 12.8,

$$n = [F : \mathbb{Z}_p] = [F : K][K : \mathbb{Z}_p].$$

In particular, if $[K : \mathbb{Z}_p] = m$, then $|K| = p^m$ and $m \mid n$.

Let m be a divisor of n and let $K = \{a \in F : a^{p^m} = a\}$. By Lemma 12.7, K is a subfield of F. Furthermore, the preceding theorem tells us that $U(F)$ is cyclic of order $p^n - 1$. In addition,

$$p^n - 1 = (p^m - 1)(1 + p^m + p^{2m} + p^{3m} + \cdots + p^{n-m}).$$

Thus, $(p^m - 1) \mid (p^n - 1)$. By Corollary 3.3, $U(F)$ has a subgroup G of order $p^m - 1$. But every element a of G satisfies $a^{p^m - 1} = 1$, and hence $a^{p^m} = a$. That is $G \subseteq K$. Also, $0 \in K$, and therefore K has at least p^m elements. But every element of K is a root of $x^{p^m} - x$, and therefore K can have at most p^m elements.

To prove the uniqueness of this subfield, suppose that L is another subfield of F with p^m elements. Then $U(K)$ and $U(L)$ are both subgroups of order $p^m - 1$ in $U(F)$. However, Corollary 3.3 tells us that $U(F)$ has only one such subgroup. Therefore, $U(K) = U(L)$. As the unit group of a field consists of everything except 0, we have $K = L$, as required.   □

Exercises

12.39. Find the smallest field containing exactly 3 proper subfields.

12.40. Let F be a field and $f(x), g(x) \in F[x]$. Show that $(f(x)g(x))' = f'(x)g(x) + f(x)g'(x)$.

12.41. Let $f(x) \in \mathbb{Z}_5[x]$ be an irreducible polynomial of degree 3. If K is a splitting field of $f(x)$ over $\mathbb{Z}_5$, show that $|K| = 5^3$ or 5^6.

12.42. Let K be a field of order p^n for some prime p and positive integer n, having subfields F and L of orders p^m and p^r, respectively. Find the order of $F \cap L$.

12.43. Let F be a field and $f(x) \in F[x]$ an irreducible polynomial having a multiple root in some extension field of F. Show that char $F = p$ for some prime p, $f(x) = a_0 + a_p x^p + a_{2p} x^{2p} + \cdots + a_{mp} x^{mp}$ for some $a_i \in F$, and that at least one of the a_i is transcendental over the prime subfield of F.

12.44. Let $\mathbb{Z}_2[t]$ be a polynomial ring over $\mathbb{Z}_2$ and $F = \mathbb{Z}_2(t)$ its field of fractions. Show that the polynomial $x^2 - t \in F[x]$ is irreducible over F, but that it has a multiple root in some extension field of F. In particular, conclude that F is not a perfect field.

12.45. Theorem 12.19 tells us that the unit group of a finite field is cyclic. If char $F \neq 2$, show that the unit group of an infinite field is not cyclic.

12.46. Suppose char $F = 2$. Let us prove that the preceding exercise still holds. Suppose, to the contrary, that $U(F)$ is cyclic. Let $U(F) = \langle a \rangle$.

1. Show that $F = \mathbb{Z}_2(a)$.
2. If a is algebraic over $\mathbb{Z}_2$, show that F is finite, and we are done.
3. If a is transcendental over $\mathbb{Z}_2$, show that there exists an integer n such that $a^n = a + 1$, and obtain a contradiction.

12.47. Suppose we wrote $x^{125} - x$ as a product of irreducibles over $\mathbb{Z}_5$. Show that each of these irreducible polynomials has degree 1 or 3. (Please do not actually write the polynomials!)

12.48. Show that for every prime p and positive integer n, there exists an irreducible polynomial of degree n in $\mathbb{Z}_p[x]$.

Reference

1. Baker, A.: Transcendental Number Theory, 2nd edn. Cambridge University Press, Cambridge (1990)

Part V
Applications

Chapter 13
Public Key Cryptography

In this short chapter, we talk a bit about cryptography. First, we discuss some classical sorts of private key methods, and their limitations in the modern world. We then look at the first public key cryptographic method.

13.1 Private Key Cryptography

Countless methods of encrypting messages have been invented over the centuries, and we will not attempt to give an exhaustive list here. Let us discuss a few well-known codes.

One ancient method is known as the **Caesar cipher**. It could not be much simpler! Each letter in the alphabet is shifted forward three letters. Thus, A becomes D, B becomes E, and Z becomes C. If we wish to send the message HOWDY[1] then our encrypted message is KRZGB. Decrypting the message is equally simple; the recipient shifts each letter back three positions.

This is not a particularly good code. An opponent who knew that we were using a Caesar cipher could read any intercepted message instantly. We could complicate things a bit by selecting a positive integer k as a **key**. Instead of shifting letters 3 positions ahead, we would shift them k positions ahead. Decryption would then be a matter of shifting back k positions. We call this an **additive cipher**. This is better, but not much. There are only 26 possible keys (really 25, as one of them will just leave the message unencrypted). It would not take an opponent long to try all of the possible keys on an intercepted encrypted message, and see which one gives sensible text.

But we can be more sophisticated than that. Let ρ be any permutation of the set of letters of the alphabet. We can then encrypt text by replacing each letter with its

---

[1] The author acknowledges that the circumstances under which this message would need to be sent secretly are few and far between.

© Springer International Publishing AG, part of Springer Nature 2018
G. T. Lee, *Abstract Algebra*, Springer Undergraduate Mathematics Series,
https://doi.org/10.1007/978-3-319-77649-1_13

image under ρ. This is called a **simple substitution cipher**. For instance, suppose
we use Table 13.1.

Table 13.1 Encryption table for a simple substitution cipher

original text	A B C D E F G H I J K L M N O P Q R S T U V W X Y Z
encrypted text	R V C X N O A Y W B K U J E T D I L Q M Z F S P G H

Encrypting HOWDY, we obtain YTSXG. To decrypt, we apply the inverse of ρ.
To put this another way, we flip the rows of the table and, for the sake of convenience,
sort by the encrypted letter rather than the original letter, as in Table 13.2.

Table 13.2 Decryption table for the simple substitution cipher from Table 13.1

encrypted text	A B C D E F G H I J K L M N O P Q R S T U V W X Y Z
original text	G J C P N V Y Z Q M K R T E F X S A W O L B I D H U

Thus, YTSXG decrypts to HOWDY. This is a vast improvement over the additive
cipher in terms of security, because the number of possible keys is 26!. Even for a
computer, that is a huge number of permutations to consider. It does come at the cost
of having a larger key to exchange. Also, if a substantial amount of text is intercepted,
this cipher is vulnerable to frequency analysis. That is, in English text, some letters
occur much more frequently than others. For instance, E is by far the most common
letter, T is the second most common, and so forth. An opponent could look for the
most common letters in our text and make an educated guess that those represent
T and E. If another moderately common letter occurs between them frequently, it
might just be THE. Proceeding in this way, the code could be cracked.

Can anything be done about this? There is always the **one time pad**. This is, in
fact, an unbreakable cipher. It is also quite simple. The key is a string of random
letters, at least as long as our message to be encrypted. We then assign a numerical
value to each letter. We let A be 0, B be 1, and so on, letting Z be 25. To encrypt,
we add the value of the first letter of our message to the value of the first letter of
our key in $\mathbb{Z}_{26}$. We then do the same with the second letter of our message and the
second letter of our key, until we reach the end of our message. Each of our sums is
then converted back to a letter.

For instance, say we wanted to encrypt HOWDY, and our randomly selected key
was NCVBT. Now, H is 7 and N is 13, so the sum is 20, which is U. Similarly, O
and C are 14 and 2 respectively, giving a sum of 16, which is Q. Next, W is 22 and
V is 21, giving a sum of 17, which is R. Now, D and B give $3 + 1 = 4$, and hence E,
and Y and T give $24 + 19 = 17$, which is R. Thus, our encrypted text is UQRER.
To decrypt, we subtract the value of the key letter from the corresponding encrypted

letter value. For this message, $20 - 13 = 7$, $16 - 2 = 14$, $17 - 21 = 22$, $4 - 1 = 3$ and $17 - 19 = 24$, and we obtain HOWDY.

Assuming that the key is truly random, is only used once and is kept secret, an opponent who intercepts an encrypted message will be unable to determine anything more than the length of the message. The difficulty with this cipher is that the participants must have the ability to exchange a very large key secretly. In general, anyone who can do that may not need a code! For certain purposes, though, it is ideal. For example, if two people are able to meet once, exchange a briefcase full of random letters, and then leave for distant cities, they will be able to exchange messages while they are apart.

In the internet age, the problem is that most encrypted messages are sent between two distant computers, and the computers can never meet secretly to exchange information. None of the schemes we have discussed are suitable. These are all **private key** methods. That is, the key used must be kept secret. Any opponent who discovered it could easily decrypt an intercepted message, since the ability to encrypt implies the ability to decrypt. Modern codes use **public key** schemes. The key can be released to an opponent without fear, because in these methods, it is quite possible to be able to encrypt and not be able to decrypt.

The next section is devoted to a discussion of the first such scheme.

Exercises

Spaces and punctuation have been deleted from all messages to be encrypted or decrypted.

13.1. Encrypt the following message using a Caesar cipher:
THETREASUREISBURIEDTWENTYPACESNORTHOFTHEPALMTREE

13.2. A message in English has been encrypted using an additive cipher. Decrypt it.
BPMBQUMPIAKWUMBPMEITZCAAIQLBWBITSWNUIVGBPQVOA

13.3. Let us define a multiplicative cipher as follows. Assign the letters of the alphabet numerical values as usual (A is 0, B is 1, Z is 25), and choose a positive integer k as a key. Then if a letter with value v appears in the text, encrypt it as kv, with the multiplication taking place in $\mathbb{Z}_{26}$. Which values of k will produce a valid cipher?

13.4. A message was encrypted using a multiplicative cipher, as in the preceding problem, with $k = 7$. Decrypt it.
WUGCDEGCWERCHCZECRCVAWGANMAWWE
FEGBUWWEHZCDXENQWHCJUPCHPCASJAWD

13.5. We establish a simple substitution cipher using the following table.

original text	A B C D E F G H I J K L M N O P Q R S T U V W X Y Z
encrypted text	V Y Z X E N A W R I O P C S B D F G H J K L M Q T U

Encrypt the following message:
TRANSFERTENMILLIONDOLLARSONTUESDAY

13.6. Using the same simple substitution cipher as in the preceding problem, a message is encrypted. Decrypt it.

YEJJEGJWGEEWBKGHJBBHBBSJWVSVCRSKJEJBBPVJE

13.7. Making use of the key

RQPKDFOCMWODKFJDKSKDFVKQUYCHTISOXETX,

encrypt the following message with a one time pad.

THEDOCUMENTSAREHIDDENBEHINDTHERENOIR

13.8. Making use of the key

ICOWLDIFNSXZIEEOWPAMWRUSDMFJEJFJBAWUQH,

the following message was encrypted with a one time pad. Decrypt it.

EJSJLQOWLULTVXXCBDUDSYYFYQWHEWLAZSSYQY

13.2   The RSA Scheme

The **RSA Scheme** is a public key cipher first described by Ronald L. Rivest, Adi Shamir and Leonard N. Adleman in 1977. In fact, an equivalent system was created by Clifford C. Cocks in 1973, but his work was classified and not made public for more than two decades.

For convenience, let us say that Bob will be sending messages to June. In this case, it is June who creates the cipher. She selects two large distinct primes p and q, and lets $n = pq$. By Theorem 3.19, $\varphi(n) = (p - 1)(q - 1)$. June chooses a number e, with $1 < e < \varphi(n)$, such that $(e, \varphi(n)) = 1$. The public key consists of the numbers e and n. She sends these to Bob, without worrying about whether they are intercepted. She does not, however, tell anyone what p and q are!

Bob prepares to send a message m which must be an integer with $0 \leq m < n$. (We will discuss how to convert text to this format shortly.) Bob calculates

$$m^e \equiv a \pmod{n},$$

where $0 \leq a < n$. He then sends the encrypted message a to June.

How does June decrypt? The number e was chosen so that $e \in U(\varphi(n))$. Let d be the inverse of e in this group. To put that another way,

$$de \equiv 1 \pmod{\varphi(n)}.$$

But now we have the following theorem.

Theorem 13.1. *Let p and q be distinct primes and $n = pq$. If $k \equiv 1$ (mod $\varphi(n)$), then for any integer b, we have*

$$b^k \equiv b \pmod{n}.$$

Proof. First suppose that $(b, n) = 1$. Then $b \in U(n)$. But by Theorem 3.17, $|U(n)| = \varphi(n)$. Thus, by Corollary 3.5, $b^{\varphi(n)} \equiv 1 \pmod{n}$. Now, $\varphi(n) \mid (k - 1)$, and hence $b^{k-1} \equiv 1 \pmod{n}$. Thus, $b^k \equiv b \pmod{n}$.

Now, suppose that exactly one of $\{p, q\}$ divides b. Without loss of generality, say $p \mid b$ but $q \nmid b$. Then $b \in U(q)$. As $|U(q)| = \varphi(q) = q - 1$, we see that $b^{q-1} \equiv 1 \pmod{q}$. Thus, $b^{(p-1)(q-1)} \equiv 1 \pmod{q}$ and hence, as above, $b^k \equiv b \pmod{q}$. Also, $b \equiv b^k \equiv 0 \pmod{p}$. Thus, p and q both divide $b^k - b$. Since p and q are relatively prime, Corollary 2.3 tells us that $n = pq \mid (b^k - b)$, as required.

Finally, if both p and q divide b, then $b \equiv b^k \equiv 0 \pmod{n}$. $\square$

Therefore, to decrypt, June calculates

$$a^d \equiv (m^e)^d \equiv m \pmod{n}.$$

The scheme works because June is the only one who knows d. In order to calculate d, an opponent would need to find $\varphi(n)$. But knowing that means being able to calculate p and q (see Exercise 13.10). And it is precisely upon the difficulty of this problem that the security of the system rests. To be sure, if Bob and June were foolish enough to use $n = 143$, an opponent would be able to find p and q instantly. But what if n had 300 digits? Factoring that into two primes of roughly 150 digits each is certainly beyond human abilities, and even for a computer, it is going to take a very long time. In theory, the cipher would be breakable. But any system that will take a fast computer a trillion years to crack is good enough for most purposes.

How do we create our messages? Suppose that n has d digits. Then we will create a message m that is at most $d - 1$ digits long, so that $m < n$. If $d - 1$ is even, we will do this by grouping our message into blocks of $\frac{d-1}{2}$ letters; if it is odd, then the blocks will have size $\frac{d-2}{2}$. Use the same values for letters introduced in the previous section, but make each letter have two digits. Thus, A is 00, B is 01, and Z is 25. Put the numbers from one block together to form a message m. (We have to use the two-digit method. If we dropped the leading zeroes, we would not know if 123 meant 1, 2 and 3 (BCD), 1 and 23 (BX) or 12 and 3 (MD).) If the length of our text message does not split evenly into blocks of the appropriate length, then pad out the last block with random letters.

Example 13.1. June decides to create an RSA scheme using the primes $p = 113$ and $q = 137$. (Yes, these are much too small to produce a secure system. However, the author is far too lazy to perform calculations using 300-digit numbers, and these will suffice for an illustration.) Then $n = pq = 15481$ and $\varphi(n) = (p - 1)(q - 1) = 15232$. How can June find a suitable e? A prime larger than both p and q will certainly be relatively prime to $\varphi(n)$. June selects $e = 151$. She then sends the values of n and e to Bob.

As n has five digits, Bob knows that he must break his message into two-letter blocks. Suppose he wishes to send the message HOWDY. As the length is not a multiple of 2, he pads it out by adding a Q to the end. Now, HO is 0714, WD is 2203 and YQ is 2416. To encrypt the first message, Bob calculates

$$714^{151} \equiv 14628 \quad (\text{mod } 15481).$$

Next, he calculates

$$2203^{151} \equiv 2494 \quad (\text{mod } 15481)$$

and

$$2416^{151} \equiv 8498 \quad (\text{mod } 15481).$$

Bob sends June the messages 14628, 2494 and 8498.

June must calculate d. Since $(e, \varphi(n)) = 1$, we know that there exist $u, v \in \mathbb{Z}$ such that $eu + \varphi(n)v = 1$. Then d will be u (modulo $\varphi(n)$). The Euclidean algorithm shows us how to calculate d, and we find that $d = 807$. Thus, June calculates

$$14628^{807} \equiv 714 \quad (\text{mod } 15481),$$

$$2494^{807} \equiv 2203 \quad (\text{mod } 15481)$$

and

$$8498^{807} \equiv 2416 \quad (\text{mod } 15481).$$

The original message was, therefore, 071422032416, which converts to HOWDYQ.

We should mention a couple of practical points. First, as any power of 0 is 0 and any power of 1 is 1, the messages 0 and 1 will not be encrypted using any RSA scheme. For that matter, since e will always be odd, $n - 1$ (which is -1 modulo n), will not change either. Given our method of encrypting English text, $n - 1$ will not arise, but 0 and 1 might. Can we do anything about it? Keep in mind that in the preceding example, we had $n = 15481$ but the possible messages would have been in the range of 0 to 2525. Would there be any harm in pushing them into the range of 2 to 2527? Surely not! Thus, we can agree to add 2 to every message before encrypting, and then subtract 2 after decrypting. (Do not do this in the exercises!)

Another point worth mentioning is that e should be reasonably large. To see why, note that in the example above, we could have used $e = 3$. This would be a problem if we sent a relatively small message. For example, if we had $m = 6$, the encrypted message would be $6^3 = 216$. No reduction modulo n takes place! An opponent who intercepted the message could simply take the cube root of 216 and recover the original message, without knowing anything about p and q. If e is large, we can ensure that this is avoided.

While modern ciphers are more complex than the RSA scheme, their security invariably rests upon the fact that it is very difficult to factor large numbers.

Exercises

Spaces have been deleted from all messages to be encrypted or decrypted. Where a letter is needed to pad out a block, use Q.

13.9. Suppose that someone foolishly used numbers as small as $n = 1961$ and $e = 43$ to create an RSA scheme. Crack the code by determining d.

13.10. If n is a product of two distinct primes, and both n and $\varphi(n)$ are known, show how to determine the two primes quickly. Illustrate the method using $n = 10961$, $\varphi(n) = 10752$.

13.11. Encrypt the message ALGEBRA using an RSA scheme with $n = 17399$ and $e = 149$.

13.12. Encrypt the message ABELIANGROUP using an RSA scheme with $n = 18203$ and $e = 191$.

13.13. Having set up an RSA scheme using $p = 103$, $q = 179$ and $e = 151$, we receive the following message: 2469, 7093, 14773, 10900, 143. Decrypt it.

13.14. Having set up an RSA scheme using $p = 89$, $q = 167$ and $e = 181$, we receive the following message: 13962, 8768, 7864, 4297, 12341. Decrypt it.

Chapter 14
Straightedge and Compass Constructions

We now apply our knowledge of field extensions in order to answer three questions posed by the ancient Greeks.

14.1 Three Ancient Problems

More than 2000 years ago, the ancient Greeks performed many geometric constructions using a straightedge and compass. For our purposes, a **straightedge** is an infinitely long ruler having no markings on it. If we have constructed two points, then we can use the straightedge to construct the line passing through those points. Furthermore, if we have constructed two points A and B, then for any point C we have constructed (which may or may not be distinct from A and B), we can use the **compass** to draw a circle centred at C with radius equal to the distance between A and B.

Next, we can take any two lines, any two circles, or one of each, that we have constructed, and construct their points of intersection. Then we repeat! The general question is, what can we construct in finitely many steps?

Let us discuss a few simple examples that will be of use.

Example 14.1. If we have constructed points A and B, let us construct a perpendicular bisector to the line segment AB. To this end, construct a circle centred at A with radius AB, and a circle centred at B with radius AB. Call the intersection points of these circles C and D. Then construct the line through C and D. It is a perpendicular bisector of AB, as illustrated in Figure 14.1.

Example 14.2. Suppose that we have constructed points A and B, and the line passing through them. Let us say that we have constructed point C as well, although we do not insist that $C \notin \{A, B\}$. We claim that we can construct a line through C that is perpendicular to the line through A and B. Without loss of generality, we may assume that C and A are distinct points. Construct the circle centred at C with radius

© Springer International Publishing AG, part of Springer Nature 2018
G. T. Lee, *Abstract Algebra*, Springer Undergraduate Mathematics Series,
https://doi.org/10.1007/978-3-319-77649-1_14

Fig. 14.1 Construction of a
perpendicular bisector of AB

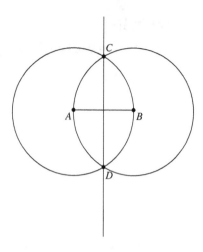

AC. If it intersects the line through A and B at a single point (which must necessarily be A), then the line through A and C will suffice, as illustrated in Figure 14.2.

Otherwise, suppose that the circle meets the line at points A and D. Then the line we are looking for is the perpendicular bisector of AD, which the preceding example allows us to construct. See Figure 14.3.

Example 14.3. Suppose that we have constructed three points A, B and C, that are not collinear. The three points must lie on a circle. Let us construct the centre of the circle, and hence the circle itself. Using Example 14.1, construct the perpendicular bisector of the chord AB. It must pass through the centre of the circle. Similarly, we can construct the perpendicular bisector of BC, and it too passes through the centre of the circle. Therefore, the point of intersection D of the two lines we have just constructed is the centre of the circle, and we can construct the circle itself, as it is centred at D and has radius AD. See Figure 14.4.

For all the remarkable geometric constructions that were performed in antiquity, some problems could not be solved at the time.

Question 14.1. (Squaring the Circle). Given an arbitrary circle, can we construct a square having the same area?

As we shall see, if we are given a square, we can construct another square whose area is twice that of the first square. If we extend our constructions into three dimensions, we have the following.

Question 14.2. (Doubling the Cube). Given an arbitrary cube, can we construct another cube having twice the volume of the first cube?

If we are given three distinct points A, B and C, then we can construct a bisector of the angle $\angle ABC$. That is, we can construct a point D such that $\angle DBC = \frac{1}{2}\angle ABC$. See Exercise 14.5. This naturally led to the following question.

Fig. 14.2 Construction of a perpendicular to AB passing through C (first case)

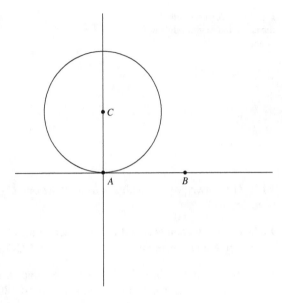

Fig. 14.3 Construction of a perpendicular to AB passing through C (second case)

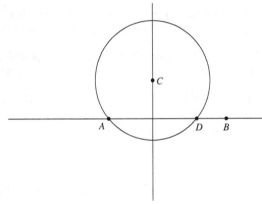

Question 14.3. (Trisecting the Angle). Given any three distinct points A, B and C, can we construct a point D such that $\angle DBC = \frac{1}{3}\angle ABC$?

In fact, all three questions have a negative answer, but it was not until the nineteenth century that the tools of modern algebra allowed a proof to be given.

Exercises

14.1. Suppose that we have points A and B, and the distance from A to B is 1. Construct points C and D such that the distance from C to D is 1.5.

14.2. Suppose that we have points A and B, and the distance from A to B is 1. Construct points C and D such that the distance from C to D is $\sqrt{2}$.

Fig. 14.4 Construction of
the circle passing through A,
B and C

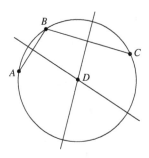

14.3. Given two points A and B, construct a point C such that ABC is an equilateral triangle.

14.4. Given two points A and B, construct points C, D and E all lying on the circle centred at A and passing through B, such that $BCDE$ is a square.

14.5. Given three points A, B and C, construct a point D such that $\angle DBC = \frac{1}{2}\angle ABC$ (where $\angle ABC$ is assumed to be at most $180°$).

14.6. Given three points A, B and C, construct a point D such that $\angle DBC = 2\angle ABC$ (where $\angle ABC$ is assumed to be at most $90°$).

14.7. Given two points A and B, construct points C and D on the circle centred at A and passing through B, such that BCD is an equilateral triangle.

14.8. Suppose we are given three points A, B and C, not collinear. Construct the inscribed circle for the triangle ABC; that is, construct a circle that lies inside the triangle but intersects each side at a single point.

14.2 The Connection to Field Extensions

In order to tackle these problems, we need to be able to discuss them in algebraic terms. Let us formalize our procedure.

We will begin with two points. (Without those, we cannot construct any lines or circles, and so we get nowhere.) Let us identify these with the points $(0, 0)$ and $(1, 0)$ in the plane. We let $P_1 = \{(0, 0), (1, 0)\}$. Then we proceed as follows. For every positive integer i, take all pairs of distinct points A and B in P_i, and draw the line through A and B. Also, for every pair of distinct points A and B in P_i, and for every point C in P_i (where C may or may not be in $\{A, B\}$), draw the circle centred at C with radius equal to the distance between A and B. Let Q_i be the set of all lines and circles obtained in this way. Then let P_{i+1} be the set of all points of intersection of any two distinct lines, any two distinct circles, or any line and any circle in Q_i.

We note that each P_i and Q_i is a finite set, with $P_i \subseteq P_{i+1}$ and $Q_i \subseteq Q_{i+1}$ for all i.

Definition 14.1. A line or circle in the plane is **constructible** if it is in some Q_i. A point in the plane is **constructible** if it is in some P_i. A real number r is **constructible** if the point $(r, 0)$ is constructible.

Let us start by proving what numbers we can construct, and then see what limits there are upon constructibility.

Lemma 14.1. *Let $r \in \mathbb{R}$. Then the following are equivalent:*

1. *r is constructible;*
2. *$-r$ is constructible;*
3. *the point $(0, r)$ is constructible; and*
4. *the point $(0, -r)$ is constructible.*

Proof. If $r = 0$, there is nothing to do, so assume that $r \neq 0$.

Suppose that (1) holds. Then let $A = (0, 0)$ and $B = (r, 0)$. We can construct the circle centred at A with radius AB and the line through A and B (namely, the x-axis). They intersect at $C = (-r, 0)$, giving (2). As in Example 14.1, construct the perpendicular bisector of BC, which is the y-axis. The circle we constructed above intersects it at $(0, r)$ and $(0, -r)$, giving (3) and (4). By symmetry, (2) implies (1) as well.

If we assume (3), then again, we can construct a circle centred at $(0, 0)$ with radius $|r|$. As we are given $(0, 0)$ and $(1, 0)$, we can construct the x-axis, which intersects the circle at $(r, 0)$, giving (1). By symmetry, (4) implies (1) as well. $\square$

Note from the proof that since $(0, 0)$ and $(1, 0)$ are constructible, so are the x- and y-axes.

Lemma 14.2. *Let $a, b \in \mathbb{R}$. Then the point (a, b) is constructible if and only if the numbers a and b are constructible.*

Proof. Suppose that (a, b) is constructible. As in Example 14.2, construct a line through (a, b) perpendicular to the x-axis. It intersects the x-axis at $(a, 0)$, and so a is constructible. Then construct the line through (a, b) perpendicular to the y-axis. It intersects the y-axis at $(0, b)$. Hence, by the preceding lemma, b is constructible.

Conversely, let a and b be constructible. Then the points $(a, 0)$ and $(0, b)$ are constructible. Construct the line perpendicular to the x-axis through $(a, 0)$. Similarly, construct the line perpendicular to the y-axis through $(0, b)$. These two lines meet at (a, b). $\square$

Theorem 14.1. *The constructible numbers form a subfield of $\mathbb{R}$.*

Proof. By definition, 1 is constructible. Suppose that a and b are constructible. We would like to show that $a + b$ and $a - b$ are constructible. If $b = 0$, there is nothing to do. Otherwise, construct the circle centred at $(a, 0)$, the radius of which is the distance from $(0, 0)$ to $(b, 0)$. It intersects the x-axis at the points $(a + b, 0)$ and $(a - b, 0)$. See Figure 14.5.

Fig. 14.5 Construction of
$a + b$ and $a - b$

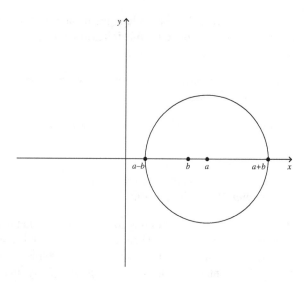

If $b \neq 0$, we also need to construct ab^{-1}. As it is sufficient to construct $-ab^{-1}$, we may assume that $a \geq 0$ and $b > 0$. But in view of the preceding lemma, we can construct the points $(-b, 0)$, $(0, a)$ and $(0, -1)$. As in Example 14.3, we can construct the circle passing through these points. Either geometrically or through algebraic manipulation (see Exercise 14.10), we can prove that this circle intersects the x-axis at $(ab^{-1}, 0)$. See Figure 14.6.

Thus, ab^{-1} is constructible. Theorem 8.12 completes the proof. $\square$

As $\mathbb{Q}$ is the prime subfield of $\mathbb{R}$, we now know that every rational number is constructible. But we can say more.

Theorem 14.2. *If a is a positive constructible number, then so is $\sqrt{a}$.*

Proof. As a and 1 are constructible, Lemma 14.1 tells us that we can construct the points $A = (0, a)$ and $B = (0, -1)$. By Example 14.1, we can construct the perpendicular bisector of AB and, hence, its midpoint $C = (0, \frac{a-1}{2})$. Now construct the circle with centre C and radius AC. This circle intersects the x-axis at $(d, 0)$ and $(-d, 0)$ for some $d > 0$. Again using Exercise 14.10, we see that $d = \sqrt{a}$. Thus, $\sqrt{a}$ is constructible. $\square$

Corollary 14.1. *Suppose there exist fields $\mathbb{Q} = F_0 \subseteq F_1 \subseteq \cdots \subseteq F_k$, where each F_{i+1} is a quadratic extension of F_i and $F_k \subseteq \mathbb{R}$. Then every element of F_k is constructible.*

Proof. We noted above that every element of F_0 is constructible. Thus, by induction, it suffices to show that if every element of a field F is constructible, and $[K : F] = 2$, then every element of K is constructible. Now, if $a \in K$, but $a \notin F$, then $\{1, a\}$ is linearly independent over F and hence, in view of Theorem 12.6, a basis for K. In

Fig. 14.6 Construction of ab^{-1}

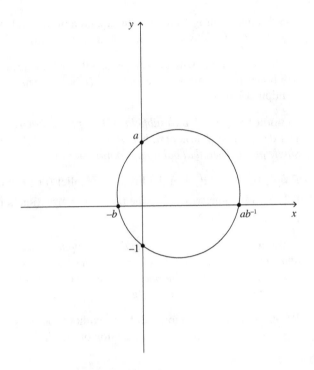

particular, $K = F(a)$. By Theorem 12.9, a is algebraic over F and, in particular, Theorem 12.11 tells us that the minimal polynomial has degree 2. Say that it is $x^2 + bx + c$, with $b, c \in F$. But then $a = \frac{-b \pm \sqrt{b^2 - 4c}}{2}$. By the preceding theorem, $\sqrt{b^2 - 4c}$ is constructible. But now Theorem 14.1 tells us that a is constructible, and hence so is every element of $F(a) = K$, as required. $\qquad \square$

Example 14.4. The number $\sqrt{3} + \sqrt[4]{2 + 5\sqrt{15}}$ is constructible. Let $F_0 = \mathbb{Q}$, $F_1 = F_0(\sqrt{3})$, $F_2 = F_1(\sqrt{15})$, $F_3 = F_2(\sqrt{2 + 5\sqrt{15}})$ and $F_4 = F_3(\sqrt{\sqrt{2 + 5\sqrt{15}}})$. It is clear that each extension is of degree at most 2, since if $a \in F$, then either $\sqrt{a} \in F$ or the minimal polynomial of a over F is $x^2 - a$. Furthermore, $\sqrt{3} + \sqrt[4]{2 + 5\sqrt{15}} \in F_4$.

Now, let us try to restrict the sorts of numbers that can be constructed.

Lemma 14.3. *Let F be a subfield of $\mathbb{R}$. Suppose that we have two distinct points A and B such that the coordinates of both points lie in F. Then the line through A and B has an equation of the form $ax + by = c$, for some $a, b, c \in F$. If C is any point with coordinates in F, then the circle centred at C with radius equal to the distance between A and B has an equation of the form $(x - d)^2 + (y - e)^2 = f$, for some $d, e, f \in F$.*

Proof. Let us say that $A = (a_1, a_2)$, $B = (b_1, b_2)$ and $C = (c_1, c_2)$. Then the equation of the line is $(b_2 - a_2)x + (a_1 - b_1)y = a_1 b_2 - a_2 b_1$, and we can see that the coefficients are in F. Similarly, the equation of the circle is $(x - c_1)^2 + (y - c_2)^2 = (a_1 - b_1)^2 + (a_2 - b_2)^2$, which is of the correct form. $\square$

Readers familiar with linear algebra will not find the next lemma surprising, as the solution to a system of linear equations can be found using only addition, subtraction, multiplication and division.

Lemma 14.4. *Let F be a subfield of* $\mathbb{R}$. *Suppose that we have two lines,* $ax + by = c$ *and* $dx + ey = f$, *where* $a, b, c, d, e, f \in F$, *and that the two lines intersect at a single point. Then that point has coordinates in F.*

Proof. If $ae = bd$, then the lines are parallel (or identical), which is not permitted. Assume otherwise. Then the point of intersection is $\left(\frac{ce - bf}{ae - bd}, \frac{af - cd}{ae - bd} \right)$, and these coordinates lie in F. $\square$

Lemma 14.5. *Let F be a subfield of* $\mathbb{R}$. *Suppose that we have a line* $ax + by = c$ *and a circle* $(x - d)^2 + (y - e)^2 = f$, *with* $a, b, c, d, e, f \in F$. *If the line and circle intersect, then there is a nonnegative number* $g \in F$ *such that the coordinates of the intersection point(s) lie in* $F(\sqrt{g})$.

Proof. As a and b cannot both be 0, without loss of generality, say $a \neq 0$. Then $x = \frac{c - by}{a}$. Substituting into the equation of the circle, we obtain

$$\left(\frac{c - by}{a} - d \right)^2 + (y - e)^2 = f.$$

Simplifying, we obtain an equation of the form $uy^2 + vy + w = 0$, for some $u, v, w \in F$. Furthermore, $u = \frac{b^2}{a^2} + 1 > 0$. Then $y = \frac{-v \pm \sqrt{v^2 - 4uw}}{2u}$. If $v^2 - 4uw < 0$, then the line and circle do not intersect, contradicting our assumption. Thus, let $g = v^2 - 4uw \geq 0$. Then $y \in F(\sqrt{g})$, and as $x = \frac{c - by}{a}$, we see that $x \in F(\sqrt{g})$ as well. $\square$

Lemma 14.6. *Let F be a subfield of* $\mathbb{R}$. *Suppose that we have two distinct circles* $(x - a)^2 + (y - b)^2 = c$ *and* $(x - d)^2 + (y - e)^2 = f$, *with* $a, b, c, d, e, f \in F$. *If these circles intersect, then there is a nonnegative* $g \in F$ *such that the coordinates of the intersection point(s) lie in* $F(\sqrt{g})$.

Proof. Subtracting one equation from the other, we obtain

$$(2a - 2d)x + (2b - 2e)y = f - c + a^2 + b^2 - d^2 - e^2.$$

This is the equation of a line unless $2a - 2d = 2b - 2e = 0$. But in the latter case, the circles have the same centre, meaning that they are identical or do not intersect, so we may assume otherwise. Thus, we are now looking at the intersection of a circle and a line (with coefficients in F), and Lemma 14.5 applies. $\square$

Time to put it all together!

Theorem 14.3. *A real number a is constructible if and only if there exist subfields F_i of $\mathbb{R}$ such that $\mathbb{Q} = F_0 \subseteq F_1 \subseteq \cdots \subseteq F_k$, where each F_{i+1} is a quadratic extension of F_i and $a \in F_k$.*

Proof. One direction of the theorem is given by Corollary 14.1. Let us prove the other. Suppose that a is constructible. Referring to the sets P_i and Q_i from the definition of constructibility, let K_i be the intersection of all subfields of $\mathbb{R}$ containing all of the coordinates of the points in P_i. Then each K_{i+1} is an extension field of K_i.

We claim that for each i, there exist fields F_j with $\mathbb{Q} = F_0 \subseteq F_1 \subseteq \cdots \subseteq F_m = K_i$, where each F_{j+1} is a quadratic extension of F_j. Our proof is by induction on i. If $i = 1$, then $P_i = \{(0,0),(1,0)\}$ and hence $K_i = \mathbb{Q}$ and there is nothing to do. Assume that our claim holds for i. Then we have $\mathbb{Q} = F_0 \subseteq F_1 \subseteq \cdots \subseteq F_m = K_i$, where each F_{j+1} is a quadratic extension of F_j. Now, by Lemma 14.3, every line and circle in Q_i has coefficients in K_i. Furthermore, Lemmas 14.4, 14.5 and 14.6 tell us that every possible intersection point of two distinct lines, two distinct circles or a line and a circle in P_i has coordinates either in K_i or in $K_i(\sqrt{b})$, for some nonnegative $b \in K_i$. But P_i is a finite set, so there are only finitely many values of b being used. Let them be $b_1, \ldots, b_n$. For each k, $1 \le k \le n$, let $F_{m+k} = F_{m+k-1}(\sqrt{b_k})$. If $\sqrt{b_k} \in F_{m+k-1}$, then $F_{m+k} = F_{m+k-1}$, and we can discard F_{m+k}. Otherwise, $x^2 - b_k$ is irreducible over F_{m+k-1}, and so F_{m+k} is a quadratic extension of F_{m+k-1}, by Theorem 12.11. But the last of the F_{m+k} is, by definition, K_{i+1}, establishing the claim.

Since a is constructible, $(a, 0) \in P_i$, for some i, hence $a \in K_i$ and the proof is complete. $\qquad\square$

Corollary 14.2. *Let $a \in \mathbb{R}$. If a is constructible, then a is algebraic over $\mathbb{Q}$ and, in fact, its minimal polynomial over $\mathbb{Q}$ has degree 2^m for some nonnegative integer m.*

Proof. Using F_i as in the statement of the preceding theorem, we have $a \in F_k$, and by Theorem 12.8,

$$[F_k : \mathbb{Q}] = [F_k : F_{k-1}][F_{k-1} : F_{k-2}] \cdots [F_1 : \mathbb{Q}] = 2^k.$$

Thus, by Lemma 12.1, the numbers $1, a, a^2, \ldots, a^{2^k}$ are linearly dependent over $\mathbb{Q}$. That is, a satisfies a nonzero polynomial over $\mathbb{Q}$. By Theorem 12.11, $[\mathbb{Q}(a) : \mathbb{Q}] < \infty$ and therefore, by Theorem 12.8,

$$2^k = [F_k : \mathbb{Q}] = [F_k : \mathbb{Q}(a)][\mathbb{Q}(a) : \mathbb{Q}].$$

That is, $[\mathbb{Q}(a) : \mathbb{Q}]$ divides 2^k, and so it is 2^m, for some m. By Theorem 12.11, the degree of the minimal polynomial of a over $\mathbb{Q}$ is 2^m. $\qquad\square$

Please note that while the condition given in Theorem 14.3 is necessary and sufficient for a to be constructible, the condition given in the corollary is not. It is possible to find a real number a whose minimal polynomial over $\mathbb{Q}$ has degree 4, but such that a is not constructible.

Exercises

14.9. Let a and b be nonzero real numbers. If a is constructible and b is not, show that neither $a + b$ nor ab is constructible. If c is not constructible, show by example that $b + c$ may or may not be constructible.

14.10. Let a, b, c and d be positive real numbers and suppose that a circle in the plane passes through the points $(a, 0)$, $(-b, 0)$, $(0, c)$ and $(0, -d)$. Show that $ab = cd$.

14.11. Are the following numbers constructible?

1. $\sqrt[4]{2 + \sqrt{5} - \sqrt{3}}$
2. $\sqrt{3} + \sqrt[3]{3}$

14.12. Is $\sqrt[3]{2} + \sqrt[4]{4}$ constructible?

14.13. Let a be a real root of the polynomial $x^6 - 15x^5 + 27x^4 - 12x^3 + 30x^2 - 21x + 87$. Is a constructible?

14.14. Let a be a real root of the polynomial $x^6 - 6x^4 + 12x^2 - 8$. Is a constructible?

14.3 Proof of the Impossibility of the Problems

We now have the machinery necessary to answer the three questions from Section 14.1. First, let us look at squaring the circle. We may as well assume that our two initial points are the centre of the circle, $(0, 0)$, and a point on the circle, $(1, 0)$. Thus, we can construct the unit circle immediately. Its area is π. If we were to construct a square with area π, we would need to construct an edge of length $\sqrt{\pi}$. The following theorem tells us that we cannot.

Theorem 14.4. *The number $\sqrt{\pi}$ is not constructible.*

Proof. If $\sqrt{\pi}$ were constructible, then by Theorem 14.1, π would be constructible as well. But as we mentioned in Section 12.3, π is transcendental over $\mathbb{Q}$. This contradicts Corollary 14.2.   □

As we discussed in Section 14.1, doubling a square is possible. Indeed, if we can construct a side with length s, then by Theorems 14.1 and 14.2, the number $s\sqrt{2}$ is also constructible, and this will be the side length of a square with twice the area. To deal with the problem of doubling the cube, without worrying about going into the third dimension, we may simply suppose that one edge extends between our two initial points, and thus has length 1. This would lead to a cube with volume 1. To obtain a cube with volume 2, we would need an edge with length $\sqrt[3]{2}$. This is not going to happen.

Theorem 14.5. *The number $\sqrt[3]{2}$ is not constructible.*

Proof. By Example 12.27, the minimal polynomial of $\sqrt[3]{2}$ over $\mathbb{Q}$ is $x^3 - 2$. But then Corollary 14.2 tells us that $\sqrt[3]{2}$ is not constructible. ☐

Finally, what about trisecting an angle? Some angles can be trisected (see Exercise 14.15). But not all. Indeed, we will show that an angle of $60°$ (or $\frac{\pi}{3}$, as we will be doing a bit of trigonometry) cannot be trisected. In view of Theorems 14.1 and 14.2, the numbers 0, 1, $\frac{1}{2}$ and $\frac{\sqrt{3}}{2}$ are all constructible. Thus, by Lemma 14.2, we can construct the points $A = \left(\frac{1}{2}, \frac{\sqrt{3}}{2}\right)$, $B = (0,0)$ and $C = (1,0)$. Then $\angle ABC = \frac{\pi}{3}$. Thus, we do not need to assume anything extra to obtain the angle. If we could find a point D such that $\angle DBC = \frac{\pi}{9}$, then we could draw the line through D and B, and then intersect with the unit circle centred at B. An intersection point would be $\left(\cos\left(\frac{\pi}{9}\right), \sin\left(\frac{\pi}{9}\right)\right)$. By Lemma 14.2, this would require $\cos\left(\frac{\pi}{9}\right)$ to be constructible. However, the following theorem dashes any hopes of that.

Theorem 14.6. *The number* $\cos\left(\frac{\pi}{9}\right)$ *is not constructible.*

Proof. For any $\theta \in \mathbb{R}$, note that

$$
\begin{aligned}
\cos(3\theta) &= \cos(2\theta)\cos(\theta) - \sin(2\theta)\sin(\theta) \\
&= (\cos^2(\theta) - \sin^2(\theta))\cos(\theta) - 2\sin^2(\theta)\cos(\theta) \\
&= \cos^3(\theta) - 3\sin^2(\theta)\cos(\theta) \\
&= \cos^3(\theta) - 3(1 - \cos^2(\theta))\cos(\theta) \\
&= 4\cos^3(\theta) - 3\cos(\theta).
\end{aligned}
$$

Let $\theta = \frac{\pi}{9}$. Then $\cos(3\theta) = \frac{1}{2}$, and so

$$
\frac{1}{2} = 4\cos^3(\theta) - 3\cos(\theta).
$$

That is, $\cos\left(\frac{\pi}{9}\right)$ satisfies the polynomial $8x^3 - 6x - 1$. If this polynomial were reducible over $\mathbb{Q}$, then by Corollary 11.2, it would have a rational root. By Theorem 11.3, the only possible roots are $\pm 1, \pm\frac{1}{2}, \pm\frac{1}{4}, \pm\frac{1}{8}$. But none of these work. Thus, $8x^3 - 6x - 1$ is irreducible over $\mathbb{Q}$. Therefore, the minimal polynomial of $\cos\left(\frac{\pi}{9}\right)$ over $\mathbb{Q}$ is $x^3 - \frac{3}{4}x - \frac{1}{8}$. Corollary 14.2 completes the proof. ☐

Exercises

All angles are expressed in radians.

14.15. Show that it is possible to trisect a right angle using straightedge and compass.

14.16. Show that the angles $\pi/6$ and $2\pi/3$ cannot be trisected using straightedge and compass.

14.17. In the next two problems, we will show that it is impossible to construct an angle of $\theta = 2\pi/7$. If it were possible, then as we are given the points $(0,0)$ and $(1,0)$, we would also be able to construct $(\cos(\theta), \sin(\theta))$. In particular, the number $\cos(\theta)$ would be constructible. Let us show that it is not. To this end, for each n, $1 \le n \le 6$, express $\cos(n\theta)$ as a linear combination over $\mathbb{Q}$ of $\cos^k(\theta)$, $0 \le k \le 3$.

14.18. Let $\theta = 2\pi/7$.

1. Show that $\cos(\theta) + \sin(\theta)i$ is a complex root of $1 + x + x^2 + x^3 + x^4 + x^5 + x^6$.
2. Show that $1 + \cos(\theta) + \cos(2\theta) + \cos(3\theta) + \cos(4\theta) + \cos(5\theta) + \cos(6\theta) = 0$.
3. Use the answer to the preceding exercise to show that $\cos(\theta)$ is a root of $8x^3 + 4x^2 - 4x - 1$.
4. Conclude that $\cos(\theta)$ is not constructible.

14.19. Suppose we are given the vertices $A = (0,0)$, $B = (1,0)$ and $C = (c_1, c_2)$ of a triangle. Show that we can construct points D, E and F such that the triangles ABC and DEF are similar, but DEF has twice the area of ABC.

14.20. Suppose that we are forced to perform our constructions using a straightedge and a collapsing compass. That is, any time we lift the compass, it collapses. In particular, we cannot directly use it to construct a circle centred at A with radius equal to the distance between B and C. All we can do is take two points A and B that we have constructed, and construct a circle centred at A and passing through B. Show that this does not change the set of constructible numbers in any way. That is, show that any number that was constructible before is still constructible using a straightedge and collapsing compass.

Appendix A
The Complex Numbers

The complex numbers are an extension of the real numbers.

Definition A.1. A **complex number** is a formal expression $a + bi$, with $a, b \in \mathbb{R}$. The set of all complex numbers is denoted $\mathbb{C}$.

We define addition and multiplication on $\mathbb{C}$ via

$$(a + bi) + (c + di) = (a + c) + (b + d)i$$

and

$$(a + bi)(c + di) = (ac - bd) + (ad + bc)i$$

for all $a, b, c, d \in \mathbb{R}$.

Example A.1. Observe that $(2+3i)+(5-9i) = 7-6i$ and $(2+3i)(5-9i) = 37-3i$.

We identify the real number a with the complex number $a+0i$. A complex number $0 + bi$, with $b \in \mathbb{R}$, is said to be **purely imaginary**. We simply write bi for such a number. In particular, note that $i^2 = -1$. Also, if $u = a + bi$, write $-u = -a - bi$.

Let us summarize a few properties concerning complex addition.

Theorem A.1. *Let $u, v, w \in \mathbb{C}$. Then*

1. $u + v \in \mathbb{C}$;
2. $u + v = v + u$;
3. $(u + v) + w = u + (v + w)$;
4. $u + 0 = u$; *and*
5. $u + (-u) = 0$.

Proof. The calculations are all straightforward. For instance, to show (2), we note that

$$(a+bi)+(c+di) = (a+c)+(b+d)i = (c+a)+(d+b)i = (c+di)+(a+bi).$$

The remaining parts are left to the reader. □

© Springer International Publishing AG, part of Springer Nature 2018
G. T. Lee, *Abstract Algebra*, Springer Undergraduate Mathematics Series,
https://doi.org/10.1007/978-3-319-77649-1

Similarly, we can list some properties of complex multiplication.

Theorem A.2. *Let $u, v, w \in \mathbb{C}$. Then*

1. $uv \in \mathbb{C}$;
2. $uv = vu$;
3. $(uv)w = u(vw)$;
4. $u(v + w) = uv + uw$;
5. $1u = u$; and
6. *if $u \neq 0$, then there exists a $z \in \mathbb{C}$ such that $uz = 1$.*

Proof. Again, all of the calculations in (2) through (5) are straightforward. For instance, to prove (3), let $u = a + bi$, $v = c + di$ and $w = e + fi$, with $a, b, c, d, e, f \in \mathbb{R}$. Then

$$
\begin{aligned}
(uv)w &= ((ac - bd) + (ad + bc)i)(e + fi) \\
&= (ace - bde - adf - bcf) + (acf - bdf + ade + bce)i \\
&= (a + bi)((ce - df) + (cf + de)i) \\
&= u(vw).
\end{aligned}
$$

(6) If $u = a + bi$, then let $z = \frac{a}{a^2+b^2} - \frac{b}{a^2+b^2}i$. $\square$

(Readers who have finished Chapter 3 will realize that Theorem A.1 shows that $\mathbb{C}$ is an abelian group under addition. Those who have completed Chapter 8 will understand that the two theorems combined show that $\mathbb{C}$ is a field.)

Let us discuss a simple example of a way in which the complex numbers differ from the real numbers.

Definition A.2. If $z \in \mathbb{C}$ and n is a positive integer, then we say that z is a **primitive nth root of unity** if $z^n = 1$ but $z^m \neq 1$ for any positive integer $m < n$.

In $\mathbb{R}$, the only roots of unity are 1 and -1. But we see immediately that in $\mathbb{C}$, we have a primitive fourth root of unity, i. We can say more, however. We will need this well-known theorem due to Abraham de Moivre.

Theorem A.3 (De Moivre's Theorem). *Let $\theta \in \mathbb{R}$. Then*

$$
(\cos(\theta) + \sin(\theta)i)^n = \cos(n\theta) + \sin(n\theta)i,
$$

for any positive integer n.

Proof. We proceed by induction on n. If $n = 1$, there is nothing to do. Assume that the theorem holds for n. Then

$$
\begin{aligned}
(\cos(\theta) + \sin(\theta)i)^{n+1} &= (\cos(n\theta) + \sin(n\theta)i)(\cos(\theta) + \sin(\theta)i) \\
&= (\cos(n\theta)\cos(\theta) - \sin(n\theta)\sin(\theta)) \\
&\quad + (\cos(n\theta)\sin(\theta) + \sin(n\theta)\cos(\theta))i \\
&= \cos((n+1)\theta) + \sin((n+1)\theta)i,
\end{aligned}
$$

as required. $\qquad$ □

Corollary A.1. *Let n be a positive integer. Then* $\cos\left(\frac{2\pi}{n}\right) + \sin\left(\frac{2\pi}{n}\right) i$ *is a primitive nth root of unity in* $\mathbb{C}$.

Proof. By Theorem A.3,

$$\left(\cos\left(\frac{2\pi}{n}\right) + \sin\left(\frac{2\pi}{n}\right) i\right)^m = \cos\left(\frac{2m\pi}{n}\right) + \sin\left(\frac{2m\pi}{n}\right) i,$$

for any positive integer m. If $m = n$, then we obtain $\cos(2\pi) + \sin(2\pi)i = 1$. On the other hand, if $1 \leq m < n$, then $0 < \frac{m}{n} < 1$, and hence $\cos\left(\frac{2m\pi}{n}\right) \neq 1$. □

Example A.2. Letting $n = 3$, we obtain a primitive cube root of unity, namely, $\frac{-1}{2} + \frac{\sqrt{3}}{2}i$.

Appendix B
Matrix Algebra

Let us discuss a few definitions and basic properties of matrices. The entries in the matrices will come from rings and fields. Readers who are not yet familiar with these terms can simply assume that the entries are real numbers.

Definition B.1. Let R be a ring and m and n positive integers. Then an $m \times n$ **matrix** over R is an array of elements of R with m rows and n columns. If our matrix is A, then we write a_{ij} for the (i, j)-**entry** of A; that is,

$$A = \begin{pmatrix} a_{11} & a_{12} & \cdots & a_{1n} \\ a_{21} & a_{22} & \cdots & a_{2n} \\ \vdots & \vdots & \ddots & \vdots \\ a_{m1} & a_{m2} & \cdots & a_{mn} \end{pmatrix}.$$

Example B.1. If we let

$$A = \begin{pmatrix} 4 & 6 & 7 \\ 3 & 8 & 5 \end{pmatrix},$$

then A is a 2×3 matrix over $\mathbb{R}$. Furthermore, $a_{12} = 6$ and $a_{21} = 3$.

Definition B.2. Let A and B be $m \times n$ matrices over a ring R. Then their **sum** $A + B$ is the $m \times n$ matrix C such that $c_{ij} = a_{ij} + b_{ij}$ for all i and j.

Example B.2. Working with 2×2 matrices over $\mathbb{R}$, we have

$$\begin{pmatrix} 3 & 6 \\ 2 & 5 \end{pmatrix} + \begin{pmatrix} 4 & 6 \\ 1 & 2 \end{pmatrix} = \begin{pmatrix} 7 & 12 \\ 3 & 7 \end{pmatrix}.$$

For any $m \times n$ matrix A, we also let $-A$ be the $m \times n$ matrix B such that $b_{ij} = -a_{ij}$ for all i and j. Furthermore, the $m \times n$ **zero matrix** has every entry 0. We denote this matrix by 0.

Let us list a few properties of matrix addition.

© Springer International Publishing AG, part of Springer Nature 2018
G. T. Lee, *Abstract Algebra*, Springer Undergraduate Mathematics Series,
https://doi.org/10.1007/978-3-319-77649-1

Theorem B.1. *Let R be a ring, and m and n positive integers. If A, B and C are $m \times n$ matrices over R, then*

1. *$A + B$ is an $m \times n$ matrix over R;*
2. *$A + B = B + A$;*
3. *$(A + B) + C = A + (B + C)$;*
4. *$A + 0 = A$; and*
5. *$A + (-A) = 0$.*

Proof. The first part is contained in the definition. The other parts are all obtained by calculating the (i, j)-entry of each side. For instance, to prove (3), we note that the (i, j)-entry of $(A+B)+C$ is $(a_{ij}+b_{ij})+c_{ij}$, whereas the (i, j)-entry of $A+(B+C)$ is $a_{ij} + (b_{ij} + c_{ij})$, and these are equal. The rest of the proof is left to the reader. □

Anyone who has read Chapter 3 will note that Theorem B.1 implies that the $m \times n$ matrices over a ring form an abelian group under addition.

Definition B.3. Let A be an $m \times n$ matrix over a ring R. If $r \in R$, then the **scalar multiple** rA is the $m \times n$ matrix B such that $b_{ij} = ra_{ij}$ for all i and j.

Example B.3. Working with 3×2 matrices over $\mathbb{R}$, we have

$$5 \begin{pmatrix} 1 & 3 \\ 12 & 6 \\ 1 & 2 \end{pmatrix} = \begin{pmatrix} 5 & 15 \\ 60 & 30 \\ 5 & 10 \end{pmatrix}.$$

Here are a few properties of scalar multiplication.

Theorem B.2. *Let R be a ring and $m, n \in \mathbb{N}$. If A and B are $m \times n$ matrices over R and $r, s \in R$, then*

1. *rA is an $m \times n$ matrix over R;*
2. *$r(A + B) = rA + rB$;*
3. *$(r + s)A = rA + sA$; and*
4. *$r(sA) = (rs)A$.*

Proof. (1) is clear from the definition. Each of the other parts is proved by calculating the (i, j)-entry of both sides of the equation. For instance, the (i, j)-entry of $r(A+B)$ is $r(a_{ij} + b_{ij})$, whereas that of $rA + rB$ is $ra_{ij} + rb_{ij}$, but these are the same, establishing (2). The rest of the proof is left to the reader. □

If F is a field, then Theorems B.1 and B.2, when combined with the obvious fact that $1A = A$, show us that the $m \times n$ matrices over F form a vector space over F, as discussed in Chapter 12.

Matrix multiplication is a bit different.

Definition B.4. Let R be a ring, and let A be a $k \times m$ matrix over R, and B an $m \times n$ matrix over R. Then the **product** AB is the $k \times n$ matrix C such that

$$c_{ij} = a_{i1}b_{1j} + a_{i2}b_{2j} + \cdots + a_{im}b_{mj},$$

for all i and j.

Example B.4. Let

$$A = \begin{pmatrix} 1 & 3 & 2 \\ 2 & 0 & 3 \end{pmatrix} \quad \text{and } B = \begin{pmatrix} 3 & 4 & 1 & 5 \\ 2 & 1 & 2 & 0 \\ 1 & 2 & 6 & 1 \end{pmatrix}$$

be matrices over $\mathbb{R}$. Then

$$AB = \begin{pmatrix} 11 & 11 & 19 & 7 \\ 9 & 14 & 20 & 13 \end{pmatrix}.$$

If R is a ring with identity, then we also have the $n \times n$ **identity matrix** I_n, which is the $n \times n$ matrix A such that $a_{ii} = 1$ for all i and $a_{ij} = 0$ if $i \neq j$. For instance,

$$I_3 = \begin{pmatrix} 1 & 0 & 0 \\ 0 & 1 & 0 \\ 0 & 0 & 1 \end{pmatrix}.$$

For any positive integer n, write $M_n(R)$ for the set of $n \times n$ matrices over a ring R.

Theorem B.3. *Let R be a ring, n a positive integer, and $A, B, C \in M_n(R)$. Then*

1. *$AB \in M_n(R)$;*
2. *$(A + B)C = AC + BC$;*
3. *$A(B + C) = AB + AC$;*
4. *$(AB)C = A(BC)$; and*
5. *if R is a ring with identity, then $I_n A = A I_n = A$.*

Proof. (1) follows from the definition.

(2) The (i, j)-entry of $(A + B)C$ is

$$(a_{i1} + b_{i1})c_{1j} + (a_{i2} + b_{i2})c_{2j} + \cdots + (a_{in} + b_{in})c_{nj},$$

whereas the (i, j)-entry of $AB + AC$ is

$$(a_{i1}c_{1j} + a_{i2}c_{2j} + \cdots + a_{in}c_{nj}) + (b_{i1}c_{1j} + b_{i2}c_{2j} + \cdots + b_{in}c_{nj}),$$

and these are equal.

(3) is similar to (2).

(4) Through repeated applications of (2) and (3), we can reduce (4) to the case where each of A, B and C has at most one nonzero entry. But then it is trivial.

(5) Let $D = I_n$. Then the (i, j)-entry of DA is

$$d_{i1}a_{1j} + d_{i2}a_{2j} + \cdots + d_{in}a_{nj} = a_{ij}.$$

Thus, $I_n A = A$. The proof that $A I_n = A$ is similar.   □

As discussed in Chapter 8, we have now proved that if R is a ring, then so is $M_n(R)$, for any positive integer n. Furthermore, if R is a ring with identity, then so is $M_n(R)$. It is, however, worth mentioning, that $M_n(R)$ need not be commutative, even if R is. For instance, in $M_2(\mathbb{R})$,

$$\begin{pmatrix} 1 & 1 \\ 0 & 1 \end{pmatrix} \begin{pmatrix} 1 & 0 \\ 1 & 1 \end{pmatrix} = \begin{pmatrix} 2 & 1 \\ 1 & 1 \end{pmatrix} \neq \begin{pmatrix} 1 & 1 \\ 1 & 2 \end{pmatrix} = \begin{pmatrix} 1 & 0 \\ 1 & 1 \end{pmatrix} \begin{pmatrix} 1 & 1 \\ 0 & 1 \end{pmatrix}.$$

Definition B.5. Let F be a field and n a positive integer. Then a matrix $A \in M_n(F)$ is said to be **invertible** if there exists a $B \in M_n(F)$ such that $AB = BA = I_n$. In this case, we call B the **inverse** of A and write $B = A^{-1}$.

Example B.5. In $M_2(\mathbb{R})$, the matrix

$$A = \begin{pmatrix} 3 & 2 \\ 7 & 5 \end{pmatrix}$$

is invertible, as

$$A^{-1} = \begin{pmatrix} 5 & -2 \\ -7 & 3 \end{pmatrix}.$$

In most linear algebra courses, a couple of different methods of finding the inverse of a matrix are presented (often just in $M_n(\mathbb{R})$, but the same methods work in $M_n(F)$, for any field F). There is, however, a shortcut for determining if a matrix is invertible.

Definition B.6. Let F be a field and n a positive integer. If $A \in M_n(F)$, then the **determinant** of A, $\det(A)$, is an element of F defined recursively as follows. If $n = 1$, then $\det((a_{11})) = a_{11}$. If $n > 1$, then for any $1 \leq i, j \leq n$, let $A_{ij} \in M_{n-1}(F)$ be the matrix obtained by discarding row i and column j of A. Then

$$\det(A) = a_{11} \det(A_{11}) - a_{12} \det(A_{12}) + a_{13} \det(A_{13}) - \cdots + (-1)^{n+1} a_{1n} \det(A_{1n}).$$

Example B.6. In $M_2(F)$, we have

$$\det \left(\begin{pmatrix} a_{11} & a_{12} \\ a_{21} & a_{22} \end{pmatrix} \right) = a_{11}a_{22} - a_{12}a_{21}.$$

Example B.7. In $M_3(\mathbb{R})$, let

$$A = \begin{pmatrix} 2 & 5 & 3 \\ 1 & 4 & 6 \\ 8 & 9 & 7 \end{pmatrix}.$$

Then

$$\det(A) = 2 \det\left(\begin{pmatrix} 4 & 6 \\ 9 & 7 \end{pmatrix}\right) - 5 \det\left(\begin{pmatrix} 1 & 6 \\ 8 & 7 \end{pmatrix}\right) + 3 \det\left(\begin{pmatrix} 1 & 4 \\ 8 & 9 \end{pmatrix}\right)$$

$$= 2(-26) - 5(-41) + 3(-23)$$

$$= 84.$$

We conclude with the following result.

Theorem B.4. *Let F be a field and n a positive integer. If $A, B \in M_n(F)$, then*

1. $\det(AB) = \det(A)\det(B)$; *and*
2. A *is invertible if and only if* $\det(A) \neq 0$.

Proof. We will prove the $n = 2$ case. The general case can be found in standard introductory linear algebra textbooks.

(1) Observe that

$$AB = \begin{pmatrix} a_{11}b_{11} + a_{12}b_{21} & a_{11}b_{12} + a_{12}b_{22} \\ a_{21}b_{11} + a_{22}b_{21} & a_{21}b_{12} + a_{22}b_{22} \end{pmatrix}.$$

Thus,

$$\det(AB) = (a_{11}b_{11} + a_{12}b_{21})(a_{21}b_{12} + a_{22}b_{22}) - (a_{11}b_{12} + a_{12}b_{22})(a_{21}b_{11} + a_{22}b_{21}).$$

On the other hand

$$\det(A)\det(B) = (a_{11}a_{22} - a_{12}a_{21})(b_{11}b_{22} - b_{12}b_{21}),$$

and these are equal.

(2) If $\det(A) \neq 0$, then let

$$B = (\det(A))^{-1} \begin{pmatrix} a_{22} & -a_{12} \\ -a_{21} & a_{11} \end{pmatrix}.$$

It is easy to verify that $AB = BA = I_2$; thus, $B = A^{-1}$. Suppose, on the other hand, that $\det(A) = 0$. If $AB = I_2$, then by (1), $\det(A)\det(B) = \det(I_2) = 1$, which is impossible. $\qquad\qquad\square$

Solutions

Solutions to the odd-numbered problems.

Problems of Chapter 1

1.1 $S \cap T = \{3\}$, $S \cup T = \{1, 2, 3, 4\}$, $S \backslash T = \{1, 2\}$, $T \backslash S = \{4\}$ and $S \times T = \{(1, 3), (1, 4), (2, 3), (2, 4), (3, 3), (3, 4)\}$.

1.3 Let $a \in R \cup T$. Then $a \in R$ or $a \in T$. If $a \in R$, then as $R \subseteq S$, we have $a \in S$, and hence $a \in S \cup T$. If $a \in T$, then $a \in S \cup T$.

1.5 Take $a \in R \cup (S \cap T)$. Then $a \in R$ or $a \in S \cap T$. If $a \in R$, then $a \in R \cup S$ and $a \in R \cup T$, so $a \in (R \cup S) \cap (R \cup T)$. If $a \in S \cap T$, then $a \in S$ and $a \in T$. Therefore, $a \in R \cup S$ and $a \in R \cup T$, so $a \in (R \cup S) \cap (R \cup T)$. Thus, $R \cup (S \cap T) \subseteq (R \cup S) \cap (R \cup T)$. Conversely, suppose that $a \in (R \cup S) \cap (R \cup T)$. If $a \in R$, then $a \in R \cup (S \cap T)$. If $a \notin R$, then as $a \in R \cup S$, we must have $a \in S$ and, similarly, $a \in T$. Thus, $a \in S \cap T$, and hence $a \in R \cup (S \cap T)$. That is, $(R \cup S) \cap (R \cup T) \subseteq R \cup (S \cap T)$.

1.7 $(2, 3)$, $(2, 4)$, $(2, 5)$, $(3, 8)$.

1.9 Reflexive? Yes. If $a \in \mathbb{R}$, then $a - a = 0 \in \mathbb{Q}$, so $a\rho a$. Symmetric? Yes. If $a\rho b$, then $a - b \in \mathbb{Q}$, so $b - a = -(a - b) \in \mathbb{Q}$, and hence $b\rho a$. Transitive? Yes. If $a\rho b$ and $b\rho c$, then $a - b, b - c \in \mathbb{Q}$. But then $a - c = (a - b) + (b - c) \in \mathbb{Q}$, so $a\rho c$.

1.11 (1) A relation is a subset of $\{1, 2, 3\} \times \{1, 2, 3\}$. This Cartesian product has 9 elements, and therefore $2^9 = 512$ subsets. (See Exercise 1.4.)

(2) A relation ρ is symmetric provided $1\rho 2$ if and only if $2\rho 1$, $1\rho 3$ if and only if $3\rho 1$ and $2\rho 3$ if and only if $3\rho 2$. In short, we do not get to decide if $2\rho 1$, $3\rho 1$ or $3\rho 2$, once all the other possibilities are decided. Thus, only 6 of the 9 possible pairs remain to be determined, so the total number is $2^6 = 64$.

© Springer International Publishing AG, part of Springer Nature 2018
G. T. Lee, *Abstract Algebra*, Springer Undergraduate Mathematics Series,
https://doi.org/10.1007/978-3-319-77649-1

1.13 Reflexivity: As $a - a = 3 \cdot 0$, we have $a \sim a$ for all $a \in \mathbb{N}$. Symmetry: If $a \sim b$, then $a - b = 3k$, and hence $b - a = 3(-k)$; thus, $b \sim a$. Transitivity: Suppose $a \sim b$ and $b \sim c$. Then $a - b = 3k$, $b - c = 3l$, for some $k, l \in \mathbb{Z}$. Thus, $a - c = (a - b) + (b - c) = 3(k + l)$; that is, $a \sim c$. It is an equivalence relation. As for the classes, $[1] = \{1, 4, 7, \ldots\}$, $[2] = \{2, 5, 8, \ldots\}$ and $[3] = \{3, 6, 9, \ldots\}$.

1.15 Reflexivity: As $|a| = |a|$, we have $a \sim a$ for all $a \in \mathbb{Z}$. Symmetry: If $a \sim b$, then $|a| = |b|$. Therefore, $|b| = |a|$ and hence $b \sim a$. Transitivity: If $a \sim b$ and $b \sim c$ then $|a| = |b| = |c|$, and hence $a \sim c$. It is an equivalence relation. The classes are $[0] = \{0\}$, $[1] = \{1, -1\}$, $[2] = \{2, -2\}$, and so on.

1.17 Note that $\{1\}$ is a subset of $\{1, 2\}$, but $\{1, 2\}$ is not a subset of $\{1\}$. Therefore, $\sim$ is not symmetric, and hence not an equivalence relation.

1.19 Reflexivity: If $(a, b) \in \mathbb{R}^2$, then $3a - b = 3a - b$, so $(a, b) \sim (a, b)$. Symmetry: If $(a, b) \sim (c, d)$, then $3a - b = 3c - d$, so $3c - d = 3a - b$ and hence $(c, d) \sim (a, b)$. Transitivity: If $(a, b) \sim (c, d)$ and $(c, d) \sim (e, f)$, then $3a - b = 3c - d = 3e - f$, and hence $(a, b) \sim (e, f)$. Also, $(a, b) \in [(4, 2)]$ if and only if $3a - b = 3 \cdot 4 - 2 = 10$; that is, if and only if $b = 3a - 10$. Thus, $[(4, 2)] = \{(a, 3a - 10) : a \in \mathbb{R}\}$.

1.21 Let $a \sim b$ if and only if either both or neither of a and b lie in $\{1, 2, 3\}$. Reflexivity and symmetry are clear. Suppose $a \sim b$ and $b \sim c$. If $a \in \{1, 2, 3\}$, then $b \in \{1, 2, 3\}$ and hence $c \in \{1, 2, 3\}$. Similarly if $a \notin \{1, 2, 3\}$. Thus, $\sim$ is an equivalence relation. The classes are $[1] = \{1, 2, 3\}$ and $[4] = \{4, 5, 6, \ldots\}$.

1.23 If $\alpha(a) = \alpha(b)$, then $2a - 1 = 2b - 1$, and hence $a = b$. Thus, α is one-to-one. But there is no $a \in \{1, 2, 3, 4\}$ such that $\alpha(a) = 2$, so α is not onto.

1.25 If $\alpha(a) = \alpha(b)$, then $2^{3a-5} = 2^{3b-5}$. Taking the base 2 logarithm, we have $3a - 5 = 3b - 5$, and hence $a = b$. Thus, α is one-to-one. If $c \in T$, then $\alpha(((\log_2 c) + 5)/3) = c$. Therefore, α is onto as well. In fact, we will use $\beta : T \to S$ given by $\beta(c) = ((\log_2 c) + 5)/3$. We have $\beta(\alpha(a)) = \beta(2^{3a-5}) = (\log_2(2^{3a-5}) + 5)/3 = a$, for all $a \in S$.

1.27 (1) and (3) are binary operations, as $ab \in \mathbb{N}$ for all $a, b \in \mathbb{N}$, and $3 \in \mathbb{N}$. But (2) is not, as $1 * 2 = -1 \notin \mathbb{N}$.

1.29 Surely β is onto. If $t \in T$, then there exists an $r \in R$ such that $(\beta\alpha)(r) = t$. But then $\beta(\alpha(r)) = t$. However, α need not be. To see this, let R and S be the set of real numbers and let T be the set of nonnegative real numbers. Let $\alpha(r) = r^2$ and $\beta(s) = s^2$. Then α is not onto, as there is no $r \in R$ such that $\alpha(r) = -1$. However, if $t \in T$, then $\beta(\alpha(\sqrt[4]{t})) = \beta(\sqrt{t}) = t$; thus, $\beta\alpha$ is onto.

1.31 (1) For each of the m elements a of S, there are n possible choices for $\alpha(a)$, so n^m.

(2) If $n < m$, the answer is 0, as the m elements of S need to map to m different places. Suppose $n \geq m$ and let $S = \{a_1, \ldots, a_m\}$. Then there are n choices for $\alpha(a_1)$, leaving $n - 1$ choices for $\alpha(a_2)$, and so on. The answer is $n(n - 1) \cdots (n - m + 1)$.

Problems of Chapter 2

2.1 Apply induction. When $n = 1$, both sides are 1. Assume the result for n, then prove it for $n+1$: $1 + \cdots + n + (n+1) = n(n+1)/2 + (n+1) = (n+1)(n+2)/2 = (n+1)((n+1)+1)/2$, as required.

2.3 (1) This is the Binomial Theorem with $a = b = 1$.
   (2) This is the Binomial Theorem with $a = 1, b = -1$.

2.5 (1) By induction on n. We have nothing to prove for $n = 1$, so we begin with $n = 2$. Here, $(1 + a)^2 = 1 + 2a + a^2 > 1 + 2a$, as a is positive. Assume the result for n, and prove it for $n + 1$. But $(1+a)^{n+1} = (1+a)^n(1+a) > (1+na)(1+a) = 1 + (n+1)a + na^2 > 1 + (n+1)a$, as $a > 0$.
   (2) Apply (1) with $a = (n-1)/n$, then take nth roots.

2.7 By strong induction on n. If $n = 1$ or 2, the result is obvious, so assume that $n > 2$ and the result is true for smaller n. Then $f_n = f_{n-1} + f_{n-2} \leq (7/4)^{n-2} + (7/4)^{n-3} = (7/4)^{n-3}(7/4 + 1) < (7/4)^{n-1}$, since $11/4 < (7/4)^2$.

2.9 By strong induction on the area $a = rc$ of the bar. If the area is 1, then $r = c = 1$ and no actions are necessary. Suppose the area is $a > 1$ and the result is true for bars of smaller area. Then a break turns the bar into two bars with areas b and c, both less than a. By our inductive hypothesis, it will take $b - 1$ and $c - 1$ actions, respectively, to break down these two bars. We have already used 1 action, so the total is $1 + (b-1) + (c-1) = (b+c) - 1 = a - 1$, as required. Alternative solution: We must turn 1 bar into rc bars. Each action adds one bar. So we need $rc - 1$ actions.

2.11 (1) $57 = 20(2) + 17$; $20 = 17(1) + 3$; $17 = 3(5) + 2$; $3 = 2(1) + 1$; $2 = 1(2) + 0$. Thus, $(57, 20) = 1$.
   (2) $117 = 51(2) + 15$; $51 = 15(3) + 6$; $15 = 6(2) + 3$; $6 = 3(2) + 0$. Thus, $(117, 51) = 3$.

2.13 Let us write $b = ac$ and $a = bd$, with $c, d \in \mathbb{Z}$. Then $a = acd$; that is, $a(1 - cd) = 0$. If $a = 0$, then as $b = ac$, we have $b = 0$ as well. Otherwise, $1 - cd = 0$, so $cd = 1$. Thus, $d \in \{1, -1\}$, so $a \in \{b, -b\}$.

2.15 Let $d = (b, c)$. Then $d|c$ and $c|a$, so $d|a$. But also $d|b$. As $(a, b) = 1$, we must have $d = 1$.

2.17 If a and n are relatively prime, write $au + nv = 1$. Then $n(-v) = au - 1$. On the other hand, if $(a, n) = d > 1$ and $au - 1 = nb$, for some $b \in \mathbb{Z}$, then $1 = au - nb$. Now, $d|a$ and $d|n$, so $d|1$, which is impossible.

2.19 By strong induction on n. It is clear if $n \leq 4$. So let $n > 4$ and suppose that it is true for smaller n. Then $f_n = f_{n-1} + f_{n-2} = (f_{n-2} + f_{n-3}) + f_{n-2} = 2f_{n-2} + f_{n-3} = 2(f_{n-3} + f_{n-4}) + f_{n-3} = 3f_{n-3} + 2f_{n-4}$. If $4|n$, then $4|(n-4)$, so $3|f_{n-4}$, and hence $3|f_n$. Suppose that $4 \nmid n$. Then $4 \nmid (n-4)$, so $3 \nmid f_{n-4}$. If $3|f_n$, then $3|(f_n - 3f_{n-3}) = 2f_{n-4}$. As $(3, 2) = 1$, we see that $3|f_{n-4}$, giving us a contradiction.

2.21 $3528 = 2^3 \cdot 3^2 \cdot 7^2$, $30030 = 2 \cdot 3 \cdot 5 \cdot 7 \cdot 11 \cdot 13$ and $220000 = 2^5 \cdot 5^4 \cdot 11$.

2.23 Let $d = (p, n)$. As d is a positive integer and $d | p$, we can only have $d = 1$ or p. If $d = 1$, we are done. If $d = p$, then as $d | n$, we are done.

2.25 If $p_i | (p_1 \cdots p_k + 1)$, then as $p_i | p_1 \cdots p_k$, we have $p_i | (p_1 \cdots p_k + 1 - p_1 \cdots p_k) = 1$, which is impossible.

2.27 By Corollary 2.4, $p | a$. Let us say $a = pb$, with $b \in \mathbb{Z}$. Then $a^n = p^n b^n$, so $p^n | a^n$.

2.29 (1) and (2) are clearly commutative, but (3) is not, since $1 * 2 = 1$ but $2 * 1 = 2$.

2.31 (1) No. There is no $e \in \mathbb{Z}$ such that $2e + 1 = 2$.
  (2) Yes, let $e = 0$. Then $a * e = a = e * a$ for all $a \in \mathbb{Z}$.

2.33 (1) 4.
  (2) $(4 \cdot 5)^{25} = (-1)^{25} = -1 = 6$.

2.35 (2) We have $[a] + ([b] + [c]) = [a] + [b + c] = [a + (b + c)] = [(a + b) + c] = [a + b] + [c] = ([a] + [b]) + [c]$.
  (4) Note that $[a] + [0] = [a + 0] = [a]$.
  (5) Observe that $[a] + [-a] = [a + (-a)] = [0]$.

2.37 $2 \cdot 10 = 4 \cdot 5 = 6 \cdot 10 = 8 \cdot 15 = 12 \cdot 5 = 14 \cdot 10 = 16 \cdot 5 = 18 \cdot 10 = 0$. If $a \in \{1, 3, 7, 9, 11, 13, 17, 19\}$, there is no such b.

2.39 If $a^2 = 1$ in $\mathbb{Z}_p$, then $a^2 \equiv 1 \pmod{p}$; that is, $p | (a^2 - 1) = (a - 1)(a + 1)$. By Euclid's lemma, $p | (a - 1)$ or $p | (a + 1)$. That is, $a \equiv 1 \pmod{p}$ or $a \equiv -1 \equiv p - 1 \pmod{p}$. If $p = 8$, then 1, 3, 5 and 7 are solutions.

2.41 Proceeding as in the proof of Theorem 2.13, we have $d_1 = 70$, $d_2 = 30$ and $d_3 = 21$. Solving $3u_1 + 70v_1 = 1$ using the Euclidean algorithm, we get $u_1 = -23$ and $v_1 = 1$. Solving $7u_2 + 30v_2 = 1$, we get $u_2 = 13$ and $v_2 = -3$. Solving $10u_3 + 21v_3 = 1$, we get $u_3 = -2$ and $v_3 = 1$. Thus, our answer is $a = 70 \cdot 1 \cdot 2 + 30(-3)(4) + 21 \cdot 1 \cdot 3 = -157$. (As our answer is only unique modulo $3 \cdot 7 \cdot 10 = 210$, we can also add 210 and get 53.)

Problems of Chapter 3

3.1 (1) $\begin{pmatrix} 1 & 2 & 3 & 4 \\ 4 & 2 & 3 & 1 \end{pmatrix}$.

  (2) $\begin{pmatrix} 1 & 2 & 3 & 4 \\ 1 & 3 & 2 & 4 \end{pmatrix}$.

  (3) $\begin{pmatrix} 1 & 2 & 3 & 4 \\ 2 & 4 & 1 & 3 \end{pmatrix}$.

3.3 There are n places to map 1, then $n - 1$ to map 2, and so on. So there are $n!$ permutations. If we fix 2, then the other four numbers can be arranged at will, so there are $4! = 24$ possibilities.

3.5 Closure: Yes, the composition of two functions is a function. Associativity: Yes, the composition of functions is associative. Identity: Yes, we have the identity function sending each element of $\{1, 2, 3, 4, 5\}$ to itself. Inverses: No. Define α : $\{1, 2, 3, 4, 5\} \to \{1, 2, 3, 4, 5\}$ via $\alpha(i) = 1$ for all i. There is no possible function $\beta : \{1, 2, 3, 4, 5\} \to \{1, 2, 3, 4, 5\}$ such that $\alpha\beta$ is the identity function.

3.7 (1)

	0	1	2
0	0	1	2
1	1	2	0
2	2	0	1

(2)

	(0, 0)	(1, 0)	(2, 0)	(0, 1)	(1, 1)	(2, 1)
(0, 0)	(0, 0)	(1, 0)	(2, 0)	(0, 1)	(1, 1)	(2, 1)
(1, 0)	(1, 0)	(2, 0)	(0, 0)	(1, 1)	(2, 1)	(0, 1)
(2, 0)	(2, 0)	(0, 0)	(1, 0)	(2, 1)	(0, 1)	(1, 1)
(0, 1)	(0, 1)	(1, 1)	(2, 1)	(0, 0)	(1, 0)	(2, 0)
(1, 1)	(1, 1)	(2, 1)	(0, 1)	(1, 0)	(2, 0)	(0, 0)
(2, 1)	(2, 1)	(0, 1)	(1, 1)	(2, 0)	(0, 0)	(1, 0)

3.9 Take $g_i \in G$, $h_i \in H$. Now, $(g_1, h_1)(g_2, h_2) = (g_2, h_2)(g_1, h_1)$ if and only if $(g_1 g_2, h_1 h_2) = (g_2 g_1, h_2 h_1)$; that is, if and only if $g_1 g_2 = g_2 g_1$ and $h_1 h_2 = h_2 h_1$.

3.11 (1) Division is not associative; for instance, $(1/2)/3 \neq 1/(2/3)$.

(2) There is no inverse for 2 (or anything else other than 1).

3.13 Yes. Let G be our set. If $a + bi, c + di \in G$, then $(a + bi)(c + di) = (ac - bd) + (ad + bc)i$. Now, $(ac - bd)^2 + (ad + bc)^2 = a^2 c^2 + b^2 d^2 + a^2 d^2 + b^2 c^2 = (a^2 + b^2)(c^2 + d^2) = 1 \cdot 1 = 1$, so $(a + bi)(c + di) \in G$. Complex multiplication is associative. Clearly $1 \in G$, and it will serve as the identity. If $a + bi \in G$, then $(a + bi)(a - bi) = a^2 + b^2 = 1$. Now, $a^2 + (-b)^2 = 1$, so $a - bi \in G$ and $(a + bi)^{-1} = a - bi$.

3.15 Yes. To show closure, note that $(a/p^n) + (b/p^m) = (ap^m + bp^n)/p^{m+n} \in G$. Addition of rational numbers is certainly associative, and $0 = 0/p$ is the additive identity. The additive inverse of a/p^n is $-a/p^n \in G$.

3.17 (1) $aca^{-1}b^{-1}c^{-1}$.

(2) $a^{-1}c^{-1}b^{-1}a$.

3.19

	a	b	c	d
a	d	a	b	c
b	a	b	c	d
c	b	c	d	a
d	c	d	a	b

3.21 If (1) holds, then for any $g, h \in G$, let $a = g, b = hg, c = gh$. Then $ab = ca = ghg$, so by assumption, $hg = gh$, and G is abelian. If (2) holds, then whenever $ab = ca$, we have $ab = ac$, so by cancellation, $b = c$.

3.23 (1) $|\mathbb{Z}_{12}| = 12$. Also, $|0| = 1, |1| = |5| = |7| = |11| = 12, |2| = |10| = 6, |3| = |9| = 4, |4| = |8| = 3$ and $|6| = 2$.

(2) $|\mathbb{Z}_2 \times \mathbb{Z}_4| = 8$. Also, $|(0, 0)| = 1, |(1, 0)| = |(1, 2)| = |(0, 2)| = 2$ and every other element has order 4.

3.25 $|a^3| = 20/(3, 20) = 20/1 = 20, |a^{12}| = 20/(12, 20) = 20/4 = 5$ and $|a^{15}| = 20/(15, 20) = 20/5 = 4$.

3.27 We are looking for the smallest positive integer n such that $(a, b)^n = (e, e)$; that is, such that $a^n = e$ and $b^n = e$. But $a^n = e$ if and only if $12|n$ and $b^n = e$ if and only if $18|n$. Thus, we want the smallest positive integer n divisible by both 12 and 18. The order is 36.

3.29 (1) Note that $a^n = e$ if and only if $(a^n)^{-1} = e^{-1}$; that is, if and only if $(a^{-1})^n = e$.

(2) Recall that conjugates have the same order. Also, $ab = b^{-1}(ba)b$.

3.31 First, note that $U(8)$ has exactly three elements of order 2, namely 3, 5 and 7. Suppose that a and b are distinct elements of order 2. Now, $(ab)^2 = a^2b^2 = e^2 = e$, since G is abelian. Furthermore, if $ab = e$, then $a = b^{-1} = b$, since b has order 2. But this is impossible. Thus, $|ab| = 2$. Furthermore, if $ab = a$, then $b = e$ and if $ab = b$, then $a = e$. Thus, a, b and ab are distinct elements of order 2. Let c be a fourth distinct element of order 2. By the same argument, ac has order 2. If $ac = a$, then $c = e$. If $ac = b$, then $c = a^{-1}b = ab$. If $ac = ab$, then $c = b$. None of these are true. Thus, ac is a fifth distinct element of order 2.

3.33 (1) Yes. Clearly H contains the identity matrix. If $A, B \in H$, then $\det(AB) = \det(A)\det(B) = 1 \cdot 1 = 1$, so $AB \in H$. Furthermore, $\det(A^{-1}) = 1/\det(A) = 1$, so $A^{-1} \in H$.

(2) No. H does not contain the identity.

(3) Yes. First, we see that $0 = 0/1 \in H$. Next, if $a/b, c/d \in H$, then $(a/b) + (c/d) = (ad + bc)/(bd) \in H$, and $-(a/b) = (-a)/b \in H$.

3.35 Let F be any flip and R any rotation. Drawing out the effects of each operation, we find that $FR = R^{-1}F$. This is RF if and only if $R = R^{-1}$. Letting $R = R_{360/n}$, we find that $R \neq R^{-1}$. Thus, no flip is central. In fact, $R = R^{-1}$ if and only if

$R = R_0$ or $R = R_{180}$. If n is odd, there is no R_{180}, so $Z(D_{2n}) = \{R_0\}$. If n is even, we see that R_{180} commutes with every flip, and surely with every rotation, so $Z(D_{2n}) = \{R_0, R_{180}\}$.

3.37 Let H and K be subgroups of G. If $a, b \in H \cap K$, then $a, b \in H$, so $ab \in H$. Similarly, $ab \in K$, and therefore $ab \in H \cap K$. By the same argument, $a^{-1} \in H$, $a^{-1} \in K$, so $a^{-1} \in H \cap K$. Finally, as H and K are subgroups, $e \in H$ and $e \in K$, so $e \in H \cap K$. The argument for an arbitrary intersection is similar.

3.39 (1) $\langle 0 \rangle = \{0\}$, $\langle 1 \rangle = \langle 3 \rangle = \langle 7 \rangle = \langle 9 \rangle = \langle 11 \rangle = \langle 13 \rangle = \langle 17 \rangle = \langle 19 \rangle = \mathbb{Z}_{20}$, $\langle 2 \rangle = \langle 6 \rangle = \langle 14 \rangle = \langle 18 \rangle = \{0, 2, 4, 6, 8, 10, 12, 14, 16, 18\}$, $\langle 4 \rangle = \langle 8 \rangle = \langle 12 \rangle = \langle 16 \rangle = \{0, 4, 8, 12, 16\}$, $\langle 5 \rangle = \langle 15 \rangle = \{0, 5, 10, 15\}$, $\langle 10 \rangle = \{0, 10\}$.

   (2) $\langle 1 \rangle = \{1\}$, $\langle 3 \rangle = \langle 11 \rangle = \{1, 3, 9, 11\}$, $\langle 5 \rangle = \langle 13 \rangle = \{1, 5, 9, 13\}$, $\langle 7 \rangle = \{1, 7\}$, $\langle 9 \rangle = \{1, 9\}$, $\langle 15 \rangle = \{1, 15\}$.

3.41 Label the vertices of the regular n-gon from 1 to n, counterclockwise. Then notice that a rotation leaves the vertices in counterclockwise order, whereas a flip changes them to clockwise. This makes clear what must happen in each case.

3.43 (1) We have $\langle a^1 \rangle = G$, $\langle a^2 \rangle = \{e, a^2, a^4, a^6, a^8, a^{10}\}$, $\langle a^3 \rangle = \{e, a^3, a^6, a^9\}$, $\langle a^4 \rangle = \{e, a^4, a^8\}$, $\langle a^6 \rangle = \{e, a^6\}$, $\langle a^{12} \rangle = \{e\}$.

   (2) As $\mathbb{Z}_{12}$ is cyclic of order 12 with generator 1, we have $\langle 1 \rangle = \mathbb{Z}_{12}$, $\langle 2 \rangle = \{0, 2, 4, 6, 8, 10\}$, $\langle 3 \rangle = \{0, 3, 6, 9\}$, $\langle 4 \rangle = \{0, 4, 8\}$, $\langle 6 \rangle = \{0, 6\}$, $\langle 0 \rangle = \{0\}$.

3.45 A positive integer is not relatively prime to p^n if and only if it is divisible by p. Thus, we are excluding $p, 2p, 3p, \ldots, p^n$. There are p^{n-1} such numbers.

3.47 It does follow, as $H \cap K$ is a subgroup of H, and every subgroup of a cyclic group is cyclic.

3.49 Let $G = \langle a \rangle$ be a cyclic group of order n. Now, $|a^i| = n/(n, i)$, for every integer i. In particular, each element of G has order dividing n. Now, for every k dividing n, the number of elements of order k is $\varphi(k)$. Thus, the sum of the $\varphi(k)$ is the number of elements in G, namely, n.

3.51 (1) If $a \in H \cap K$ has order n, then $\langle a \rangle$ is a subgroup of order n in both H and K. As $|H| = |K| = n$, this means that $H = K = \langle a \rangle$, which is impossible.

   (2) If G has no elements of order n, we are done. Otherwise, take $a \in G$ of order n. Then $\langle a \rangle$ has $\varphi(n)$ elements of order n. If those are all of the elements in G, we are done. Otherwise, find $b \notin \langle a \rangle$ of order n. Then $\langle b \rangle$ contains $\varphi(n)$ elements of order n, and by (1), $\langle a \rangle$ and $\langle b \rangle$ have no elements of order n in common. Thus, we now have $2\varphi(n)$ elements of order n. Repeat. If the process stops, we have a multiple of $\varphi(n)$. If not, we have infinitely many.

3.53 (1) Left cosets: $0+H = \{\ldots, -4, 0, 4, 8, \ldots\}$, $1+H = \{\ldots, -3, 1, 5, 9, \ldots\}$, $2 + H = \{\ldots, -2, 2, 6, 10, \ldots\}$, $3 + H = \{\ldots, -1, 3, 7, 11, \ldots\}$. As G is abelian, the right cosets are the same.

   (2) Left cosets: $R_0 H = \{R_0, F_2\}$, $R_{90} H = \{R_{90}, F_4\}$, $R_{180} H = \{R_{180}, F_1\}$, $R_{270} H = \{R_{270}, F_3\}$. Right cosets: $H R_0 = \{R_0, F_2\}$, $H R_{90} = \{R_{90}, F_3\}$, $H R_{180} = \{R_{180}, F_1\}$, $H R_{270} = \{R_{270}, F_4\}$.

3.55 Let $G = pq$. If $H \leq G$, then $|H|$ divides $|G|$, so $|H| \in \{1, p, q, pq\}$. As H is a proper subgroup, the order is not pq. But the trivial group is cyclic, as is any group of prime order.

3.57 As $H \cap K$ is a subgroup of H and K, its order divides both 28 and 65. But $(28, 65) = 1$, so we can only have $H \cap K = \{e\}$.

3.59 By Exercise 3.30, $a_1 \cdots a_k$ has order 1 or 2. But a group of odd order has no elements of order 2.

3.61 Suppose otherwise, and let $h_1(H \cap K), \ldots, h_{n+1}(H \cap K)$ be distinct left cosets of $H \cap K$ in H. Then if $i \neq j$, we have $h_i^{-1} h_j \notin H \cap K$. Since $h_i^{-1} h_j \in H$, we have $h_i^{-1} h_j \notin K$. That is, $h_1 K, \ldots, h_{n+1} K$ are distinct left cosets of K in G, contradicting the assumption that $[G : K] = n$.

Problems of Chapter 4

4.1 (1) Yes. Clearly H contains the identity matrix. If $A, B \in H$, then $\det(AB) = \det(A)\det(B) \in \mathbb{Q}$, and $\det(A^{-1}) = 1/\det(A) \in \mathbb{Q}$; thus, $AB, A^{-1} \in H$, and H is a subgroup. Also, if $C \in GL_2(\mathbb{R})$, then $\det(C^{-1}AC) = \det(C^{-1})\det(A)\det(C) = \det(A)\det(C^{-1})\det(C) = \det(AC^{-1}C) = \det(A) \in \mathbb{Q}$; thus, $C^{-1}AC \in H$, so H is normal.

(2) No, $\begin{pmatrix} 1 & 1 \\ 0 & 1 \end{pmatrix}^{-1} \begin{pmatrix} 1 & 0 \\ 0 & 2 \end{pmatrix} \begin{pmatrix} 1 & 1 \\ 0 & 1 \end{pmatrix} = \begin{pmatrix} 1 & -1 \\ 0 & 2 \end{pmatrix}$, which is not diagonal.

4.3 Let $N = \{e, a\}$. If $b \in G$, then $b^{-1}ab \in N$. But if $b^{-1}ab = e$, then $a = beb^{-1} = e$; impossible. Thus, $b^{-1}ab = a$, and a is central; naturally, e is always central.

4.5 Let N and K be normal subgroups of G. By Exercise 3.37, $N \cap K$ is a subgroup. Let $a \in N \cap K$ and $g \in G$. Since $a \in N$, we have $g^{-1}ag \in N$. Similarly, $g^{-1}ag \in K$, so $g^{-1}ag \in N \cap K$. The proof of the generalization is similar.

4.7 If $a \in G$, then $a^{-1}Ha$ is a subgroup of order n, so $a^{-1}Ha = H$.

4.9 If $g \in G$, $n \in N$, then $n^{-1}g^{-1}ng \in N$; thus, $g^{-1}ng = n(n^{-1}g^{-1}ng) \in N$.

4.11 We know that $|a|$ is divisible by 5. Also, $(aN)^5 = eN$, so $a^5 \in N$. By Lagrange's theorem, $(a^5)^{14} = e$. Thus, the order of a divides 70. So $|a| \in \{5, 10, 35, 70\}$. To see that these are all possible, let $G = \mathbb{Z}_{70}$, $N = \langle 5 \rangle$ and let a be 14, 7, 2 and 1, respectively.

4.13 For both parts, $G = D_8 \times \mathbb{Z}$ will suffice. We have $Z(G) = \langle R_{180} \rangle \times \mathbb{Z}$. As $G/Z(G)$ has order 4, it clearly satisfies (2). As for (1), it remains to show that $D_8/\langle R_{180} \rangle$ is abelian. But this can be seen by examining the group table from Exercise 4.12.

4.15 Let $|aN| = 42$. As G is finite, we know that a has finite order, and so its order is a multiple of 42, say $42n$. But then a^n has order 42. It need not hold for infinite groups. Indeed, let $G = \mathbb{Z}$, $N = \langle 42 \rangle$ and $a = 1$. We see that $|1 + N| = 42$, but every nonidentity element of G has infinite order.

4.17 By Exercise 4.16, $a^{-1}b^{-1}ab \in K$, for all $a, b \in G$. Similarly, $a^{-1}b^{-1}ab \in N$. But $K \cap N = \{e\}$, so $a^{-1}b^{-1}ab = e$, and hence $ab = ba$.

4.19 Clearly $e \in N$. If $a, b \in N$, say $a^k = b^l = e$, for some $k, l \in \mathbb{N}$, then $(ab)^{kl} = (a^k)^l(b^l)^k = e^l e^k = e$; thus, $ab \in N$. Also, $|a| = |a^{-1}|$, so $a^{-1} \in N$. Thus, N is a subgroup. As G is abelian, it is normal. Take any $c \in G$. If, for some $n \in \mathbb{N}$, we have $(cN)^n = eN$, then $c^n \in N$; that is, c^n has finite order, so $c^{nm} = e$ for some $m \in \mathbb{N}$. In other words, $c \in N$, so $cN = eN$.

4.21 (1) It is a homomorphism. Indeed, $\alpha(ab) = \log_{10}(ab) = \log_{10} a + \log_{10} b = \alpha(a) + \alpha(b)$. It is one-to-one, as if $\alpha(a) = 0$, then $\log_{10} a = 0$, so $a = 1$; that is, $\ker(\alpha) = \{1\}$. It is also onto, as if $b \in \mathbb{R}$, then $\alpha(10^b) = b$.
  (2) It is not a homomorphism, as $\beta(0 + 0) = 1 \neq 2 = \beta(0) + \beta(0)$.

4.23 We have $\alpha((a, b)(c, d)) = \alpha((ac, bd)) = ac(bd)^{-1} = ab^{-1}cd^{-1}$ (since $U(16)$ is abelian), and this is $\alpha((a, b))\alpha((c, d))$. Thus, α is a homomorphism. Also, $\langle 7 \rangle = \{1, 7\}$. Now, $\alpha((a, b)) = 1$ if and only if $ab^{-1} = 1$; that is, we have the pairs $(1, 1), (3, 3), \ldots, (15, 15)$. Similarly, $\alpha((a, b)) = 7$ if and only if $a = 7b$, so we have the pairs $(7, 1), (5, 3), (3, 5), (1, 7), (15, 9), (13, 11), (11, 13), (9, 15)$.

4.25 (1) Not necessarily. For instance, H could be the trivial group.
  (2) Yes. Let $h \in H$ have order n. As α is onto, say $\alpha(g) = h$. Since G is finite, $|g| < \infty$, so $|h|$ divides $|g|$. Let us say that $|g| = mn$. Then $|g^m| = n$.

4.27 Note that $gh = \alpha((g, h)) = \alpha((e, h)(g, e)) = \alpha((e, h))\alpha((g, e)) = hg$.

4.29 Let $H = G/N$. Define $\alpha : G \to H$ via $\alpha(g) = gN$. It is a homomorphism, as $\alpha(g_1 g_2) = g_1 g_2 N = g_1 N g_2 N = \alpha(g_1)\alpha(g_2)$, and $g \in \ker(\alpha)$ if and only if $gN = eN$; that is, if and only if $g \in N$.

4.31 (1) Count the elements of order 2.
  (2) One is abelian and the other is not.
  (3) We know that $\mathbb{Z}$ is cyclic. Suppose that $\mathbb{Z} \times \mathbb{Z} = \langle (a, b) \rangle$. Then there exists an $n \in \mathbb{Z}$ such that $(1, 0) = n(a, b)$. Since n cannot be 0, we see that $b = 0$. Similarly, $a = 0$. But this is impossible.

4.33 Define $\alpha : \mathbb{Z} \to GL_2(\mathbb{R})$ via $\alpha(a) = \begin{pmatrix} 1 & 0 \\ a & 1 \end{pmatrix}$. Note that $\alpha(a + b) = \begin{pmatrix} 1 & 0 \\ a+b & 1 \end{pmatrix} = \begin{pmatrix} 1 & 0 \\ a & 1 \end{pmatrix}\begin{pmatrix} 1 & 0 \\ b & 1 \end{pmatrix} = \alpha(a)\alpha(b)$. Thus, α is a homomorphism. In particular, $\alpha(\mathbb{Z}) = G$ is a subgroup of $GL_2(\mathbb{R})$. Furthermore, if $\alpha(a)$ is the identity matrix, then $a = 0$; thus, α is one-to-one. Therefore, $\mathbb{Z}$ is isomorphic to $\alpha(\mathbb{Z}) = G$.

4.35 Define $\alpha : H \to a^{-1}Ha$ via $\alpha(h) = a^{-1}ha$. Then for any $h, k \in H$, we have $\alpha(hk) = a^{-1}hka = a^{-1}haa^{-1}ka = \alpha(h)\alpha(k)$; thus, α is a homomorphism. By definition, it is onto. Also, if $h \in \ker(\alpha)$, then $a^{-1}ha = e$; therefore, $h = aa^{-1} = e$, and α is one-to-one.

4.37 If $n > 1$ is a positive integer, then $n\mathbb{Z}$ is a proper subgroup which is infinite cyclic, and therefore isomorphic to $\mathbb{Z}$.

4.39 Define $\alpha : G \to G$ via $\alpha((a_1, a_2, \ldots)) = (0, a_1, a_2, \ldots)$. It is a homomorphism; indeed, if $a = (a_1, a_2, \ldots)$ and $b = (b_1, b_2, \ldots)$, then $\alpha(a + b) = \alpha(a) + \alpha(b) = (0, a_1 + b_1, a_2 + b_2, \ldots)$. Furthermore, it is one-to-one; if $\alpha(a) = (0, 0, \ldots)$, then clearly $a = (0, 0, \ldots)$. Thus, G is isomorphic to $\alpha(G)$, which is a proper subgroup of G.

4.41 Define $\alpha : G \to \mathbb{Z}$ via $\alpha((a, b)) = a - b$. Now, α is a homomorphism, since $\alpha((a, b) + (c, d)) = \alpha((a + c, b + d)) = (a + c) - (b + d) = (a - b) + (c - d) = \alpha((a, b)) + \alpha((c, d))$. Also, α is onto, since for any $a \in \mathbb{Z}$, $\alpha((a, 0)) = a$. Finally, the kernel is the set of all (a, b) such that $a - b = 0$; that is, $\ker(\alpha) = N$. Apply the First Isomorphism Theorem.

4.43 Define $\alpha : \mathbb{R} \to H$ via $\alpha(r) = \cos(2\pi r) + \sin(2\pi r)i$ (where we are working in radians). As $\cos^2(\theta) + \sin^2(\theta) = 1$ for any $\theta \in \mathbb{R}$, we see that $\alpha(\mathbb{R}) \subseteq H$. Furthermore, for any $a, b \in \mathbb{R}$ such that $a^2 + b^2 = 1$, we can surely find $r \in \mathbb{R}$ such that $\cos(2\pi r) = a$ and $\sin(2\pi r) = b$; thus, $\alpha(\mathbb{R}) = H$. To show that α is a homomorphism, calculate $\alpha(r + s)$ and $\alpha(r)\alpha(s)$ and use trigonometric identities. Finally, the kernel is the set of all $r \in \mathbb{R}$ such that $\cos(2\pi r) = 1$ and $\sin(2\pi s) = 0$; that is, $\ker(\alpha) = \mathbb{Z}$. Apply the First Isomorphism Theorem.

4.45 (1) If $\begin{pmatrix} 1 & a & b \\ 0 & 1 & c \\ 0 & 0 & 1 \end{pmatrix}$ commutes with $\begin{pmatrix} 1 & 1 & 0 \\ 0 & 1 & 0 \\ 0 & 0 & 1 \end{pmatrix}$, then $c = 0$. Similarly, $a = 0$. But matrices with $a = c = 0$ are easily seen to commute with everything in G, so those matrices form the centre.

(2) Define $\alpha : G \to \mathbb{Z} \times \mathbb{Z}$ via $\alpha\left(\begin{pmatrix} 1 & a & b \\ 0 & 1 & c \\ 0 & 0 & 1 \end{pmatrix}\right) = (a, c)$. We can see that

$$\alpha\left(\begin{pmatrix} 1 & a & b \\ 0 & 1 & c \\ 0 & 0 & 1 \end{pmatrix}\begin{pmatrix} 1 & d & e \\ 0 & 1 & f \\ 0 & 0 & 1 \end{pmatrix}\right) = \alpha\left(\begin{pmatrix} 1 & a+d & e+af+b \\ 0 & 1 & c+f \\ 0 & 0 & 1 \end{pmatrix}\right)$$

$$= (a+d, c+f) = \alpha\left(\begin{pmatrix} 1 & a & b \\ 0 & 1 & c \\ 0 & 0 & 1 \end{pmatrix}\right) + \alpha\left(\begin{pmatrix} 1 & d & e \\ 0 & 1 & f \\ 0 & 0 & 1 \end{pmatrix}\right),$$

so α is a homomorphism. It is clearly onto. Furthermore, its kernel is precisely $Z(G)$, as we found in the first part. Now apply the First Isomorphism Theorem.

4.47 As $\alpha(ab) = (ab)^m = a^m b^m = \alpha(a)\alpha(b)$ (since G is abelian), we know that α is a homomorphism. If $\alpha(a) = e$, then $a^m = e$, and hence $|a|$ divides m. But by Lagrange's theorem, $|a|$ divides n as well. Since $(m, n) = 1$, we can only have $|a| = 1$. Thus, α is one-to-one. As G is finite, it must be onto as well.

4.49 Let α be an automorphism of $\mathbb{Z}_2 \times \mathbb{Z}_2$. Now, any homomorphism sends the identity to the identity. As α is one-to-one, $\alpha((1, 0)) \in \{(1, 0), (0, 1), (1, 1)\}$. Furthermore, once $\alpha((1, 0))$ is chosen, that leaves only two possibilities for $\alpha((0, 1))$. Once both of these are decided, there is only one option left for $\alpha((1, 1))$. So there are only $3 \cdot 2 = 6$ possible automorphisms. This does not mean that all of them are necessarily automorphisms, but as it happens, they are. To see this, note that every group of prime order is cyclic, and groups of order 4 are abelian. Thus, since Example 4.26 shows us that there are noncommuting automorphisms, no order less than 6 is possible. So all of the functions we have considered are actually automorphisms. Also, every group of order 6 is isomorphic to $\mathbb{Z}_6$ or D_6. As the automorphism group is nonabelian, it must be D_6.

4.51 As $\alpha(e) = e$, we see that e is in our set. If $\alpha(a) = a$ and $\alpha(b) = b$, then $\alpha(ab) = \alpha(a)\alpha(b) = ab$, so ab is in our set. Also, $\alpha(a^{-1}) = (\alpha(a))^{-1} = a^{-1}$, so a^{-1} is in our set.

4.53 As $\alpha(\langle a \rangle \times \{e\}) \subseteq \langle a \rangle \times \{e\}$, let us say that $\alpha((a, e)) = (a^i, e)$. Similarly, $\alpha((e, b)) = (e, b^j)$ and $\alpha((a, b)) = (a, b)^k$. But then $(a^i, b^j) = (a^k, b^k)$. That is, $\alpha((a, e)) = (a^k, e)$ and $\alpha((e, b)) = (e, b^k)$. Then for any $r, s \in \mathbb{Z}$, $\alpha((a^r, b^s)) = \alpha((a, e))^r \alpha((e, b))^s = (a^k, e)^r (e, b^k)^s = (a^r, b^s)^k$.

4.55 If m is an integer, then we know that $\alpha(m) = \alpha(m \cdot 1) = m \cdot \alpha(1)$. If n is a nonzero integer, then $\alpha(m) = \alpha\left(n \cdot \frac{m}{n}\right) = n\alpha\left(\frac{m}{n}\right)$. Thus, $\alpha\left(\frac{m}{n}\right) = \frac{1}{n}\alpha(m) = \frac{m}{n}\alpha(1)$.

Problems of Chapter 5

5.1 Let $H = \langle 3 \rangle$ and $K = \langle 31 \rangle$. We see that $|H| = 8$, $|K| = 2$. As the group is abelian, both subgroups are normal, and $H \cap K = \{1\}$. Thus $|HK| = 8 \cdot 2/1 = 16 = |U(32)|$, so $U(32) = H \times K$.

5.3 Note that $5(a, b) = (0, 0)$ if and only if $5a = 0$ and $5b = 0$. But $5a = 0$ for all a, whereas $5b = 0$ if and only if $b \in \{0, 5, 10, 15, 20\}$. Thus, $5 \cdot 5 = 25$ elements satisfy $5(a, b) = 0$. Now, these elements have order dividing 5, so we need only exclude the identity, which has order 1; thus, there are 24 elements of order 5. As $25(a, b) = (0, 0)$ for all a and b, we see that every element has order 1, 5 or 25. We have found 25 elements not having order 25, which means that $5 \cdot 25 - 25 = 100$ elements have order 25.

5.5 If $D_8 = H \times K$, then $|H||K| = 8$. As the subgroups are both proper, $|H| = 4$ and $|K| = 2$ (or vice versa). By Corollaries 4.2 and 4.3, H and K are abelian, so D_8 is abelian, giving us a contradiction.

5.7 Let $G = \mathbb{Z}_2 \times \mathbb{Z}_2$, $N_1 = \langle(1,0)\rangle$, $N_2 = \langle(0,1)\rangle$ and $N_3 = \langle(1,1)\rangle$. As G is abelian, normality is not an issue. We can see that $G = N_1 N_2$, so surely $G = N_1 N_2 N_3$. Also, each $N_i \cap N_j = \{(0,0)\}$. But we cannot have $G = N_1 \times N_2 \times N_3$, since the order is wrong.

5.9 It does not follow. Let $G = \mathbb{Z}_2 \times \mathbb{Z}_2$ and $H = \mathbb{Z}_2$. Define $\alpha : G \to H$ via $\alpha((a,b)) = a+b$. We have $\alpha((a,b)+(c,d)) = a+b+c+d = \alpha((a,b))+\alpha((c,d))$, so α is a homomorphism. As $\alpha((0,0)) = 0$ and $\alpha((1,0)) = 1$, we see that α is onto. Now, $G = \langle(1,0)\rangle \times \langle(0,1)\rangle$, but $\alpha(\langle(1,0)\rangle) = \alpha(\langle(0,1)\rangle) = H$; thus, the intersection of the images is not trivial, so we do not have a direct product in H.

5.11 (1) $\mathbb{Z}_3 \times \mathbb{Z}_7$.

(2) $\mathbb{Z}_{81}$, $\mathbb{Z}_{27} \times \mathbb{Z}_3$, $\mathbb{Z}_9 \times \mathbb{Z}_9$, $\mathbb{Z}_9 \times \mathbb{Z}_3 \times \mathbb{Z}_3$, $\mathbb{Z}_3 \times \mathbb{Z}_3 \times \mathbb{Z}_3 \times \mathbb{Z}_3$.

(3) $\mathbb{Z}_8 \times \mathbb{Z}_{25} \times \mathbb{Z}_{49}$, $\mathbb{Z}_4 \times \mathbb{Z}_2 \times \mathbb{Z}_{25} \times \mathbb{Z}_{49}$, $\mathbb{Z}_2 \times \mathbb{Z}_2 \times \mathbb{Z}_2 \times \mathbb{Z}_{25} \times \mathbb{Z}_{49}$, $\mathbb{Z}_8 \times \mathbb{Z}_5 \times \mathbb{Z}_5 \times \mathbb{Z}_{49}$, $\mathbb{Z}_4 \times \mathbb{Z}_2 \times \mathbb{Z}_5 \times \mathbb{Z}_5 \times \mathbb{Z}_{49}$, $\mathbb{Z}_2 \times \mathbb{Z}_2 \times \mathbb{Z}_2 \times \mathbb{Z}_5 \times \mathbb{Z}_5 \times \mathbb{Z}_{49}$, $\mathbb{Z}_8 \times \mathbb{Z}_{25} \times \mathbb{Z}_7 \times \mathbb{Z}_7$, $\mathbb{Z}_4 \times \mathbb{Z}_2 \times \mathbb{Z}_{25} \times \mathbb{Z}_7 \times \mathbb{Z}_7$, $\mathbb{Z}_2 \times \mathbb{Z}_2 \times \mathbb{Z}_2 \times \mathbb{Z}_{25} \times \mathbb{Z}_7 \times \mathbb{Z}_7$, $\mathbb{Z}_8 \times \mathbb{Z}_5 \times \mathbb{Z}_5 \times \mathbb{Z}_7 \times \mathbb{Z}_7$, $\mathbb{Z}_4 \times \mathbb{Z}_2 \times \mathbb{Z}_5 \times \mathbb{Z}_5 \times \mathbb{Z}_7 \times \mathbb{Z}_7$, $\mathbb{Z}_2 \times \mathbb{Z}_2 \times \mathbb{Z}_2 \times \mathbb{Z}_5 \times \mathbb{Z}_5 \times \mathbb{Z}_7 \times \mathbb{Z}_7$.

5.13 As $|U(56)| = \varphi(56) = 24$, the possibilities are $\mathbb{Z}_8 \times \mathbb{Z}_3$, $\mathbb{Z}_4 \times \mathbb{Z}_2 \times \mathbb{Z}_3$ and $\mathbb{Z}_2 \times \mathbb{Z}_2 \times \mathbb{Z}_2 \times \mathbb{Z}_3$. But running through the elements of $U(56)$, we see that none have order larger than 6. As $\mathbb{Z}_8 \times \mathbb{Z}_3$ and $\mathbb{Z}_4 \times \mathbb{Z}_2 \times \mathbb{Z}_3$ both have elements of order 12, it must be $\mathbb{Z}_2 \times \mathbb{Z}_2 \times \mathbb{Z}_2 \times \mathbb{Z}_3$.

5.15 We see that G is isomorphic to a direct product of groups of the form $\mathbb{Z}_{p^{n_i}}$, for various $n_i \in \mathbb{N}$. But if $n_i > 1$, then such a group has elements of order p^2.

5.17 Solving $5u + 7v = 1$ in $\mathbb{Z}$, one possible solution is $u = 3$, $v = -2$. Then $a = a^{5u+7v} = (a^5)^3(a^7)^{-2}$. Now $|a^{15}| = 7$ and $|a^{-14}| = |a^{21}| = 5$.

5.19 We proceed by strong induction on $|G|$. There is nothing to do if $|G| = 1$, so we start the induction with $|G| = 2$. In this case, $p = 2$ and G has an element of order 2. Let $|G| > 2$ and assume the result for groups of smaller order. If $e \neq b \in G$, then choose some prime q dividing $|b|$. Let $a = b^{|b|/q}$. Then $|a| = q$. If $q = p$, we are done. Otherwise $|G/\langle a\rangle| = |G|/q$, and this is still divisible by p. By our inductive hypothesis, $G/\langle a\rangle$ has an element $c\langle a\rangle$ of order p. Thus, $c^p \in \langle a\rangle$, so $c^{pq} = e$. Hence, $|c^q| = 1$ or p. But if $c^q = e$, then $(c\langle a\rangle)^q = e\langle a\rangle$. As $|c\langle a\rangle| = p$, this is impossible.

5.21 (1) 8, 2, 3, 3, 25, 7, 7.

(2) 2, 2, 2, 3, 9, 27.

5.23 p^3, q^2, r; p^2, p, q^2, r; p, p, p, q^2, r; p^3, q, q, r; p^2, p, q, q, r; p, p, p, q, q, r.

5.25 It is obviously the case for $n = 1$. For larger n, we claim that it is true if and only if n is a product of distinct primes. If $n = p_1 \cdots p_k$, where the p_i are all distinct, then the only possible list of elementary divisors is $p_1, \ldots, p_k$, so the groups are all isomorphic. On the other hand, if $p^2 | n$ for some prime p, then we have the cyclic group of order n and $\mathbb{Z}_p \times \mathbb{Z}_{n/p}$. Since $(p, n/p) = p > 1$, we see that this group is not cyclic.

5.27 The list of elementary divisors of $G_1 \times G_2$ is obtained by combining the lists of elementary divisors of G_1 and G_2. Similarly for $G_1 \times G_3$. If these lists are the same, then deleting the elementary divisors of G_1 from each list, we see that G_2 and G_3 have the same elementary divisors, and hence are isomorphic.

5.29 We know that G is isomorphic to $\mathbb{Z}_{2^{n_1}} \times \cdots \times \mathbb{Z}_{2^{n_k}}$. If $2(a_1, \ldots, a_k) = (0, \ldots, 0)$, then $2a_i = 0$ for all i; that is, each a_i has order 1 or 2. But a cyclic group of order 2^{n_i} has only one element of order 2, so there are only two such a_i, for each i. In total, we get 2^k elements. But we must exclude the identity, so our number is $2^k - 1$.

5.31 (1) Remember that $q + \mathbb{Z} = r + \mathbb{Z}$ if and only if $q - r \in \mathbb{Z}$. This is basically the same as Example 1.19, using $\mathbb{Q}$ instead of $\mathbb{R}$.
   (2) We have $b(a/b + \mathbb{Z}) = a + \mathbb{Z} = 0 + \mathbb{Z}$. Thus, $|a/b + \mathbb{Z}| \le b$. But if $c \in \mathbb{N}$ and $c(a/b + \mathbb{Z}) = 0 + \mathbb{Z}$, then $ca/b \in \mathbb{Z}$; that is, $b|ac$. Since $(a, b) = 1$, this means that $b|c$. In particular, $c \ge b$, so the order is b.

5.33 We have $\alpha(a+b) = n(a+b) = na+nb = \alpha(a)+\alpha(b)$, so α is a homomorphism. If $a \in G$, then since G is divisible, there exists a $b \in G$ such that $nb = a$. Thus, $\alpha(b) = a$, and α is onto. But it is not necessarily an isomorphism. Let G be the Prüfer p-group and take $n = p$. Then we see that $1/p + \mathbb{Z} \in \ker(\alpha)$.

5.35 If N is a subgroup of G, take $a + N \in G/N$. Then for any $n \in \mathbb{N}$, there exists a $b \in G$ such that $nb = a$. Therefore, $n(b + N) = a + N$, and G/N is divisible. However, $\mathbb{Q}$ is divisible but $\mathbb{Z}$ is not, as there is no $b \in \mathbb{Z}$ such that $2b = 1$.

5.37 Let $G/N = \langle aN \rangle$. If $gN \in G/N$, then $gN = (aN)^k$, for some $k \in \mathbb{Z}$. That is, $g = a^k n$, for some $n \in N$. In other words, $G = \langle a \rangle N$. If $e \ne b \in \langle a \rangle \cap N$, then $b = a^l \in N$, for some $0 \ne l \in \mathbb{Z}$. If $l < 0$, then we may replace b with b^{-1}, so let $l > 0$. Then $(aN)^l = eN$, which means that aN has finite order in G/N. As $G/N = \langle aN \rangle$ and G/N is infinite cyclic, this is impossible. Therefore, $G = \langle a \rangle \times N$. It remains only to show that $\langle a \rangle$ is infinite cyclic. It is surely cyclic, and if $|a| = k$, then again, $(aN)^k = eN$ gives us a contradiction.

Problems of Chapter 6

6.1 (1) $(2\ 4\ 7\ 6)(3\ 5)$
   (2) $(1\ 2\ 5)(3\ 6\ 4)(7\ 8)$

6.3 (1) $(1\ 4\ 2)(3\ 6\ 7\ 5)$.
   (2) Writing the permutation as a product of disjoint cycles, we get $(1\ 2)(3\ 5\ 4)$, so the inverse is $(1\ 2)(3\ 4\ 5)$.

6.5 An element of order 3 must be a product of one or more disjoint 3-cycles. Let us count the 3-cycles $(a\ b\ c)$. There are 9 choices for a, 8 for b and 7 for c. But

$(a\ b\ c) = (b\ c\ a) = (c\ a\ b)$, so we must divide by 3, giving $9 \cdot 8 \cdot 7/3 = 168$. For pairs of disjoint 3-cycles, we get $9 \cdot 8 \cdot 7 \cdot 6 \cdot 5 \cdot 4/(3 \cdot 3 \cdot 2) = 3360$, using the same argument and the fact that the order of the two cycles is irrelevant. Finally, to get three disjoint 3-cycles, we have $9!/(3 \cdot 3 \cdot 3 \cdot 3!) = 2240$, again noting that the three cycles can be permuted as we please. Our total is 5768.

6.7 If τ exists, then it has order k. Thus, if k is even, then τ^2 has order $k/2$, and therefore it cannot be a k-cycle. So suppose that k is odd. Let $\tau = \sigma^{(k+1)/2}$. Then $\tau^2 = \sigma^{k+1} = \sigma$, as σ has order k. Furthermore, as $2((k + 1)/2) + (-1)k = 1$, we know that $((k + 1)/2, k) = 1$. The preceding exercise tells us that τ is a k-cycle.

6.9 Let $|\sigma| = 105 = 3 \cdot 5 \cdot 7$. We know that $|\sigma|$ is the least common multiple of the lengths of its cycles in the disjoint cycle decomposition. The product of a 3-cycle, a 5-cycle and a 7-cycle would work, so $m = 15$ is a possibility. Can there be a smaller value? There must surely be a cycle whose length is a multiple of 7 and a divisor of 105. If it is smaller than 15, it can only be 7. Similarly for 3 and 5. Thus, $m = 15$ is the smallest possible value.

Let $|\tau| = 125$. The only way to make this happen is for the disjoint cycle decomposition for τ to include a 125-cycle. We see that $n = 125$.

6.11 (1) even

(2) odd

6.13 Without loss of generality, the possible products are $(1\ 2)(1\ 2) = (1)$, having order 1, $(1\ 2)(1\ 3) = (1\ 3\ 2)$, having order 3 and $(1\ 2)(3\ 4)$, having order 2.

6.15 It is certainly impossible if n is 2 or 3, as the groups are too small. But if $n \geq 4$, then A_n has the subgroup $\{(1), (1\ 2)(3\ 4), (1\ 3)(2\ 4), (1\ 4)(2\ 3)\}$. (It contains the identity, and closure is easily checked.) This subgroup is not cyclic. Indeed, if σ is a permutation of order 4, then its disjoint cycle decomposition is a product of one or more 4-cycles and, possibly, some 2-cycles. But a 4-cycle by itself is odd, so we need $n \geq 6$ to get something like $(1\ 2\ 3\ 4)(5\ 6) \in A_n$.

6.17 The order of an element in S_n is the least common multiple of the lengths of its disjoint cycles. If this order is odd, then these cycles all have odd length. But a cycle of odd length is even.

6.19 We see by inspection that $n = 1$ falls into the second category and $n = 2$ and 3 fall into the third category. Let $n \geq 4$. By Exercise 6.17, all elements of odd order lie in A_n. As A_n contains half the elements of S_n, we see that there are at least as many elements of even order as of odd order, and they can only be equal if every element of A_n has odd order. However, $(12)(34) \in A_n$ has order 2. So if $n \geq 4$, we are in the first category.

6.21 Such a subgroup would have index 2, and therefore be normal, by Theorem 4.1. But A_5 is simple.

6.23 It can. Note that A_6 has an isomorphic copy of A_5 as a proper subgroup. (Just use the exact same permutations as in A_5, assuming that each fixes the number 6.)

6.25 Let N be a nontrivial proper normal subgroup of A_4. By the preceding exercise, N contains no 3-cycles, so N is a subgroup of the group exhibited in Example 6.11. If it is not the same group, then it can only have order 2. But by Exercise 4.3, a normal subgroup of order 2 is central. However, the elements of order 2 in A_4 are the products of two disjoint transpositions, and these are not central. For instance, $(1\ 2)(3\ 4)(1\ 2\ 3) \neq (1\ 2\ 3)(1\ 2)(3\ 4)$.

6.27 In view of the preceding exercise, it suffices to show that each $(1\ i), 2 \leq i \leq n$, is the product of such transpositions. We proceed by induction on i, beginning with $i = 2$. There is nothing to do there, so assume the result for i and prove it for $i + 1$, when $1 < i < n$. However, $(i\ (i+1))(1\ i)(i\ (i+1)) = (1\ (i+1))$, completing the proof.

Problems of Chapter 7

7.1 Let $A = \begin{pmatrix} a & b \\ c & d \end{pmatrix} \in GL_2(\mathbb{R})$. Then A commutes with $\begin{pmatrix} 1 & 1 \\ 0 & 1 \end{pmatrix}$ if and only if $a = a + c, a + b = b + d$ and $c + d = d$; that is, if and only if $c = 0$ and $a = d$. Thus, the matrices in the centralizer have the form $\begin{pmatrix} a & b \\ 0 & a \end{pmatrix}$, where $a, b \in \mathbb{R}$ (and $a \neq 0$, so that the matrix is invertible).

7.3 We always have $C(H) \subseteq N(H)$. Let $A = \begin{pmatrix} 2 & 3 \\ 5 & 6 \end{pmatrix}$, and suppose that $B \in N(H)$. Then $B^{-1}AB \in H$, so $B^{-1}AB = A^n$, for some integer n. However, $\det(B^{-1}AB) = \det(A) = -3$, whereas $\det(A^n) = (\det(A))^n = (-3)^n$. We conclude that $n = 1$, and hence $B^{-1}AB = A$.

7.5 As always, $C(H) \subseteq N(H)$. Let $H = \{e, a\}$. If $b \in N(H)$, then $b^{-1}eb = e$ and we must have $b^{-1}ab \in H$. If $b^{-1}ab = e$, then $a = e$, which is impossible. Therefore, $b^{-1}ab = a$, and $b \in C(H)$.

7.7 Take $b \in C(a)$. As $C(a)$ is a subgroup, $b^{-1} \in C(a)$, so $b^{-1}a = ab^{-1}$. Inverting, we get $a^{-1}b = ba^{-1}$; thus, $b \in C(a^{-1})$. This means that $C(a) \subseteq C(a^{-1}) \subseteq C((a^{-1})^{-1}) = C(a)$.

7.9 (1) As H has prime order, it is abelian, so $H \leq C(H)$. In particular, $|C(H)|$ is divisible by 11 and divides 77, so it is 11 or 77. If it is 11, we must have $C(H) = H$. Otherwise, $H \leq Z(G)$. In the same way, we now have $|Z(G)| = 11$ or 77. Since G is not abelian, $Z(G) = H$. But this contradicts Corollary 4.1.

(2) Suppose otherwise. Combining (1) with Theorem 7.3, and noting that H is normal, we have G/H isomorphic to a subgroup of Aut(H). By Theorem 4.14, H is

isomorphic to $\mathbb{Z}_{11}$, and Theorem 4.22 tells us that Aut(H) is isomorphic to $U(11)$. But this is a group of order 10 and cannot have a subgroup of order 7.

7.11 $\{R_0\}, \{R_{180}\}, \{R_{90}, R_{270}\}, \{F_1, F_2\}, \{F_3, F_4\}$.

7.13 It does not follow. Let $G = S_3$, $H = \langle (1\ 3) \rangle$ and $K = \langle (1\ 3\ 2) \rangle$. Now consider the subgroups $\langle (1\ 2) \rangle$ and $\langle (2\ 3) \rangle$. As $(1\ 3)^{-1}(1\ 2)(1\ 3) = (2\ 3) = (1\ 3\ 2)^{-1}(1\ 2)(1\ 3\ 2)$, it follows immediately that these subgroups are both H- and K-conjugate. However, $H \cap K = \{(1)\}$, so they are not $(H \cap K)$-conjugate.

7.15 If each $[G : C(a)]$ in the class equation is divisible by p^2, then since $|G|$ is also divisible by p^2, we must have p^2 dividing the order of $|Z(G)|$, which is not the case. Thus, since each $[G : C(a)]$ divides p^n, one of them must be p. It follows that $|C(a)| = p^{n-1}$.

7.17 (1) No. Groups of order 25 are abelian, so all conjugacy classes would have just one element.

(2) Yes, S_3.

(3) No, the identity is always in a conjugacy class by itself.

7.19 Suppose that $b^{-1}ab = a^{-1}$. Then $b^{-2}ab^2 = b^{-1}a^{-1}b = (b^{-1}ab)^{-1} = a$. That is, $b^2 \in C(a)$. If G has odd order, so does b. Thus, write $|b| = 2m - 1$, for some $m \in \mathbb{N}$. Then $(b^2)^m = b$, so $b \in C(a)$. But then $b^{-1}ab = a$, so $a = a^{-1}$. That is, $a^2 = e$. As a has odd order, $a = e$, giving us a contradiction.

7.21 Sylow 2-subgroup: $\langle 25 \rangle \times \langle 7 \rangle$. Sylow 5-subgroup: $\langle (4, 0) \rangle$. Sylow 7-subgroup: $\langle (0, 2) \rangle$.

7.23 We have $|G| = 2 \cdot 3 \cdot 7^2$. The number of Sylow 7-subgroups is $1 + 7k$, for some nonnegative integer k, and divides 6. The only possible solution is $k = 0$.

7.25 Let H be a Sylow p-subgroup of G. By definition, its order is p^m, the largest power of p dividing $|G|$. Thus, $n \leq m$. By Exercise 7.16, H has a subgroup of order p^n.

7.27 By the Second Isomorphism Theorem, HN/N is isomorphic to $H/(H \cap N)$. In particular, its order divides $|H|$ and is therefore a power of p. Furthermore, $H \leq HN \leq G$, so $|G|/|HN|$ is a divisor of $|G|/|H|$. In particular, $|G|/|HN|$ is relatively prime to p. However, $[G/N : HN/N] = (|G|/|N|)/(|HN|/|N|) = |G|/|HN|$. Thus, HN/N is indeed a Sylow p-subgroup of G/N.

7.29 The number of Sylow 7-subgroups is $1 + 7k$ and divides 12. Thus, it is 1, and the Sylow 7-subgroup is normal.

7.31 The number of Sylow 17-subgroups is $1 + 17k$ and divides 256, so it is 1 or 256. If it is 1, the 17-Sylow subgroup is normal. If it is 256, then we note that each Sylow 17-subgroup is cyclic and has 16 elements of order 17. Distinct groups of prime order intersect trivially, so we have $16 \cdot 256 = 4096$ elements of order 17. This leaves only 256 other elements. But this is the size of a Sylow 2-subgroup, so there can be only one, and it is normal.

7.33 The number of Sylow p-subgroups is $1 + kp$ and divides q. If it is not 1, it is q, so $p|(q-1)$, giving us a contradiction. Thus, the Sylow p-subgroup is normal. Similarly, the number of Sylow q-subgroups is $1 + lq$ and divides p^2. Thus, it is 1, p or p^2. Suppose it is not 1. If it is p, then $q|(p-1)$, and since $(p-1)|(p^2-1)$, we have a contradiction. If it is p^2, we again obtain a contradiction. Therefore, the Sylow q-subgroup is normal as well. Thus, G is the direct product of its Sylow subgroups. Now, groups of order a prime or the square of a prime are abelian, and we are done.

7.35 The number of Sylow 3-subgroups is $1 + 3k$ and divides 19, so it is 1 or 19. If it is 1, then there are 2 elements of order 3. If it is 19, then there are 38, since subgroups of prime order intersect trivially.

7.37 Let H be a Sylow 7-subgroup and K a Sylow 17-subgroup. The number of Sylow 7-subgroups is $1 + 7k$ and divides 85, so it is 1 or 85. If it is 1, then H is normal. By Theorem 4.5, HK is a subgroup, and its order is $7 \cdot 17/1 = 119$. So assume that there are 85 Sylow 7-subgroups. Then we have $6 \cdot 85 = 510$ elements of order 7. The number of Sylow 17-subgroups is $1 + 17l$ and divides 35, so it is 1 or 35. If it is 1, then K is normal, and as above, we are done. Otherwise, we get $16 \cdot 35 = 560$ elements of order 17. But we now have too many elements.

7.39 It is not abelian, so we can rule out the two abelian groups. It has an element of order 6, namely $(R_{120}, 1)$, so we can rule out A_4. But it has no elements of order 4, unlike the group H from Example 7.14, which has $((1\ 2), 1)$. Thus, it must be D_{12}.

7.41 It suffices to show that every cyclic subgroup is normal, for then if $K \le Q_8$ and $a \in K$, $b \in Q_8$, we have $b^{-1}ab \in \langle a \rangle \le K$. As 1 and -1 are central, we need not worry about them. The remaining cases just involve checking, for instance, that $j^{-1}ij = -jij = kj = -i = i^{-1} \in \langle i \rangle$.

7.43 (1) We have $\alpha_{a,b}\alpha_{c,d}(x) = \alpha_{a,b}(cx+d) = a(cx+d)+b = acx+ad+b$. Thus, $\alpha_{a,b}\alpha_{c,d} = \alpha_{ac,ad+b} \in G$, since $p \nmid a$ and $p \nmid c$ imply that $p \nmid ac$; that is, ac is not 0 in $\mathbb{Z}_p$. Thus, we have closure. Composition of functions is always associative. The identity is $\alpha_{1,0}$. To find the inverse of $\alpha_{a,b}$, note that we want $ac = 1$ and $ad + b = 0$. Now, $(a, p) = 1$, so write $au + pv = 1$, for some $u, v \in \mathbb{Z}$. Thus, $au = 1$ in $\mathbb{Z}_p$. Let $c = u$. Similarly, letting $d = -ub$, we have $ad + b = -aub + b = 0$ in $\mathbb{Z}_p$. To see that the group is not abelian, note that $\alpha_{2,0}$ and $\alpha_{1,1}$ do not commute.

  (2) Let $H = \{\alpha_{a,b} \in G : a \in \{1, 2, 4\}\}$. Closure checks out as above, since products of 1, 2 and 4 remain in $\{1, 2, 4\}$ in $\mathbb{Z}_7$. Clearly $\alpha_{1,0} \in H$, so $H \le G$. There are 3 choices for a and 7 for b, so $|H| = 21$. Also, H is not abelian for the same reason given in the first part.

7.45 The number of Sylow p-subgroups is $1 + kp$ and divides q. As $p > q$, it is 1. The number of Sylow q-subgroups is $1 + lq$ and divides p, so it is 1 or p. But if it is p, then $q|(p-1)$, which is not allowed. Therefore, both Sylow subgroups are normal, and G is the direct product of $\mathbb{Z}_p$ and $\mathbb{Z}_q$, and hence isomorphic to $\mathbb{Z}_{pq}$, as $(p, q) = 1$.

Problems of Chapter 8

8.1 The addition table is found in Table 3.1. For multiplication, the table is as follows.

$$
\begin{array}{c|ccccc}
 & 0 & 1 & 2 & 3 & 4 \\
\hline
0 & 0 & 0 & 0 & 0 & 0 \\
1 & 0 & 1 & 2 & 3 & 4 \\
2 & 0 & 2 & 4 & 1 & 3 \\
3 & 0 & 3 & 1 & 4 & 2 \\
4 & 0 & 4 & 3 & 2 & 1 \\
\end{array}
$$

8.3 It is easy to see that R is closed under addition and contains $\{0\}$. Thus, since it is a finite set, it is an additive subgroup of $\mathbb{Z}_{15}$, which is an abelian group. Furthermore, R is closed under multiplication in $\mathbb{Z}_{15}$, and we know that this multiplication operation is associative and satisfies the distributive laws. Therefore, R is a ring. It is certainly commutative, and we can see that 6 is the identity.

8.5 It is not a ring, as it does not satisfy the distributive laws. Let $\alpha(x) = x^2$, $\beta(x) = x$ and $\gamma(x) = 2x$. Then $(\alpha \circ (\beta + \gamma))(x) = 9x^2$, but $(\alpha \circ \beta)(x) + (\alpha \circ \gamma)(x) = 5x^2$.

8.7 It is easy to see that the sum of two matrices in R also lies in R. Also, matrix addition is commutative and associative. The zero matrix is the additive identity, and negatives of matrices in R lie in R. Thus, R is an abelian group under addition. The product of two matrices in R is easily seen to be in R. Furthermore, matrix multiplication is associative and satisfies the distributive laws. Therefore, R is a ring. It contains the identity matrix, so it is a ring with identity. However, $\begin{pmatrix} 1 & 1 \\ 0 & 0 \end{pmatrix}$ and $\begin{pmatrix} 0 & 1 \\ 0 & 1 \end{pmatrix}$ do not commute, so it is not a commutative ring.

8.9 Not necessarily. Consider the additive group $\mathbb{Z}_p$, but define a multiplication operation via $ab = 0$ for all a and b. Clearly this operation is associative and the distributive laws are satisfied. Thus, we have a ring with p elements, but there is no identity.

8.11 (1) $a^2 + ba - ab - b^2$.
  (2) $a^3 - a^2b - aba - ba^2 + ab^2 + bab + b^2a - b^3$.

8.13 We have $b = b1 = bac = 1c = c$.

8.15 No, use $R = \mathbb{Z}_2 \oplus \mathbb{Z}_2$.

8.17 Note that $(a+bi) - (c+di) = (a-c) + (b-d)i \in R$ and $(a+bi)(c+di) = (ac-bd) + (ad+bc)i \in R$, for all $a, b, c, d \in \mathbb{Z}$. Also, $0 \in R$. Thus, R is a subring. In addition, R is a unital subring, as it contains $1 + 0i$, the identity of $\mathbb{C}$.

8.19 Certainly R contains the zero matrix. If $a, b \in \mathbb{R}$, then $\begin{pmatrix} 0 & 0 \\ 0 & a \end{pmatrix} - \begin{pmatrix} 0 & 0 \\ 0 & b \end{pmatrix} = \begin{pmatrix} 0 & 0 \\ 0 & a-b \end{pmatrix} \in R$, and $\begin{pmatrix} 0 & 0 \\ 0 & a \end{pmatrix}\begin{pmatrix} 0 & 0 \\ 0 & b \end{pmatrix} = \begin{pmatrix} 0 & 0 \\ 0 & ab \end{pmatrix} \in R$. Thus, R is a subring. It is a ring with identity, as $\begin{pmatrix} 0 & 0 \\ 0 & 1 \end{pmatrix}$ serves as the identity. But the identity of $M_2(\mathbb{R})$ is not there, so it is not a unital subring.

8.21 We have $(0, 0) \in T$. If $r_1, r_2 \in R$, then $(r_1, 0) - (r_2, 0) = (r_1 - r_2, 0) \in T$ and $(r_1, 0)(r_2, 0) = (r_1 r_2, 0) \in T$.

8.23 We have $0 = 0a \in S$. If $r_1, r_2 \in R$, then $r_1 a - r_2 a = (r_1 - r_2)a \in S$ and $(r_1 a)(r_2 a) = (r_1 a r_2)a \in S$.

8.25 Not necessarily. Let $R = \mathbb{Q}$, $S = \mathbb{Z}$ and $a = 2$. Then $1/2 \in T$, but $(1/2)^2 \notin T$.

8.27 We have $1 \in R$. If $a + bi, c + di \in R$, then $(a + bi) - (c + di) = (a - c) + (b - d)i \in R$. Furthermore, $(a + bi)(c + di) = (ac - bd) + (ad + bc)i \in R$. If $c + di \neq 0$, then $(c + di)(c - di) = c^2 + d^2$, which is a nonzero rational number, so the inverse of $c + di$ is $\frac{c}{c^2+d^2} - \frac{d}{c^2+d^2}i \in R$.

8.29 We have $(r, s) \in U(R \oplus S)$ if and only if there exist $r_1 \in R$, $s_1 \in S$ such that $rr_1 = r_1 r = 1$ and $ss_1 = s_1 s = 1$; that is, if and only if $r \in U(R)$ and $s \in U(S)$.

8.31 If $a^2 = a$, then $a(a - 1) = 0$, so since there are no zero divisors, $a = 0$ or 1. An integral domain must have these two elements.

8.33 By Exercise 8.20, $K \cap L$ is a subring. As $1 \in K$ and $1 \in L$, we have $1 \in K \cap L$. Also, if $0 \neq a \in K \cap L$, then $a^{-1} \in K$ and $a^{-1} \in L$, so $a^{-1} \in K \cap L$. Thus, $K \cap L$ is a subfield. The proof for an arbitrary collection of subfields is similar.

8.35 We have $(a^{10})^4 = (b^{10})^4$ and $(a^{13})^3 = (b^{13})^3$. That is, $a^{40} = b^{40}$ and $a^{39} = b^{39}$. So, $a^{39}a = b^{39}b = a^{39}b$. If $a = 0$, then since $b^{40} = 0$ and there are no zero divisors, $b = 0$. If $a \neq 0$, then cancelling a^{39}, we obtain $a = b$.

8.37 (1) 7.
    (2) 0.

8.39 As 1 cannot have infinite order in a finite additive group, we know that char $R = p$, for some prime p. Thus, $pa = 0$ for all $a \in R$, so every element of R has additive order 1 or p. If $|R|$ is divisible by some prime $q \neq p$, then by Cauchy's theorem, R has an element of additive order q, which is impossible. Thus, the only prime dividing $|R|$ is p.

8.41 (1) We have $(1 + a)(1 - a + a^2 - a^3 + \cdots + (-1)^{n-1}a^{n-1}) = 1$.
    (2) Let char $R = p$ and choose k such that $p^k > n$. Then $(1 + a)^{p^k} = 1 + a^{p^k}$ (using the Freshman's Dream). But $a^{p^k} = a^n a^{p^k - n} = 0$, so $(1 + a)^{p^k} = 1$.

Problems of Chapter 9

9.1 (1) $(0, 0)$, $(0, 1)$, $(0, 2)$, $(0, 3)$, $(0, 4)$, $(0, 5)$.
  (2) $(0, 0)$, $(2, 0)$, $(0, 3)$, $(2, 3)$.

9.3 By Exercise 8.20, $I \cap J$ is a subring. Take $a \in I \cap J$ and $r \in R$. Then $a \in I$ implies $ra, ar \in I$. Similarly, $ra, ar \in J$, so $ra, ar \in I \cap J$, and $I \cap J$ is an ideal. The argument for an arbitrary collection of ideals is similar.

9.5 (1) Let $R = 2\mathbb{Z}$, $a = 2$.
  (2) Let $R = M_2(\mathbb{R})$, $a = \begin{pmatrix} 1 & 0 \\ 0 & 0 \end{pmatrix}$. Every matrix in S is of the form $\begin{pmatrix} b & 0 \\ c & 0 \end{pmatrix}$, for some $b, c \in \mathbb{R}$. Clearly $a \in S$, but multiplying on the right by $\begin{pmatrix} 1 & 1 \\ 1 & 1 \end{pmatrix}$, we get a matrix not in S, so S is not an ideal.

9.7 Let $R = \mathbb{Z}_8$, $I = (2)$ and $J = (4)$.

9.9 We know from Exercise 3.42 that G is an additive group. It is clearly abelian. It is also closed under multiplication. Furthermore, $(a_1, a_2, \ldots)((b_1, b_2, \ldots)$ $(c_1, c_2, \ldots)) = (a_1 b_1 c_1, a_2 b_2 c_2, \ldots) = ((a_1, a_2, \ldots)(b_1, b_2, \ldots))(c_1, c_2, \ldots)$. Thus, we have associativity of multiplication. The distributive law follows similarly, and we have a ring. By Exercise 3.42, H is an additive subgroup of G, so it remains only to check absorption. If $(a_1, a_2, \ldots) \in H$, $(b_1, b_2, \ldots) \in G$, then only finitely many of the a_i are different from 0, so only finitely many of the $a_i b_i$ are different from 0, and $(a_1 b_1, a_2 b_2, \ldots) \in H$. Similarly for $(b_1 a_1, b_2 a_2, \ldots)$.

9.11 The addition table may be found in Table 4.2 (replacing each instance of "N" with "I"). The multiplication table follows.

	$0 + I$	$1 + I$	$2 + I$	$3 + I$	$4 + I$
$0 + I$	$0 + I$	$0 + I$	$0 + I$	$0 + I$	$0 + I$
$1 + I$	$0 + I$	$1 + I$	$2 + I$	$3 + I$	$4 + I$
$2 + I$	$0 + I$	$2 + I$	$4 + I$	$1 + I$	$3 + I$
$3 + I$	$0 + I$	$3 + I$	$1 + I$	$4 + I$	$2 + I$
$4 + I$	$0 + I$	$4 + I$	$3 + I$	$2 + I$	$1 + I$

9.13 Expanding, we obtain $8x^4 + 2x^3 + 7x^2 + 5x + 2 + I$. Now, $2(x^3 + 6x^2 + 2) \in I$, so $2x^3 + I = -12x^2 - 4 + I$. Also, $8x(x^3 + 6x^2 + 2) \in I$, so $8x^4 + I = -48x^3 - 16x + I$. Similarly, $48x^3 + I = -288x^2 - 96 + I$. Thus, our answer is $288x^2 + 96 - 16x - 12x^2 - 4 + 7x^2 + 5x + 2 + I = 283x^2 - 11x + 94 + I$.

9.15 By the preceding exercise, $R/(I \cap J)$ is commutative if and only if $ab - ba \in I \cap J$ for all $a, b \in R$. But this happens if and only if $ab - ba \in I$ and $ab - ba \in J$ for all $a, b \in R$; that is, if and only if R/I and R/J are commutative.

9.17 The only ideals of F are $\{0\}$ and F, so 81 and 1.

9.19 If $(a + I)^n = 0 + I$, then $a^n \in I$, and hence there exists an $m \in \mathbb{N}$ such that $(a^n)^m = 0$; that is, $a^{nm} = 0$, which means that $a \in I$, and hence $a + I = 0 + I$.

9.21 (1) No, as $\alpha(1 \cdot 1) = 2$ but $\alpha(1)\alpha(1) = 4$.

(2) Yes. If $f(x), g(x) \in \mathbb{R}[x]$, then $\alpha(f(x) + g(x)) = f(2) + g(2) = \alpha(f(x)) + \alpha(g(x))$ and $\alpha(f(x)g(x)) = f(2)g(2) = \alpha(f(x))\alpha(g(x))$.

9.23 For any $r_1, r_2 \in R$, we have $\beta(\alpha(r_1 + r_2)) = \beta(\alpha(r_1) + \alpha(r_2)) = \beta(\alpha(r_1)) + \beta(\alpha(r_2))$, and similarly for multiplication.

9.25 It is a homomorphism, as $\alpha((a, b) + (c, d)) = \alpha((a + c, b + d)) = (a + c, 0) = \alpha((a, b)) + \alpha((c, d))$, and similarly for multiplication. The kernel is $\{0\} \oplus \mathbb{Z}$. Furthermore, $\alpha^{-1}(2\mathbb{Z} \oplus 3\mathbb{Z}) = 2\mathbb{Z} \oplus \mathbb{Z}$.

9.27 Let $S = R/I$ and define $\alpha : R \to S$ via $\alpha(a) = a + I$. We have $\alpha(a + b) = a + b + I = (a + I) + (b + I) = \alpha(a) + \alpha(b)$, and similarly for multiplication, so α is a homomorphism. Also, $a \in \ker(\alpha)$ if and only if $a + I = 0 + I$; that is, if and only if $a \in I$.

9.29 Let $\alpha : F \to K$ be a homomorphism. Now, $\ker(\alpha)$ is an ideal of F. As F is a field, this means that $\ker(\alpha) = \{0\}$ or F. In the former case, α is one-to-one, which is impossible, as K has fewer elements than F. In the latter case, $\alpha(a) = 0$ for all a, so this is the only possible homomorphism.

9.31 (1) The additive groups are not isomorphic. (See Exercise 4.31.)

(2) One has an identity, the other does not.

9.33 Let $\alpha : R \to S$ be an isomorphism. We claim that $\alpha(Z(R)) \subseteq Z(S)$. But if $a \in Z(R)$, then for any $r \in R$, we have $ar = ra$, and hence $\alpha(a)\alpha(r) = \alpha(r)\alpha(a)$. As α is onto, $\alpha(a)$ commutes with everything in S. Thus, restricting α to $Z(R)$, we have a one-to-one homomorphism into $Z(S)$. But if $b \in Z(S)$, then for any $r \in R$, we have $\alpha(r)b = b\alpha(r)$. Letting $b = \alpha(c)$, this means that $\alpha(r)\alpha(c) = \alpha(c)\alpha(r)$; that is, $\alpha(rc) = \alpha(cr)$, and since α is one-to-one, $rc = cr$. In particular, $b = \alpha(c) \in \alpha(Z(R))$. Therefore, $\alpha : Z(R) \to Z(S)$ is onto as well, and hence an isomorphism.

9.35 Let K be the field of fractions, and define $\alpha : F \to K$ via $\alpha(a) = [a, 1]$ for all $a \in F$. If $a, b \in F$, then $\alpha(a + b) = [a + b, 1] = [a, 1] + [b, 1] = \alpha(a) + \alpha(b)$ and $\alpha(ab) = [ab, 1] = [a, 1][b, 1] = \alpha(a)\alpha(b)$; thus, α is a homomorphism. If $[a, 1] = [0, 1]$, then $a = 0$, so α is one-to-one. Furthermore, if $a, b \in F$, with $b \neq 0$, then $\alpha(ab^{-1}) = [ab^{-1}, 1] = [a, b]$; thus, α is onto.

9.37 No: $\mathbb{Z}$ and $\mathbb{Q}$ are certainly not isomorphic ($\mathbb{Q}$ is a field but $\mathbb{Z}$ is not), however, we already know that the field of fractions of $\mathbb{Z}$ is isomorphic to $\mathbb{Q}$, and by Exercise 9.35, the field of fractions of $\mathbb{Q}$ is isomorphic to $\mathbb{Q}$ as well.

9.39 (1) Note that

$$\alpha\left(\begin{pmatrix} a & b \\ c & d \end{pmatrix} + \begin{pmatrix} e & f \\ g & h \end{pmatrix}\right) = \begin{pmatrix} a + e & c + g \\ b + f & d + h \end{pmatrix} = \alpha\left(\begin{pmatrix} a & b \\ c & d \end{pmatrix}\right) + \alpha\left(\begin{pmatrix} e & f \\ g & h \end{pmatrix}\right).$$

Furthermore,

$$\alpha\left(\begin{pmatrix} a & b \\ c & d \end{pmatrix}\begin{pmatrix} e & f \\ g & h \end{pmatrix}\right) = \begin{pmatrix} ae+bg & ce+dg \\ af+bh & cf+dh \end{pmatrix} = \alpha\left(\begin{pmatrix} e & f \\ g & h \end{pmatrix}\right)\alpha\left(\begin{pmatrix} a & b \\ c & d \end{pmatrix}\right).$$

Also, it is clear that applying α twice returns the original matrix.

(2) requires similar computations.

9.41 Define $\alpha : R \oplus S \to S$ via $\alpha((r, s)) = s$. If $r_i \in R$, $s_i \in S$, then $\alpha((r_1, s_1) + (r_2, s_2)) = \alpha((r_1 + r_2, s_1 + s_2)) = s_1 + s_2 = \alpha((r_1, s_1)) + \alpha((r_2, s_2))$, and similarly for multiplication. Thus, α is a homomorphism. If $s \in S$, then $\alpha((0, s)) = s$, and hence α is onto. Finally, $(r, s) \in \ker(\alpha)$ if and only if $s = 0$; that is, $\ker(\alpha) = R \oplus \{0\}$. Apply the First Isomorphism Theorem.

9.43 Define $\alpha : \mathbb{Z}[x] \to \mathbb{Z}_5$ via $\alpha(f(x)) = [f(0)]$, where the square brackets denote the congruence class in $\mathbb{Z}_5$. If $f(x), g(x) \in \mathbb{Z}[x]$, then $\alpha(f(x) + g(x)) = [f(0) + g(0)] = [f(0)] + [g(0)] = \alpha(f(x)) + \alpha(g(x))$, and similarly for multiplication. Thus, α is a homomorphism. If $[a] \in \mathbb{Z}_5$, then letting $f(x)$ be the constant polynomial a, we see that $\alpha(f(x)) = a$; thus, α is onto. Furthermore, as $f(0)$ is the constant term of $f(x)$, we see that $\ker(\alpha)$ is precisely I. Now apply the First Isomorphism Theorem.

9.45 The first part is the Third Isomorphism Theorem. To see the second part, note that $3\mathbb{Z}/12\mathbb{Z} = \{0 + 12\mathbb{Z}, 3 + 12\mathbb{Z}, 6 + 12\mathbb{Z}, 9 + 12\mathbb{Z}\}$ is a commutative ring having identity $9 + 12\mathbb{Z}$. Furthermore, its characteristic is 4. Thus, it has a subring isomorphic to $\mathbb{Z}_4$. As the ring only has four elements, the ring is itself isomorphic to $\mathbb{Z}_4$.

9.47 Note that $(1 + i)(1 - i) = 2 \in (2)$, and yet neither $1 + i$ nor $1 - i$ is a multiple of 2 in R. The ideal is not prime and hence, as R is a commutative ring with identity, not maximal.

9.49 Let I be a prime ideal. Then R/I is an integral domain. But a finite integral domain is a field (see Theorem 8.10), so R/I is a field, and hence I is maximal.

9.51 Let $R = 2\mathbb{Z}_4 = \{0, 2\}$ and $I = \{0\}$. Now, I is surely maximal, since if it got any larger, it would be R. But it is not prime, as $2 \notin I$, but $2 \cdot 2 = 0 \in I$.

9.53 In a field, the only element that is not a unit is 0, and $\{0\}$ is an ideal. In $\mathbb{Z}_{p^n}$, we know (see Exercise 8.30) that the units are precisely the elements a that are relatively prime to p^n. In other words, the elements that are not units are those that are divisible by p, so (p) is the ideal in question.

9.55 Use $P = R \oplus I$. As $I \neq R$, we see that $P \neq R \oplus R$. Also, if $(a, b)(c, d) \in P$, then $bd \in I$. As I is prime, either b or d is in I, and hence (a, b) or (c, d) is in P.

Problems of Chapter 10

10.1 $f(x) - g(x) = 9x^4 + 8x^3 + 4x^2 + 6x + 4$, $f(x)g(x) = 4x^7 + 7x^6 + 2x^5 + x^4 + 7x^2 + 4x + 5$.

10.3 $q(x) = 5x^2 + 6x + 3$, $r(x) = 4x^2 + 1$.

10.5 No. According to the preceding exercise, x is not a unit.

10.7 Suppose that char $R[x] = n > 0$. Then, in particular, for every constant polynomial a, we have $na = 0$. Thus, $0 < $ char $R \le n$. On the other hand, if char $R = m > 0$, then for any $f(x) \in R[x]$, we note that for each coefficient a_i appearing in $f(x)$, we have $ma_i = 0$; thus, $mf(x) = 0$, and $0 < $ char $R[x] \le m$. The only remaining case is where char R and char $R[x]$ are both 0.

10.9 Certainly $0 \in S[x]$. Take $f(x), g(x) \in S[x]$. Then all coefficients of $f(x)$ and $g(x)$ lie in S. The coefficients of $f(x) - g(x)$ are differences of elements of S, and hence lie in S, so $f(x) - g(x) \in S[x]$. Similarly, the coefficients of $f(x)g(x)$ are sums of products of elements of S and thus lie in S. Hence, $S[x]$ is a subring. Let S be an ideal. If $f(x) \in S[x]$ and $g(x) \in R[x]$, then the coefficients of $f(x)g(x)$ are sums of products, where each term in the sum is an element of S multiplied by an element of R, and therefore lies in S. Thus, $f(x)g(x) \in S[x]$. Similarly, $g(x)f(x) \in S[x]$.

10.11 As a and ab are associates, write $ab = au$, where u is a unit. If $a \ne 0$, then cancellation gives $b = u$.

10.13 Let a be a unit. Then for any $0 \ne b \in R$, we have $b = a(a^{-1}b)$. Thus, $\varepsilon(a) \le \varepsilon(b)$, so $\varepsilon(a)$ is indeed the smallest possible value, n. Now suppose that a is not a unit. We can write $1 = aq + r$, where $q, r \in R$ and either $r = 0$ or $\varepsilon(r) < \varepsilon(a)$. In the former case, a is a unit, which is a contradiction. In the latter case, $\varepsilon(a)$ is not the smallest possible value.

10.15 We have

$$f(x) = g(x)\left(\frac{3}{2}\right) + \left(-\frac{7}{2}x^3 - \frac{13}{2}x^2 - \frac{19}{2}x + \frac{3}{2}\right)$$

$$g(x) = \left(-\frac{7}{2}x^3 - \frac{13}{2}x^2 - \frac{19}{2}x + \frac{3}{2}\right)\left(-\frac{4}{7}x - \frac{46}{49}\right) + \left(\frac{72}{49}x^2 + \frac{144}{49}x + \frac{216}{49}\right),$$

and since $\frac{72}{49}x^2 + \frac{144}{49}x + \frac{216}{49}$ divides $-\frac{7}{2}x^3 - \frac{13}{2}x^2 - \frac{19}{2}x + \frac{3}{2}$, the former is a gcd. We must make it monic, so multiplying by $49/72$, we get $(f(x), g(x)) = x^2 + 2x + 3$.

10.17 Beginning with the second of the two equations in the solution to Exercise 10.15, we see that

$$(f(x), g(x)) = \frac{49}{72}\left(g(x) - \left(-\frac{7}{2}x^3 - \frac{13}{2}x^2 - \frac{19}{2}x + \frac{3}{2}\right)\left(-\frac{4}{7}x - \frac{46}{49}\right)\right)$$

$$= \frac{49}{72}\left(g(x) - \left(f(x) - g(x)\left(\frac{3}{2}\right)\right)\left(-\frac{4}{7}x - \frac{46}{49}\right)\right)$$

$$= f(x)\left(\frac{7}{18}x + \frac{23}{36}\right) + g(x)\left(-\frac{7}{12}x - \frac{5}{18}\right).$$

10.19 We must apply the Euclidean algorithm. Let us use the notation established in Example 10.8, taking $u = 5 + 7i$ and $v = 1 + 3i$. Now, $(1 + 3i)(1 - 3i) = 10$, so $uv^{-1} = (5 + 7i)(1 - 3i)/10 = 2.6 - 0.8i$. Thus, we have $m = 3$ and $n = -1$, so $q = 3 - i$ and $r = (5 + 7i) - (1 + 3i)(3 - i) = -1 - i$. That is,

$$5 + 7i = (1 + 3i)(3 - i) + (-1 - i).$$

For the next step, we let $u = 1 + 3i$ and $v = -1 - i$. But $(-1 - i)(-1 + i) = 2$, so $uv^{-1} = (1 + 3i)(-1 + i)/2 = -2 - i$. Therefore,

$$1 + 3i = (-1 - i)(-2 - i) + 0.$$

Thus, $-1 - i$ is a gcd of $5 + 7i$ and $1 + 3i$.

10.21 Using the notation as in Example 10.15, we note that $N(1 + 2\sqrt{5}i) = 21$. If $1 + 2\sqrt{5}i = uv$, then $N(u)N(v) = 21$, and assuming without loss of generality that $N(u) \le N(v)$, we have $N(u) = 1$ or 3. As in Example 10.15, $N(u) = 3$ is impossible and $N(u) = 1$ means $u \in \{1, -1\}$. In particular, u is a unit and $1 + 2\sqrt{5}i$ is irreducible. However, $(1 + 2\sqrt{5}i)(1 - 2\sqrt{5}i) = 21 = 3 \cdot 7$. Thus, $(1 + 2\sqrt{5}i)|21$. But as $N(3) = 9$ and $N(7) = 49$, we cannot possibly have $1 + 2\sqrt{5}i$ dividing 3 or 7. Thus, $1 + 2\sqrt{5}i$ is not prime.

10.23 Combine the preceding two exercises with Theorem 10.11.

10.25 Not necessarily. We know that $\mathbb{Z}$ is a Euclidean domain, but $\mathbb{Z}[x]$ is not a PID, hence not a Euclidean domain.

10.27 This is essentially the same as Exercise 2.24.

10.29 Suppose not, and let $0 \neq a \in R$ be a nonunit. Let $I_n = (a^n)$. As $a^n|a^{n+1}$, we see that $I_{n+1} \subseteq I_n$. Suppose that $I_n = I_{n+1}$. Then $a^n \in (a^{n+1})$; that is, $a^n = a^{n+1}b$, for some $b \in R$. Cancelling a^n, we get $1 = ab$. Thus, a is a unit, giving us a contradiction.

10.31 Using the notation in Example 10.8, we have $\varepsilon(1 + i) = 2$. As we noted in that example, if u and v are in our ring, and $uv = 1 + i$, then $\varepsilon(u)\varepsilon(v) = 2$. But $\varepsilon(u)$ and $\varepsilon(v)$ are nonnegative integers, so without loss of generality, $\varepsilon(u) = 1$. This means $u \in \{\pm 1, \pm i\}$, and so u is a unit. Thus, $1 + i$ is irreducible. However, we know that R is a Euclidean domain, and hence a PID, so every irreducible is prime, by Theorem 10.11.

10.33 Not necessarily. We know that $\mathbb{Q}[x]$ is a UFD, and yet its subring R discussed in Example 10.18 is not.

10.35 Let $a = 2, b = 5, c = 2 + \sqrt{6}i$ and $d = 2 - \sqrt{6}i$. Clearly $ab = cd = 10$. Defining a norm as in Example 10.15 via $N(m + n\sqrt{6}i) = m^2 + 6n^2$, we see from the same calculation that $N(uv) = N(u)N(v)$ for all $u, v \in R$. Suppose that $uv = 2$. Then $N(u)N(v) = 4$, so either $N(u) = N(v) = 2$ (which is impossible) or $N(u) = 1$ and $N(u) = 4$ (or vice versa). But this means u is 1 or -1. In particular, u is a unit, so 2 is irreducible. Similar calculations show that b, c and d are irreducible. It is immediate that neither a nor b divides c or d. That R is not a UFD follows from the definition.

10.37 Let p be an irreducible of R. Then p is a nonzero nonunit. Suppose that $p|ab$, for some $a, b \in R$. If a is a unit, then b and ab are associates, so $p|b$. If $a = 0$, then $p|a$. Similarly if b is zero or a unit. So let a and b be nonzero nonunits. We may write $a = p_1 \cdots p_k$ and $b = q_1 \cdots q_l$, where the p_i and q_j are irreducible. By the preceding exercise, p divides some p_i or some q_j. Without loss of generality, say $p|q_1$. Since $q_1|b$, we have $p|b$.

Problems of Chapter 11

11.1 (1) As 5 is a root and the degree is greater than 1, no.

(2) Trying each possible root in $\mathbb{Z}_7$, we see that this polynomial has no root there. Thus, since the degree is 3, it is irreducible.

(3) No, since it factors as $(x^2 + 4)(x^2 + 4)$.

11.3 The possibilities are $x^3 + ax^2 + bx + c$, where $a, b, c \in \{0, 1\}$. If $c = 0$, then 0 is a root, so $c = 1$. Also, 1 is a root of $x^3 + 1$ and $x^3 + x^2 + x + 1$, so we can rule them out. The remaining polynomials are $x^3 + x^2 + 1$ and $x^3 + x + 1$. Both have degree 3, and neither has a root, so they are irreducible.

11.5 Let $h(x) = f(x) - g(x)$. If $f(x) \neq g(x)$, then $h(x)$ is not the zero polynomial. Say $\deg(h(x)) = n$. Then $h(x)$ can have at most n roots, but $h(a) = f(a) - g(a) = 0$ for all $a \in F$, giving us a contradiction.

11.7 No, take $a, b \in R$ such that $ab \neq ba$. Let $r = a$, $f(x) = x$ and $g(x) = b$. Then $\alpha(f(x)g(x)) = \alpha(bx) = ba$, whereas $\alpha(f(x))\alpha(g(x)) = ab$.

11.9 As $\deg(x^2 + 1) = 2$, the polynomial is reducible if and only if it has a root $m \in \{0, 1, \ldots, p - 1\}$. Factoring out $x - m$, we can only be left with $x - n$, for some $n \in \{0, 1, \ldots, p - 1\}$. Thus, $x^2 + 1 = x^2 - (m + n)x + mn$. That is, $x^2 + 1$ is reducible if and only if there exist $m, n \in \{0, 1, \ldots, p - 1\}$ such that $p|(m + n)$ and $p|(mn - 1)$. Given the range of values for m and n, we can only have $m + n \in \{0, p\}$. But m and n cannot possibly both be 0, so $m + n = p$.

11.11 (1) The only possible rational roots are $\pm 1, \pm 2$. But none of these work, so it has no rational roots.

(2) The possible rational roots are of the form m/n, where $m\,|\,(-2)$ and $n\,|\,6$. Trying all of the possibilities, we see that $-1/2$ and $2/3$ are roots.

11.13 (1) Looking first for rational roots, we know that they must be integers and divide 18. We find that 3 is a root, so we have $(x - 3)(x^3 - 7x^2 + 14x - 6)$. Now, 3 is also a root of $x^3 - 7x^2 + 14x - 6$, so we have $(x - 3)^2(x^2 - 4x + 2)$. By Eisenstein's criterion, we are now done.

(2) Looking first for rational roots, we see that they can only be $\pm 1, \pm 2$. In fact, -2 is a root, so we have $(x + 2)(x^3 + x + 1)$. Now, the only possible rational roots of $x^3 + x + 1$ are 1 and -1, and these do not work. A degree 3 polynomial with no roots is irreducible, so we are done.

11.15 Note that $f(x)$ is a constant polynomial if and only if $f(x + a)$ is a constant polynomial. Thus, we may assume that both have degrees larger than 1. Suppose that $f(x) = g(x)h(x)$, where $g(x)$ and $h(x)$ are nonconstant polynomials. Then $f(x+a) = g(x+a)h(x+a)$. As $g(x+a)$ and $h(x+a)$ are nonconstant polynomials, it follows that $f(x + a)$ is reducible. The converse is similar.

11.17 It is irreducible, using the preceding exercise with $p = 7$.

11.19 (1) We know that $2 - 3i$ must also be a root of $f(x)$, so $f(x)$ is divisible by $(x - (2 + 3i))(x - (2 - 3i)) = x^2 - 4x + 13$. Performing the division, we get $f(x) = (x^2 - 4x + 13)(x - 7)$, so the third root is 7.

(2) Here, $1 + i$ is also a root, so $f(x)$ is divisible by $(x - (1 - i))(x - (1 + i)) = x^2 - 2x + 2$. Performing the division, we get $f(x) = (x^2 - 2x + 2)(x^2 + 2x + 3)$. By the quadratic equation, the remaining roots are $-1 + \sqrt{2}i$ and $-1 - \sqrt{2}i$.

11.21 (1) By Eisenstein's criterion, the polynomial is irreducible over $\mathbb{Q}$. In $\mathbb{R}[x]$, we can factor it as $(x - \sqrt[4]{10})(x + \sqrt[4]{10})(x^2 + \sqrt{10})$. In $\mathbb{C}[x]$, we factor further and get $(x - \sqrt[4]{10})(x + \sqrt[4]{10})(x - \sqrt[4]{10}i)(x + \sqrt[4]{10}i)$.

(2) Using the Rational Roots Theorem, we find that 2 is a root. Thus, we can factor it as $(x - 2)(x^2 + 3x + 11)$. But $x^2 + 3x + 11$ is irreducible over $\mathbb{R}$, hence over $\mathbb{Q}$, so we are done in those two cases. For $\mathbb{C}$, we use the quadratic equation and get $(x - 2)(x - (-3 + \sqrt{35}i)/2)(x - (-3 - \sqrt{35}i)/2)$.

11.23 The roots must also include $2 + 5i$ and $4 - i$, so we can use

$$(x - (2 - 5i))(x - (2 + 5i))(x - (4 + i))(x - (4 - i))(x - 6)$$
$$= x^5 - 18x^4 + 150x^3 - 768x^2 + 2293x - 2958.$$

11.25 (1) Reducing modulo 5, we get $x^3 + 2x + 1$. We see that it has no roots in $\mathbb{Z}_5$, and since it has degree 3, the polynomial is irreducible in $\mathbb{Z}_5[x]$, and hence $f(x)$ is irreducible in $\mathbb{Q}[x]$.

(2) Reducing modulo 3, we get $x^4 + x^2 + 2$. This has no roots in $\mathbb{Z}_3$, but we must rule out the possibility of a product of two polynomials of degree 2. We may assume

that both such polynomials are monic, and we can only have something of the form $(x^2 + ax + 1)(x^2 + bx + 2) = x^4 + x^2 + 2$. Comparing coefficients, we find that $a + b = 0$ and $2a + b = 0$. Thus, $a = b = 0$. But $(x^2 + 1)(x^2 + 2) = x^4 + 2 \neq x^4 + x^2 + 2$. Thus, our polynomial is irreducible in $\mathbb{Z}_3[x]$, and $f(x)$ is irreducible in $\mathbb{Q}[x]$.

11.27 The monic polynomials of degree 2 are precisely those of the form $x^2 + ax + b$, with $a, b \in F$. There are thus n^2 of them. Such a polynomial is reducible if and only if it factors as $(x - c)(x - d)$, with $c, d \in F$. When $c = d$, there are n choices. If $c \neq d$, there are n choices for c and $n - 1$ for d. Of course, $(x - c)(x - d) = (x - d)(x - c)$, so we get $n(n - 1)/2$ possibilities, for a total of $n + n(n - 1)/2 = n(n + 1)/2$ reducible polynomials. By unique factorization, all of them are distinct. Thus, the number of irreducibles is $n^2 - n(n + 1)/2 = n(n - 1)/2$.

11.29 (1) $x^4 + 1 = (x^2 + a)(x^2 - a)$.
  (2) $x^4 + 1 = (x^2 + ax - 1)(x^2 - ax - 1)$.
  (3) $x^4 + 1 = (x^2 + ax + 1)(x^2 - ax + 1)$.

Problems of Chapter 12

12.1 No, as $x^n + 1$ and x^n lie in V but their difference, 1, does not.

12.3 We have $0 \in U$ and $0 \in W$, so $0 \in U \cap W$. If $v_1, v_2 \in U \cap W$, then $v_1, v_2 \in U$, so $v_1 + v_2 \in U$. Similarly, $v_1 + v_2 \in W$, so $v_1 + v_2 \in U \cap W$. If $a \in F$, then $av_1 \in U$ and $av_1 \in W$, so $av_1 \in U \cap W$. The argument for an arbitrary collection of subspaces is similar.

12.5 As $0 \in U$, we have $0 = \alpha(0) \in \alpha(U)$. (This follows immediately from the fact that α is, by definition, a homomorphism of additive groups.) Also, if $\alpha(u_1), \alpha(u_2) \in \alpha(U)$ and $a \in F$, then $\alpha(u_1) + \alpha(u_2) = \alpha(u_1 + u_2) \in \alpha(U)$, since $u_1 + u_2 \in U$, and $a\alpha(u_1) = \alpha(au_1) \in \alpha(U)$, since $au_1 \in U$.

12.7 It is. As $2 \cdot 0 + 3 \cdot 0 + 7 \cdot 0 = 0$, we see that $(0, 0, 0) \in W$. Suppose that $(a_1, b_1, c_1), (a_2, b_2, c_2) \in W$ and $a \in F$. Then $2(a_1 + a_2) + 3(b_1 + b_2) + 7(c_1 + c_2) = (2a_1 + 3b_1 + 7c_1) + (2a_2 + 3b_2 + 7c_2) = 0 + 0 = 0$, so $(a_1 + a_2, b_1 + b_2, c_1 + c_2) \in W$. Also, $2aa_1 + 3ab_1 + 7ac_1 = a(2a_1 + 3b_1 + 7c_1) = a \cdot 0 = 0$; thus, $(aa_1, ab_1, ac_1) \in W$.

12.9 We have $v + v + v = 1v + 1v + 1v = (1 + 1 + 1)v = 0v = 0$.

12.11 (1) As $3(1, 3, 5) + 2(2, 1, 4) - 1(7, 11, 23) = (0, 0, 0)$, they are linearly dependent.
  (2) Suppose that $a(1, 3, 4) + b(2, 2, 1) + c(3, 6, 3) = (0, 0, 0)$. Then $a + 2b + 3c = 3a + 2b + 6c = 4a + b + 3c = 0$. Thus, $3a - b = a - 2b = 0$. We see immediately that $a = b = 0$, and hence $c = 0$. Therefore, the vectors are linearly independent.

12.13 (1) No. If they did, then as $(1, 0, 2) + (2, 5, 3) = (3, 5, 5)$, the vectors are linearly dependent, which means that some proper subset would form a basis for $\mathbb{Q}^3$. But $\mathbb{Q}^3$ is 3-dimensional over $\mathbb{Q}$, so this is impossible.

(2) Yes. We claim that the vectors are linearly independent. If $a(1, 0, 2) + b(2, 3, 5) + c(0, 0, 4) = (0, 0, 0)$, we see immediately that $b = 0$, from which it follows that $a = 0$ and then $c = 0$. Thus, we can add vectors to this set to find a basis for $\mathbb{Q}^3$. But again, we are in a space with dimension 3, so no more vectors can be added. Therefore, the vectors span the space.

12.15 If the field is $\mathbb{C}$, we can see that every matrix can be written in a unique and obvious way as a linear combination of $\begin{pmatrix} 1 & 0 \\ 0 & 0 \end{pmatrix}, \begin{pmatrix} 0 & 1 \\ 0 & 0 \end{pmatrix}, \begin{pmatrix} 0 & 0 \\ 1 & 0 \end{pmatrix}, \begin{pmatrix} 0 & 0 \\ 0 & 1 \end{pmatrix}$, so these matrices form a basis and the dimension is 4. Working over $\mathbb{R}$, we would also need $\begin{pmatrix} i & 0 \\ 0 & 0 \end{pmatrix}, \begin{pmatrix} 0 & i \\ 0 & 0 \end{pmatrix}, \begin{pmatrix} 0 & 0 \\ i & 0 \end{pmatrix}, \begin{pmatrix} 0 & 0 \\ 0 & i \end{pmatrix}$, so the dimension is 8.

12.17 Let $\dim V = n$. If $n = 0$, then $V = \{0\}$ and the only possible subspace is $\{0\}$, so there is nothing to do. So assume that $n \geq 1$. If $W = \{0\}$, then again, there is nothing to do. So assume that there exists $0 \neq w_1 \in W$. Then w_1 is, by itself, linearly independent. If w_1 spans W, then we have a basis for W. If not, then there exists a $w_2 \in W$ such that w_2 is not a scalar multiple of w_1. But now w_1 and w_2 are linearly independent. If they span W, we have a basis. Otherwise, find $w_3 \in W$ such that w_3 is not a linear combination of w_1 and w_2. Repeat this procedure. We cannot possibly go beyond w_n, as V cannot have $n + 1$ linearly independent vectors. Thus, W has a basis consisting of at most n elements, so $\dim W \leq \dim V$. If $W \neq V$, then we can add to the basis for W to obtain a basis for V, which means we must have $\dim W < \dim V$.

12.19 Suppose that $a_1\alpha(v_1) + a_2\alpha(v_2) + \cdots + a_n\alpha(v_n) = 0$, for some $a_i \in F$. Then $\alpha(a_1 v_1 + \cdots + a_n v_n) = 0$. As α is one-to-one, $a_1 v_1 + \cdots + a_n v_n = 0$. But the v_i are linearly independent. Thus, $a_1 = \cdots = a_n = 0$.

12.21 Let $a = \sqrt{5} + \sqrt{7}$. Then $a^2 = 12 + 2\sqrt{35}$, so $(a^2 - 12)^2 = 140$. Thus, a satisfies $f(x) = x^4 - 24x^2 + 4$. We must show that $f(x)$ is irreducible over $\mathbb{Q}$. If it has a root in $\mathbb{Q}$, then by the Rational Roots Theorem, the root must lie in $\{\pm 1, \pm 2, \pm 4\}$. But none of these work. The only other possibility is that $f(x)$ is the product of two polynomials of degree 2. By Theorem 11.4, they may be assumed to be in $\mathbb{Z}[x]$. Up to a factor of -1, and noting that there is no x^3 term in $f(x)$, the factorization must be $(x^2 + bx + c)(x^2 - bx + d)$, for some $b, c, d \in \mathbb{Z}$. As there is no x term in $f(x)$, either $b = 0$ or $c = d$. If $b = 0$, we have $c + d = -24$ and $cd = 4$. No integers can possibly satisfy these equations. So, assume that $c = d$. We are left with the cases $(x^2 + bx + 2)(x^2 - bx + 2)$ and $(x^2 + bx - 2)(x^2 - bx - 2)$, for some integer b. These possibilities yield, respectively, $4 - b^2 = -24$ and $-4 - b^2 = -24$. Neither of these equations has a solution in $\mathbb{Z}$.

12.23 Suppose that $[K : F] = n$. If $a \in K$, then $1, a, a^2, \ldots, a^n$ are linearly dependent over F, by Lemma 12.1. Thus, there exist $b_i \in F$, not all zero, such that $b_0 + b_1 a + b_2 a^2 + \cdots + b_n a^n = 0$. That is, a is a root of $b_0 + b_1 x + \cdots + b_n x^n$.

12.25 Let $L = \bigcup_{n=1}^{\infty} F_n$. As $1 \in F_1$, we have $1 \in L$. Suppose that $r, s \in L$. Then $r \in F_m$, $s \in F_n$, for some $m, n \in \mathbb{N}$. Letting k be the larger of m and n, we have $r, s \in F_k$. Thus, $r - s \in F_k \subseteq L$ and, if $s \neq 0$, $rs^{-1} \in F_k \subseteq L$.

12.27 As $a \in F(a)$ and $F(a)$ is a field, we must have $a^2 \in F(a)$. Also, $F \subseteq F(a)$. As $F(a^2)$ is the intersection of all subfields of K containing F and a^2, it follows that $F(a^2) \subseteq F(a)$. For the second part, let $F = \mathbb{Q}$ and $a = i$. Then $\mathbb{Q}(a^2) = \mathbb{Q}(-1) = \mathbb{Q}$, but $\mathbb{Q}(a)$ contains i, so the fields are different.

12.29 The minimal polynomial of a is irreducible over $\mathbb{C}$. By the Fundamental Theorem of Algebra, this minimal polynomial has degree 1, and must therefore be $x - a \in \mathbb{C}[x]$; thus, $a \in \mathbb{C}$.

12.31 Note that $f(x) = x^3 + x + 1$ is irreducible over $\mathbb{Z}_7$. (It has degree 3 and no roots in $\mathbb{Z}_7$.) Thus, $F = \mathbb{Z}_7[x]/(f(x))$ will work. Letting $a = x + (f(x))$, we know that the elements of F are the linear combinations of 1, a and a^2 over $\mathbb{Z}_7$. Also, a is a root of $f(x)$, so $a^3 = -a - 1 = 6a + 6$ and $a^4 = (6a + 6)a = 6a^2 + 6a$. Thus, $(a^2 + 5a + 4)(3a^2 + 6) = 3a^4 + a^3 + 4a^2 + 2a + 3 = 3(6a^2 + 6a) + (6a + 6) + 4a^2 + 2a + 3 = a^2 + 5a + 2$.

12.33 If a and b are any two roots of $x^3 - 2$, then $(ab^{-1})^3 = 1$, so ab^{-1} is one of the roots of $x^3 - 1$. One such root in $\mathbb{C}$ is 1 and another is ω. Also, $(\omega^2)^3 = 1$, so ω^2 is the third complex root of $x^3 - 1$. Thus, since $\mathbb{Q}(\sqrt[3]{2}, \omega)$ must contain $\sqrt[3]{2}$, 1, ω and ω^2, we see that it contains every root of $x^3 - 2$; in particular, $x^3 - 2$ splits over $\mathbb{Q}(\sqrt[3]{2}, \omega)$. On the other hand, if $x^3 - 2$ splits over any subfield, then that subfield would have to contain all three roots, namely, $\sqrt[3]{2}$, $\omega\sqrt[3]{2}$ and $\omega^2\sqrt[3]{2}$. As it is a field, this means it must contain ω as well, so it is all of $\mathbb{Q}(\sqrt[3]{2}, \omega)$.

12.35 Note that $\mathbb{Q}(\sqrt{2})$ is a splitting field of $x^2 - 2$ over $\mathbb{Q}$. (Both roots, $\sqrt{2}$ and $-\sqrt{2}$ are in the field, and would have to be in any splitting field.) As an automorphism α must map the identity to the identity, we see immediately that $\alpha(c) = c$ for all $c \in \mathbb{Z}$. Similarly, if $m, n \in \mathbb{Z}$ with $n > 0$, then $m = \alpha(m) = \alpha(n(m/n)) = n\alpha(m/n)$. Thus, $\alpha(c) = c$ for all $c \in \mathbb{Q}$. By the preceding exercise, $\alpha(\sqrt{2})$ must be a root of $x^2 - 2$; in particular, $\alpha(\sqrt{2}) \in \{\sqrt{2}, -\sqrt{2}\}$. In the former case, α is the identity function. In the latter case, $\alpha(a + b\sqrt{2}) = a - b\sqrt{2}$ for all $a, b \in \mathbb{Q}$. By Lemma 12.4, this is an automorphism.

12.37 Let K be a splitting field for $f(x)$ over F. Say that in $K[x]$, we have $f(x) = a(x - a_1)(x - a_2) \cdots (x - a_n)$. Then $g(x) = a(x + 1 - a_1)(x + 1 - a_2) \cdots (x + 1 - a_n) = a(x - (a_1 - 1))(x - (a_2 - 1)) \cdots (x - (a_n - 1))$. Since the a_i lie in K, so do the $a_i - 1$; thus, $g(x)$ splits over K. Furthermore, for $g(x)$ to split, all of the $a_i - 1$ must be present, and hence so must all of the a_i. Thus, we cannot make K any smaller and have $g(x)$ split, so K is a splitting field for $g(x)$. Showing that splitting fields for $g(x)$ must be splitting fields for $f(x)$ involves a similar argument.

12.39 If $|F| = p^n$, for some prime p and positive integer n, then F has one proper subfield for each integer m, $1 \leq m < n$, with $m|n$. The first value n that works is 6, so the smallest such field has order $2^6 = 64$. Specifically, it is the splitting field of $x^{64} - x$ over $\mathbb{Z}_2$.

12.41 Let $a \in K$ be a root of $f(x)$. Then $[\mathbb{Z}_5(a) : \mathbb{Z}_5] = 3$. If all roots of $f(x)$ lie in $\mathbb{Z}_5(a)$, then $K = \mathbb{Z}_5(a)$, and $|K| = 5^3$. Otherwise, in $\mathbb{Z}_5(a)[x]$, we have $f(x) = (x - a)g(x)$, where $g(x)$ is an irreducible polynomial of degree 2. Letting b be a root of $g(x)$ in K, we see that $[\mathbb{Z}_5(a, b) : \mathbb{Z}_5(a)] = 2$. Furthermore, in $\mathbb{Z}_5(a, b)$, the polynomial $f(x)$ splits into linear factors, so $K = \mathbb{Z}_5(a, b)$. Now, $[K : \mathbb{Z}_5] = [K : \mathbb{Z}_5(a)][\mathbb{Z}_5(a) : \mathbb{Z}_5] = 2 \cdot 3 = 6$, and $|K| = 5^6$.

12.43 Every field of characteristic 0 is perfect, so char $F = p$, for some prime p. The fact that $f(x) = a_0 + a_p x^p + \cdots + a_{mp} x^{mp}$ follows exactly as in the proof of Theorem 12.16. Suppose that all of the a_i are algebraic over the prime subfield, (an isomorphic copy of) $\mathbb{Z}_p$. Then $[\mathbb{Z}_p(a_0) : \mathbb{Z}_p] < \infty$. Also, a_p is algebraic over $\mathbb{Z}_p$, and hence over $\mathbb{Z}_p(a_0)$, so $[\mathbb{Z}_p(a_0, a_p) : \mathbb{Z}_p(a_0)] < \infty$. Thus, $[\mathbb{Z}_p(a_0, a_p) : \mathbb{Z}_p] = [\mathbb{Z}_p(a_0, a_p) : \mathbb{Z}_p(a_0)][\mathbb{Z}_p(a_0) : \mathbb{Z}_p] < \infty$. In the same way, $[\mathbb{Z}_p(a_0, a_p, \ldots, a_{mp}) : \mathbb{Z}_p] < \infty$, which means that $\mathbb{Z}_p(a_0, a_p, \ldots, a_{mp})$ is a finite field, and hence perfect. If $f(x)$ is irreducible over F, it is surely irreducible over $\mathbb{Z}_p(a_0, \ldots, a_{mp})$. An irreducible polynomial over a perfect field cannot have multiple roots in any extension field.

12.45 If it were cyclic, it would be infinite cyclic. But note that $-1 \in U(F)$, and -1 has order 2. An infinite cyclic group has no such element.

12.47 Let F be the splitting field of $x^{125} - x$ over $\mathbb{Z}_5$. We know that it has order 125. Let $f(x) \in \mathbb{Z}_5[x]$ be an irreducible factor of $x^{125} - x$. If $a \in F$ is a root of $f(x)$, then $[\mathbb{Z}_5(a) : \mathbb{Z}_5] = \deg(f(x))$. But $\mathbb{Z}_5(a)$ is a subfield of F. A subfield of a field of order 5^3 can only have order 5 or 5^3. Thus, $\deg(f(x)) = 1$ or 3.

Problems of Chapter 13

13.1 WKHWUHDVXUHLVEXULHGWZHQWBSDFHVQRUWKRIWKHSDO-PWUHH

13.3 We need k to be relatively prime to 26. If it is not, then letting $d = (26, k)$, we see that both 0 and $26/d$ will be encrypted as 0, so decryption will be impossible. On the other hand, if $(k, 26) = 1$, then $k \in U(26)$, so we can decrypt by multiplying by k^{-1}. (If $k \equiv 1 \pmod{26}$, then multiplying by k does not change the text at all, so it would be reasonable to rule out this key as well.)

13.5 JGVSHNEGJESCRPPRBSXBPPVGHBSJKEHXVT

13.7 KXTNRHIOQJHVKWNKSVNHSWOXCLFAAMJSKSBO

13.9 Writing $n = pq$, the smaller of p and q must certainly be less than $\sqrt{n}$, so we only need to try primes up to 44. We discover that $p = 37$ and $q = 53$. Thus, $\varphi(n) = 36 \cdot 52 = 1872$. To find d, we use the Euclidean algorithm. In particular, $1872 = 43(43) + 23$; $43 = 23(1) + 20$; $23 = 20(1) + 3$; $20 = 3(6) + 2$;

$3 = 2(1)+1; 2 = 1(2)+0$. Thus, $1 = 3(1)+2(-1) = 3(1)+(20(1)+3(-6))(-1) = 20(-1) + 3(7) = 20(-1) + (23(1) + 20(-1))(7) = 23(7) + 20(-8) = 23(7) + (43(1)+23(-1))(-8) = 43(-8)+23(15) = 43(-8)+(1872(1)+43(-43))(15) = 1872(15) + 43(-653)$. Therefore, $43(-653) \equiv 1$ (mod 1872). As we need d to be positive, adding 1872, we get $d = 1219$.

13.11 We must break our message into blocks of length 2. As we have an odd number of letters, we add a Q to the end. Then AL is 0011, GE is 0604, BR is 0117 and AQ is 0016. Next, $11^{149} \equiv 5581$ (mod 17399), $604^{149} \equiv 2315$ (mod 17399), $117^{149} \equiv 4926$ (mod 17399) and $16^{149} \equiv 9527$ (mod 17399), so our encrypted message consists of the four numbers 5581, 2315, 4926 and 9527.

13.13 Note that $n = 103 \cdot 179 = 18437$ and $\varphi(n) = 102 \cdot 178 = 18156$. To find d, we apply the Euclidean algorithm. Namely, $18156 = 151(120) + 36$; $151 = 36(4) + 7$; $36 = 7(5) + 1$; $7 = 1(7) + 0$. Thus, $1 = 36(1) + 7(-5) = 36(1) + (151(1) + 36(-4))(-5) = 151(-5) + 36(21) = 151(-5) + (18156(1) + 151(-120))(21) = 18156(21) + 151(-2525)$. Therefore, $-2525e \equiv 1$ (mod $\varphi(n)$). As we need d to be positive, we add 18156 and get $d = 15631$. We now calculate $2469^{15631} \equiv 1514$ (mod 18437), $7093^{15631} \equiv 1124$ (mod 18437), $14773^{15631} \equiv 1314$ (mod 18437), $10900^{15631} \equiv 1208$ (mod 18437) and $143^{15631} \equiv 11$ (mod 18437) (which we remember to write as 0011). So, our message is 15141124131412080011, which translates to POLYNOMIAL.

Problems of Chapter 14

14.1 Construct the line through A and B. Next, construct the circle centred at A with radius AB. Say it meets the line at B and C. Construct the circle centred at B with radius AB, and say that it meets the line at A and E. Then the distance from C to E is 3, so if we construct the perpendicular bisector of CE, and it meets the line at D, then the distance from C to D is 1.5.

14.3 Construct the circle centred at A with radius AB and the circle centred at B with radius AB. Let C be either of the intersection points of these circles. By construction, the three sides of ABC have the same length.

14.5 We begin by constructing the line through B and A. Next, construct the circle centred at B with radius BC. It meets the line through B and A at two points; let E be the one of those points on the same side of B as A. Then replacing A with E, we may assume that in our original angle, A and C were equidistant from B. Construct the line through A and C, then construct the perpendicular bisector of AC. These two lines meet at the desired point, D.

14.7 Proceeding as in the solution to Exercise 14.3, construct a point E such that ABE is an equilateral triangle. It must lie on the circle. Now do the same thing with A and E; that is, construct a point C (the one that is different from B) such that AEC is an equilateral triangle. Again, C must be on the circle. Now do the same with A and C, and construct a new point F on the circle such that ACF is an equilateral triangle. Performing the same construction for A and F, we obtain a new point D on the circle such that AFD is an equilateral triangle. And then in the same way, we can construct a point G on the circle such that ADG is an equilateral triangle. But now $BECFDG$ is a regular hexagon, so BCD is an equilateral triangle.

14.9 As the constructible numbers form a field, if $a + b$ were constructible, then $a+b-a = b$ would also be constructible, giving a contradiction. Similarly, if ab were constructible, then the field of constructible numbers would include $a^{-1}ab = b$. Now, if $-b$ were constructible then b would be constructible as well, so letting $c = -b$, we see that $b + c = 0$ is constructible. On the other hand, if we let $c = b$, then we get $b + c = 2b$. If this were constructible, then since $1/2$ is also constructible, we would find that b would be constructible as well.

14.11 (1) Yes. As all integers are constructible and the field of constructible numbers is closed under the taking of square roots of its nonnegative elements, we see that $2 + \sqrt{5} - \sqrt{3}$ is constructible, and then we can take the square root twice to obtain this element.

(2) No. Once again, we know that $\sqrt{3}$ is constructible, but $\sqrt[3]{3}$ is a root of $x^3 - 3$. By Eisenstein's criterion, this polynomial is irreducible over $\mathbb{Q}$, so $\sqrt[3]{3}$ has minimal polynomial $x^3 - 3$. As the degree is not a power of 2, $\sqrt[3]{3}$ is not constructible. By Exercise 14.9, the sum of a number that is constructible with one that is not is not constructible.

14.13 By Eisenstein's criterion, this polynomial is irreducible over $\mathbb{Q}$. Thus, it is the minimal polynomial of a over $\mathbb{Q}$. As the degree is not a power of 2, a is not constructible.

14.15 We will prove a stronger statement, that an angle of $\pi/6$ can be constructed. We are given the points $(0, 0)$ and $(1, 0)$. To obtain such an angle, we only need to construct the point $(\cos(\pi/6), \sin(\pi/6)) = (\sqrt{3}/2, 1/2)$. But by Theorems 14.1 and 14.2, the numbers $\sqrt{3}/2$ and $1/2$ are constructible, so the point is constructible.

14.17 There is nothing to do for $n = 1$. When $n = 2$, we have $\cos(2\theta) = 2\cos^2(\theta) - 1$. For the $n = 3$ case, we look to the proof of Theorem 14.6, and see that $\cos(3\theta) = 4\cos^3(\theta) - 3\cos(\theta)$. To handle the remaining cases, we simply note that $\cos(\theta) = \cos(-\theta) = \cos(2\pi - \theta)$. Thus, $\cos(4\theta) = \cos(8\pi/7) = \cos(6\pi/7) = \cos(3\theta)$ and, similarly, $\cos(5\theta) = \cos(2\theta)$ and $\cos(6\theta) = \cos(\theta)$.

14.19 Let $D = A$. We know that the number $\sqrt{2}$ is constructible, which means that the point $E = (\sqrt{2}, 0)$ is constructible. Construct the circle centred at A and passing through C. It will meet the x-axis (which we can construct) at $(m, 0)$, where m is the distance from A to C. Thus, the number m is constructible, and so $m\sqrt{2}$ is constructible. In particular, we can construct the point $(m\sqrt{2}, 0)$. Now draw the circle centred at A and passing through $(m\sqrt{2}, 0)$. It intersects the line through A and C at a point F, where the distance from A to F is $m\sqrt{2}$ and F is on the same side of A as C. The triangles ABC and DEF are similar. As the side lengths are increased by a factor of $\sqrt{2}$, the area is increased by a factor of 2.

Index

© Springer International Publishing AG, part of Springer Nature 2018
G. T. Lee, *Abstract Algebra*, Springer Undergraduate Mathematics Series,
https://doi.org/10.1007/978-3-319-77649-1

Printed in the USA
CPSIA information can be obtained
at www.ICGtesting.com
LVHW011338211123
764547LV00002B/3